Optimization and Approximation

Optimization and Approximation

W. Krabs
Technische Hochschule Darmstadt

A Wiley–Interscience Publication

JOHN WILEY & SONS
Chichester · New York · Brisbane · Toronto

This edition is published by permission of Verlag B G Teubner, Stuttgart, and is the sole authorized English translation of the original German edition.

Library of Congress Cataloging in Publication Data:

Krabs, Werner, 1934–
Optimization and approximation.

'A Wiley–Interscience publication.'
Translation of Optimierung und Approximation.
Includes index.
1. Mathematical optimization. 2. Approximation theory. I. Title.
QA402.5.K713 515 78-10448
ISBN 0 471 99741 2

Typeset in Great Britain by Preface Ltd., Salisbury, Wilts.
and printed by Unwin Brothers Ltd., The Gresham Press, Old Woking, Surrey

Preface to the English Edition

The intimate connection between optimization and approximation has been recognized and exploited in recent years by a number of mathematicians and applied scientists. It is the primary purpose of this book to emphasize and draw further attention to this connection.

So I was glad not only to find colleagues who shared my opinion about the importance of this subject but who were also willing to translate the German edition into English and thus make the book accessible to a broader audience. I would therefore like to express my appreciation and my thanks to Professors Philip M. Anselone and Ronald B. Guenther for undertaking the effort of the translation.

A few errors in the German edition have been corrected and the solutions of various problems have been provided at the end of the corresponding chapter.

I am also indebted to Mrs G. Oelschlaegel for typing the final version of the manuscript. Finally, I would like to thank John Wiley & Sons Ltd for including this book in their scientific program.

Darmstadt, Summer 1978 W. Krabs

Preface to the German Edition

This book is devoted to the connection between optimization and approximation. It grew out of courses which I gave in the years from 1971 to 1973 in Aachen and Darmstadt.

Approximation theory developed first as an independent discipline. In the beginning, optimization theory had goals different from applications to approximation problems. Such applications are very natural, however, since every approximation problem can be regarded as an optimization problem. If approximation theory is included under the more general concept of optimization theory, then undoubtedly the more special properties of approximation problems are lost. So one cannot hope to answer all theoretical problems of approximation theory by means of optimization.

For questions of the characterization of best approximations and calculation or estimation of the minimal deviation, however, the use of optimization theory has turned out to be very fruitful. Even for the question of the existence of best approximations, it gives valuable information. The question of uniqueness, however, plays a minor role. Finally, various methods for the solution of optimization problems may be applied successfully to approximation problems, a circle of problems which we shall not go into in this book, however.

The goal of this book consists more of showing how numerous approximation problems of quite different types, which arise in physical and technical applications and in part from other problems of applied mathematics, can be treated in a unified way within the general framework of optimization theory.

It consists of three chapters on linear, convex and general nonlinear problems, plus an appendix on the background of functional analysis. Each of the first three chapters begins with a series of examples, which are taken up again after the development of the corresponding theoretical principles. In this way we hope to give an impression of the scope of the theory.

The inductive presentation chosen here has, of course, the disadvantage of a certain redundancy. On the other hand, it has the advantage that each of the three chapters on optimization problems can be read independently of the others. Also, the specific structure of the problems under consideration appear clearly right at the beginning and do not first have to be derived from more general classes of problems. Incidentally, the nonlinear problems are not presented in the general form from which one can derive the convex problems already treated, but from the beginning are taken in a form which is suitable for application to approximation problems.

To gain a deeper feeling for the material, problems have been inserted occasionally and small gaps in proofs have been left to the reader as exercises.

From the wealth of literature which has recently become available for infinite optimization problems, only a selection could be cited; this is collected in the references and is given in the text by the name of the author with the last two numbers in square brackets referring to the year in which the work appeared. Quotations which deal with material being presented are usually taken up directly in the text and those of a more general character are collected in special paragraphs with bibliographical remarks.

There are at present several books on infinite optimization which also contain some applications to approximation theory, for example, the book of Pschenitschny [72] and that of Luenberger [69]. The two books of Holmes [72] and Laurent [72] are even explicitly devoted to the connection of the two areas. In the German literature, there is at this time no book with a similar intention. This book is an effort to fill this gap and at the same time to emphasize especially the manifold applications. Since it deals primarily with infinite optimization problems, the use of functional analytic methods is unavoidable. In particular, the separation theorems for convex sets are of fundamental importance. These theorems have been collected in an appendix, partly without proof. A certain familiarity with the basics of the functional analysis of normed vector spaces is therefore necessary for reading the theoretical parts of this book.

Dr K. Glashoff and Dr E. Sachs have read the manuscript critically and have made numerous helpful suggestions. They, together with Mr D. Arndt, Prof. F. Lempio, and Prof. E. Bohl, have also helped in reading the corrections. I wish to thank all of them here as well as Mrs G. Oelschlaegel, who typed the manuscript and helped with all the editorial work. I wish also to thank the Teubner publishing company for taking up this book in its series of text books and for the excellent layout.

Darmstadt, Spring 1975 W. Krabs

Contents

I. Linear Problems 1
I.1 Introduction 1
I.1.1 The relation between approximation and optimization . . . 1
I.1.2 A control–approximation problem in the heating of metals . 3
I.1.3 Semi-infinite optimization in the control of air pollution . . 5
I.1.4 A preview of convex and general nonlinear problems . . . 6
I.2 Examples of linear approximation and optimization problems . . 7
I.2.1 Uniform approximation of functions 7
I.2.1.1 The general case 7
I.2.1.2 Approximation with interpolation side conditions . . . 8
I.2.1.3 The discrete case 8
I.2.2 Uniform approximation for the initial boundary value problem (IBVP) for the heat equation 9
I.2.3 A linear optimization problem derived from a boundary value problem for the potential equation 11
I.2.4 Linear boundary value problems and uniform approximation . 13
I.2.4.1 The general case 13
I.2.4.2 Problems of monotonic type and one-sided approximation 14
I.2.5 A linear control–approximation problem 15
I.3 The general linear optimization problem 16
I.3.1 Posing the problem; a weak duality theorem and simple consequences 16
I.3.2 The semi-infinite case 18
I.3.3 Semi-infinite problems in function spaces 19
I.3.3.1 Application to a boundary value problem for the potential equation 21
I.3.3.2 Application to a linear boundary value problem of monotonic type 22
I.3.4 Counter-examples to the validity of general existence and duality statements 26
I.3.4.1 Unsolvability of a semi-infinite problem 26
I.3.4.2 Occurrence of a duality gap 27
I.3.5 Bibliographical remarks 30

I.4 Existence and duality theorems 31
I.4.1 Double dualization of an optimization problem 31
I.4.2 Subconsistency and normality of an optimization problem. . 32
I.4.3 General existence and duality theorems 35
I.4.4 Existence theorems for the dual problem 39
I.4.4.1 Dualization of the existence theorem in Section I.4.3 . . 39
I.4.4.2 Application to the semi-infinite problem 40
I.4.4.3 A direct existence theorem for the dual problem . . . 41
I.4.5 Bibliographical remarks 43
I.5 Applications 45
I.5.1 Semi-infinite problems in function spaces 45
I.5.2 Uniform approximation of functions 48
I.5.3 One-sided uniform approximation 55
I.5.4 Application to a boundary value problem for the potential equation 59
I.5.5 A control–approximation problem in heat conduction . . . 61

II. Convex Problems 78
II.1 Examples of convex approximation and optimization problems . 78
II.1.1 The general linear approximation problem 78
II.1.2 Optimal error estimates for linear operator equations . . . 80
II.1.2.1 Defect estimates 80
II.1.2.2 Operator estimates 83
II.1.3 A problem of optimal control 87
II.2 Convex functions 88
II.2.1 Convex functionals 88
II.2.2 Convex mappings 91
II.2.3 Existence statements for linear control problems 94
II.3 The general convex optimization problem: existence and duality theorems 96
II.3.1 Posing the problem, existence theorems, subconsistency . . 96
II.3.2 The dual problem 99
II.3.3 General existence and duality theorems 101
II.3.4 The linear case 104
II.3.5 Bibliographical remarks 105
II.4 Min–sup, max–inf and saddle point theorems 106
II.4.1 The optimization problem as a min–sup problem 106
II.4.2 The dual problem as a max–inf problem 108
II.4.3 The equivalence of the min–sup = max–inf statement with a saddle point statement 110
II.4.4 The generalized theorem of Kuhn and Tucker 111
II.4.5 Existence theorems for the dual problem 112
II.4.6 Bibliographical remarks 113
II.5 Application to approximation problems 114
II.5.1 Existence theorems in the case of convex approximation problems in normed vector spaces 114

II.5.2 Uniform approximation of functions 115
II.5.2.1 The general convex case 115
II.5.2.2 The general linear case 118
II.5.3 An approximation problem with a mixed norm 119
II.5.4 Calculation of the minimal deviation for a convex approximation problem 122
II.6 Convex optimization problems in function spaces 124
II.6.1 Posing the problem and characterizing the optimality . . . 124
II.6.2 A mixed linear convex problem 129
II.6.3 Applications 133
II.6.3.1 Uniform linear approximation with interpolation . . . 133
II.6.3.2 A semi-infinite problem in the control of air pollution . 137

III. Nonlinear Problems 144
III.1 Some examples of nonlinear approximation and optimization problems 144
III.1.1 Nonlinear approximation in normed vector spaces 144
III.1.1.1 General remarks 144
III.1.1.2 Uniform approximation of functions 144
III.1.1.3 General rational approximation 145
III.1.2 One- and two-sided approximation in nonlinear boundary value problems 146
III.1.2.1 The general case 146
III.1.2.2 An example 149
III.1.3 Defect estimates for nonlinear boundary value problems . . 150
III.2 Minimizing convex functionals on arbitrary sets 154
III.2.1 Tangent cones in normed vector spaces 154
III.2.2 Necessary conditions for minimal points of convex functionals on arbitrary sets 156
III.2.2.1 A general theorem 156
III.2.2.2 Application to approximation in normed spaces . . . 157
III.2.2.3 Application to uniform approximation of functions . . 159
III.2.2.4 Application to L_1 approximation 162
III.2.3 Bibliographical remarks 164
III.3 Nonlinear optimization with an infinite number of side conditions 165
III.3.1 Necessary conditions for minimal points 165
III.3.1.1 The general case 165
III.3.1.2 The differentiable case 167
III.3.2 Sufficient conditions for minimal points 171
III.3.3 Application to nonlinear uniform approximation 173
III.3.4 Semi-infinite nonlinear optimization 177
III.3.5 Bibliographical remarks 180

IV. Appendix: Mathematical Aids 184
IV.1 Convex cones and linear mappings 184

IV.1.1 Convex cones and partial orderings 184
IV.1.2 The (topological) dual space 186
IV.1.3 Linear mappings 189
IV.2 Properties of convex cones and representation of positive linear forms 190
IV.2.1 Closedness of convex cones 190
IV.2.2 Adjoint cones 192
IV.2.3 Representation of positive linear forms on vector spaces of continuous functions 194
IV.2.4 Representation of continuous linear forms on vector spaces of continuous functions 197
IV.3 Convex Sets 200
IV.3.1 Algebraic and topological properties 200
IV.3.2 Separation theorems 202
IV.3.3 Weak convergence 203

References 209
Index 216

CHAPTER I

Linear Problems

I.1 INTRODUCTION

I.1.1 The relation between approximation and optimization

The development of optimization theory originated with economic problems and game theory, where optimal strategy was to be described mathematically. At about the same time, approximation theory, which was already well developed, experienced a reinvigoration brought about by the advent of electronic computers. They spurred the development of algorithms for the solution of approximation problems which previously would have been impractical because of the complexity of the calculations.

The connection between the two disciplines was first recognized in the case of the discrete linear approximation problem. Here the problem is to approximate a given real-valued function f at a finite number of points $t_1, \ldots, t_m$ of a set S (e.g. the real numbers) by functions v of a simpler type (e.g. polynomials or trigonometric sums). Often the given function values $f_i = f(t_i)$, $i = 1, \ldots, m$, are the result of measurements, so that an explicit expression for f in the form of an algebraic or analytic function defined on S is not known. In the case of a linear approximation problem, a finite dimensional vector space V is constructed consisting of linear combinations of functions $v_1, \ldots, v_n$, whose values are easily computed, and $v \in V$ is taken to be

$$v(t) = \sum_{j=1}^{n} x_j v_j(t) \qquad \text{for} \qquad t \in S \tag{I.1.1}$$

where $x_1, \ldots, x_n$ are real numbers. To measure the deviation of the functions $v \in V$ from f at the points $t_1, \ldots, t_m \in S$, we identify the function values $v(t_1), \ldots, v(t_m)$ with a vector in $\mathbb{R}^m$ (real m-tuples), introduce a suitable norm $\| \ \|$ and then determine $v \in V$ so that the deviation $\| v - f \|$ is as small as possible. In the case of the Euclidean norm in $\mathbb{R}^m$, this is the classical least squares approximation problem. It leads to a linear system of equations (the so-called normal equations) which has a closed form solution.

If one introduces the maximum norm in $\mathbb{R}^m$, one is led to the discrete linear Tchebychev approximation problem, which can be expressed as a linear optimization problem (see Section I.2.1.3). This is also possible in the case of the L_1 norm. For this purpose define the $m \times n$ matrix B by

$$B = \begin{pmatrix} v_1(t_1) & \dots & v_n(t_1) \\ \vdots & & \vdots \\ v_1(t_m) & \dots & v_n(t_m) \end{pmatrix}.$$

Then every function $v \in V$, i.e. of the form (I.1.1), can be identified with a vector $Bx \in \mathbb{R}^m$ on the set $\{t_1, \dots, t_m\}$ where x is a vector from the space $\mathbb{R}^n$ of all real n-tuples. Minimizing

$$\| v - f \| = \sum_{i=1}^{m} | v(t_i) - f(t_i) | \qquad \text{for} \qquad v \in V$$

is now equivalent to the problem of minimizing the linear functional $e^T z$, where $e = (1, \dots, 1)^T \in \mathbb{R}^m$ (where T means transpose), subject to the side conditions

$$Bx + z \geqslant f \qquad -Bx + z \geqslant -f \qquad x \in \mathbb{R}^n \qquad z \in \mathbb{R}^m$$

where $f = (f(t_1), \dots, f(t_m))^T$, and the sign $\geqslant$ is the componentwise ordering in $\mathbb{R}^m$. The proof is left as an exercise. This problem is also an example of ordinary linear optimization.

In all three cases, one can say that the problem is to minimize the convex functional (see Section II.2.1) $\varphi(v) = \| v - f \|$, for $v \in V$, which is a simple problem of convex optimization without side conditions.

Since the connection between discrete linear approximation problems and ordinary linear optimization has already been discussed in text books (e.g. Collatz and Wetterling [71], Suchowitzki and Awdejewa [69], Stiefel [65]), this topic will not be treated in any greater detail here.

We shall concern ourselves primarily with continuous approximation problems which lead to so-called infinite optimization problems in which infinitely many side conditions and also infinitely many free parameters can occur. One is led directly to such an infinite optimization problem when one seeks to approximate a continuous real-valued function f on a compact set S (e.g. a real interval) by linear combinations (I.1.1) of continuous functions on S as well as possible in the sense of the maximum norm (see Section I.2.1).

Such problems occur, for example, when one seeks simplified representations of functions for the purpose of evaluation on a computer (Hastings [55], Hart *et al.* [68]). They also occur in the approximate solution of boundary and initial-boundary value problems for ordinary and partial differential equations (see Sections I.2.2, I.2.3, and I.2.4). As a rule they can be reduced to so-called semi-infinite optimization problems (see Sections I.3.2, I.3.3, and I.5.1). They also arise in other connections. We shall give an application to the control of air pollution in Section I.1.3.

If one introduces for the approximation problem an infinite dimensional vector space V, then it can no longer be described by a finite number of parameters and the corresponding optimization problem is called completely infinite, i.e. there are infinitely many side conditions and infinitely many variables. Such types of problem do not usually come from pure problems in the approximation of given functions, but rather are approximation problems related to control theory (see Section I.2.5). We shall consider representatives of such problems.

I.1.2 A control–approximation problem in the heating of metals

In the book of Butkovskiy [69], the following problem is considered. A metal plate of uniform thickness is to be heated by applying a time dependent source to both sides of the plate. If one assumes that the plate is homogeneous, then the temperature distribution in the interior of the plate is a function of the time t and the coordinate x along which the thickness is measured. The temperature of the surrounding source (oven) should be controlled in such a way that the plate, which is initially at some constant temperature, is brought as close as possible to some desired final state within some time interval.

If one introduces dimensionless quantities and takes into account the symmetries of the problem, then the temperature $y = y(x, t)$ satisfies the heat equation in the form

$$y_t(t, x) = y_{xx}(t, x) \qquad \text{for } 0 < t \leqslant T, -1 < x < +1 \tag{I.1.2}$$

where $T > 0$ is the given duration of the heating process. Further, the heat conduction through the surface of the plate is described by Newton's law

$$\left.\begin{aligned} y_x(t, +1) &= b[u(t) - y(t, +1)] \\ -y_x(t, -1) &= b[u(t) - y(t, -1)] \end{aligned}\right. \qquad \text{for } t \in (0, T]. \tag{I.1.3}$$

where b is a suitable positive constant and $u = u(t)$ is the temperature of the surrounding oven (source). This will be assumed, for technical reasons, to be bounded above and below, which in the dimensionless formulation leads to the requirement

$$-1 \leqslant u(t) \leqslant +1 \qquad \text{for } 0 \leqslant t \leqslant T \tag{I.1.4}$$

if one chooses the zero point of the temperature properly. Let y_0 be the initial temperature, which yields the initial condition

$$y(0, x) = y_0 \qquad \text{for } -1 \leqslant x \leqslant +1. \tag{I.1.5}$$

The initial temperature should be brought as close as possible to a desired final temperature $y_T \in C[-1, 1]$ at time T which is expressed in the form

$$\sup_{-1 \leqslant x \leqslant 1} | y(T, x) - y_T(x) | = \min. \tag{I.1.6}$$

On the basis of a general existence theorem in the book of Friedman [64], for each continuous function $u = u(t)$, for $t \in [0, T]$, the initial boundary value problem given by equations (I.1.2), (I.1.3), and (I.1.5) has a unique solution $y = y(t, x, u)$

on $[0, T] \times [-1, 1]$ with $y_t, y_{xx} \in C((0, T] \times (-1, 1))$ and

$$y_x(t, +1, u) = \lim_{x \to 1-0} y_x(t, x, u)$$
$$y_x(t, -1, u) = \lim_{x \to -1+0} y_x(t, x, u) \tag{I.1.7}$$

for all $t \in (0, T]$.

This solution may be represented explicitly in the form

$$y(t, x, u) = y_0 \sum_{k=1}^{\infty} B_k \cos \mu_k x \exp(-\mu_k^2 t)$$
$$+ \sum_{k=1}^{\infty} \mu_k^2 B_k \cos \mu_k x \int_0^t \exp[-\mu_k^2 (t - \tau)] u(\tau) \, d\tau$$

where the μ_k are the positive solutions of the transcendental equation

$$\sin \mu = (b/\mu) \cos \mu$$

and

$$B_k = \frac{2 \sin \mu_k}{\mu_k + \sin \mu_k \cos \mu_k} \qquad k = 1,2, \ldots$$

(see Butkovskiy [69], and Krabs and Weck [74]).

Problem I.1.1 Show, by demonstrating the uniform convergence of the series, that $y(T, \cdot, u)$ is a continuous function on $[-1, 1]$. For this, use the fact that

$$(k-1)\pi < \mu_k < k\pi \qquad \text{for all } k = 1,2, \ldots.$$

Mathematically, the problem posed is to find a continuous control function u on $[0, T]$ satisfying (I.1.4) such that

$$\| y(T, \cdot, u) - y_T \| = \max_{-1 \leqslant x \leqslant +1} | y(T, x, u) - y_T(x) |$$

is minimized, where, for each $u \in C[0, T]$, $y(t, x, u)$ is the unique solution of the initial boundary value problem (I.1.2), (I.1.3), and (I.1.5) in the above sense.

If one defines a mapping $B : C[0, T] \to C[-1, +1]$ by

$$B(u)(x) = y(T, x, u) - y(T, x, 0)$$
$$= \sum_{k=1}^{\infty} \mu_k^2 B_k \cos \mu_k x \int_0^T \exp[-\mu_k^2 (T - t)] u(t) \, dt \tag{I.1.8}$$

for $u \in C[0, T]$, $x \in [-1, +1]$, then B is linear (see Section IV.1.3). Further if one sets

$$\hat{y}(x) = y_T(x) - y(T, x, 0)$$
$$= y_T(x) - y_0 \sum_{k=1}^{\infty} B_k \cos \mu_k x \exp(-\mu_k^2 T) \qquad \text{for } x \in [-1, +1], \tag{I.1.9}$$

then the above approximation problem is to determine a function $u \in C[0, T]$ with (I.1.4) such that

$$\| B(u) - \hat{y} \| = \max_{-1 \leqslant x \leqslant +1} | B(u)(x) - \hat{y}(x) |$$

is minimized.

This is a linear approximation problem with an infinite number of linear side conditions (I.1.4), in which the linear subspace

$$V = \{B(u) : u \in C[0, T]\}$$

of $C[-1, +1]$ of the approximating functions is of infinite dimension.

If one introduces the further side conditions

$$-\gamma \leqslant B(u)(x) - \hat{y}(x) \leqslant \gamma \qquad \text{for all } x \in [-1, +1] \tag{I.1.10}$$

then the approximation problem is to determine a pair $(\gamma, u) \in \mathbb{R} \times C[0, T]$ for which γ is minimized under the side conditions (I.1.4), and (I.1.10). This is a typical problem of infinite linear optimization, which we will examine further in Sections I.5.5 and II.5.4.

I.1.3 Semi-infinite optimization in the control of air pollution

As already mentioned in Section I.1.1, semi-infinite optimization problems (with infinitely many variables and infinitely many side conditions) occur not only in connection with linear approximation problems, but also have other applications. Thus, recently, the problem of air pollution has been treated mathematically by Gustafson [72], and Gustafson and Kortanek [73a], [73b] and the following model was constructed. In a given (two dimensional) control region S, a certain air quality is to be guaranteed. At the same time, the yearly average concentration of a pollutant (e.g. sulphur dioxide or carbon monoxide) is to be kept below a prescribed standard, which is described by a real-valued function φ on S. The concentration observed in S is assumed to come from two kinds of sources: (*a*) sources which can be controlled, and hence regulated, and (*b*) sources which cannot be controlled.

If, say, n controllable sources are present, then it is assumed that these contribute a yearly average $u_1, \ldots, u_n$ to the air pollution. We denote by u_0 the contribution from the uncontrollable sources. Then $u_0, u_1, \ldots, u_n$ are again real-valued functions on S, whose actual determination as a rule is naturally a very difficult problem. From the requirement that the concentration φ is not to be exceeded, we are led first to the side conditions

$$\sum_{j=1}^{n} u_j(s) + u_0(s) \leqslant \varphi(s) \qquad \text{for all } s \in S.$$

If any of these conditions are not satisfied, then the contributions of the controllable sources are reduced and, in fact, the jth source is reduced by a factor x_j,

$0 \leqslant x_j \leqslant 1$ for $j = 1, \ldots, n$, so that the side conditions

$$\sum_{j=1}^{n} (1 - x_j)u_j(s) + u_0(s) \leqslant \varphi(s) \qquad \text{for all } s \in S \tag{I.1.11}$$

are satisfied. Obviously these will be satisfied if, for all $s \in S$, the condition $u_0(s) \leqslant \varphi(s)$ is satisfied.

If an $x_j \neq 0$ must be chosen, then, in general, costs naturally arise (e.g. by changing production plans, introducing air purifiers, etc.). In the simplest case these costs are assumed to be proportional to x_j. Let c_j be the proportionality constant, so that with the choice of reduction factors $x_1, \ldots, x_n \in [0, 1]$ the total costs are

$$c(x_1, \ldots, x_n) = \sum_{j=1}^{n} c_j x_j. \tag{I.1.12}$$

One now tries to choose the factors $x_1, \ldots, x_n$ subject to the side conditions (I.1.11) so that the cost $c(x_1, \ldots, x_n)$ is as small as possible.

In summary, one therefore obtains the problem of minimizing the linear functional $c(x_1, \ldots, x_n)$ defined by (I.1.12), subject to the side conditions

$$\sum_{j=1}^{n} u_j(s)x_j \geqslant \sum_{j=0}^{n} u_j(s) - \varphi(s) \qquad \text{for all } s \in S \tag{I.1.11$'$}$$

and

$$0 \leqslant x_j \leqslant 1 \qquad \text{for } j = 1, \ldots, n. \tag{I.1.13}$$

This is a typical problem of semi-infinite optimization, which we will explore in more detail in Section II.6.3.2.

I.1.4 A preview of convex and general nonlinear problems

The theory of linear optimization treated in this chapter encompasses, of course, numerous applications. However, problems often occur which either cannot be put into such a form or can be put so only with great difficulty. That happens, for example, for the general convex approximation problem in a normed vector space treated in Sections II.5.1 and II.5.4, which cannot be treated within the framework of general linear optimization. In special cases of linear approximation problems, it is often a matter of convenience whether they are considered as linear or convex optimization problems. Thus, for example, the control problem described in Section I.1.2 can be treated both as a linear problem (see Section I.5.5) and also as a convex problem (see Section II.5.4), where the formula (II.5.52) for the minimal deviation (II.5.51) can be derived in the convex theory with less effort. One encounters typical problems of convex optimization in approximation problems with mixed norm which are investigated in Section II.5.3. They occur in connection with optimal error estimates in the case of linear operator equations (see Section II.1.2.1). Numerous approximation problems, such as those of rational approximation (Section III.1.1.3) or those which arise in connection with nonlinear boundary value problems (Sections III.1.2, III.1.3) can be described only as nonlinear optimization problems, to which Section III is devoted. Because of the complexity of this type of problem, there is naturally no well developed and polished theory as in the linear and convex cases. In the nonlinear case, for example, there are only weak duality results, but no duality theory which is valid in general. Developed furthest are necessary and sufficient conditions for minimal points, which is a principal topic in Section III.

I.2 EXAMPLES OF LINEAR APPROXIMATION AND OPTIMIZATION PROBLEMS

I.2.1 Uniform approximation of functions

I.2.1.1 The general case

Let M be a compact subset of a normed space and $C(M)$ be the vector space of continuous real-valued functions on M equipped with the maximum norm

$$\| g \|_\infty = \max_{t \in M} | g(t) | \qquad \text{for } g \in C(M). \tag{I.2.1}$$

Let $f \in C(M)$ be a given fixed function and let n linear independent functions $v_1, \ldots, \in C(M)$ be given. Let V be the n-dimensional linear subspace spanned by $v_1, \ldots, v_n$ which consists of all linear combinations

$$v = \sum_{j=1}^{n} v_j x_j \qquad \text{for } x_j \in \mathbb{R}, j = 1, \ldots, n.$$

We seek an element $\hat{v} \in V$ with

$$\| \hat{v} - f \|_\infty \leqslant \| v - f \|_\infty \qquad \text{for } v \in V \tag{I.2.2}$$

that is, an element $\hat{v} \in V$ which is a best uniform approximation of f, among all elements of V, i.e. $\hat{v}$ is the best approximation of f in V in the sense of the maximum norm (I.2.1). The quantity

$$\rho_\infty(f, V) = \inf_{v \in V} \| v - f \|_\infty \tag{I.2.3}$$

is called the minimal deviation (or also the distance) of the function f from V. We shall show later that this approximation problem is solvable (see Theorem II.5.2).

An important special case is the problem of polynomial approximation with $M = [a, b]$, $a < b$, $v_j(t) = t^{j-1}$, $j = 1, \ldots, n$. (Why are the v_j linearly independent?) In order to recast the approximation problem (I.2.2) as an optimization problem, we define for each $t \in M$ two (linear) functionals $\varphi_t^1, \varphi_t^2 : \mathbb{R}^{n+1} \to \mathbb{R}$ by

$$\left.\begin{aligned} \varphi_t^1(x, \gamma) &= \sum_{j=1}^{n} v_j(t) x_j + \gamma \\ \varphi_t^2(x, \gamma) &= \sum_{j=1}^{n} -v_j(t) x_j + \gamma \end{aligned}\right\} \qquad \text{for } x \in \mathbb{R}^n, \gamma \in \mathbb{R}. \tag{I.2.4}$$

If one sets

$$v = \sum_{j=1}^{n} v_j x_j$$

then obviously

$$\left.\begin{array}{l}\varphi_t^1(x,\gamma) \geqslant f(t)\\ \varphi_t^2(x,\gamma) \geqslant -f(t)\end{array}\right\} \quad \text{for all } t \in M \Leftrightarrow \| v - f \|_\infty \leqslant \gamma.$$

Further if one defines a (linear) functional $\varphi : \mathbb{R}^{n+1} \to \mathbb{R}$ by $\varphi(x, \gamma) = \gamma$, then the approximation problem (I.2.2) is equivalent to the problem of minimizing the functional $\varphi = \varphi(x, \gamma)$ subject to the side conditions

$$\varphi_t^1(x,\gamma) \geqslant f(t) \qquad \text{and} \qquad \varphi_t^2(x,\gamma) \geqslant -f(t) \qquad \text{for all } t \in M. \tag{I.2.5}$$

The proof is left as an exercise.

In this case we are dealing with a problem of linear optimization in general with infinitely many side conditions.

I.2.1.2 Approximation with interpolation side conditions

We imagine $r < n$ distinct points $[t_1, \ldots, t_r] \in M$ to be given and try to minimize $\| v - f \|_\infty$ with $v \in V$, subject to the additional side conditions

$$v(t_i) - f(t_i) = 0 \qquad \text{for } i = 1, \ldots, r.$$

If one assumes that

$$V_0 = \{v \in V : v(t_i) = f(t_i), i = 1, \ldots, r\} \tag{I.2.6}$$

is non-empty, then a $\hat{v} \in V_0$ is sought such that

$$\| \hat{v} - f \|_\infty = \rho_\infty(f, V_0) = \inf_{v \in V_0} \| v - f \|_\infty. \tag{I.2.7}$$

For the corresponding linear optimization problem, besides (I.2.5), one has to impose

$$\sum_{j=1}^{n} v_j(t_i) x_j = f(t_i) \qquad \text{for } i = 1, \ldots, r.$$

Problem I.2.1 Show, for the case of polynomial approximation, that V_0 is not empty.

I.2.1.3 The discrete case

Let M be a finite point set, say $M = \{t_1, \ldots, t_m\}$. The vector space $C(M)$ can then be identified with $\mathbb{R}^m$. Since we assume $V \subseteq C(M)$ with $\dim V = n$, $m \geqslant n$. We assume further that $m \geqslant (n + 1)$, since otherwise every $f \in C(M)$ would be uniquely representable as

$$f = \sum_{j=1}^{n} v_j x_j$$

and consequently $\rho_\infty(f, V) = 0$, and the approximation problem would be a representation problem. Define a $2m \times (n + 1)$ matrix A, a $2m$ vector b and an $(n + 1)$ vector c by

$$A = \begin{pmatrix} v_j(t_i)e_i \\ -v_j(t_i)e_i \end{pmatrix} \qquad b = \begin{pmatrix} f(t_i) \\ -f(t_i) \end{pmatrix} \qquad c = \begin{pmatrix} \Theta_n \\ 1 \end{pmatrix} \tag{I.2.8}$$

with $e_i = 1$ for all i, and Θ_n the null vector of $\mathbb{R}^n$. Define $z^1 \geqslant z^2 \Leftrightarrow z_i^1 \geqslant z_i^2$ for all i. Then the side conditions (I.2.5) become

$$Ay \geqslant b \qquad y \in \mathbb{R}^{n+1}. \tag{I.2.5$'$}$$

On the basis of these considerations, the discrete problem of approximating $f \in C(M) = \mathbb{R}^m$ by an element $\hat{v} \in V$ as closely as possible in the sense of the maximum norm (I.2.1) is equivalent to the problem of minimizing the linear functional $\varphi(y) = c^{\mathrm{T}} y$ (as before the superscript T means transpose) subject to the side conditions (I.2.5′).

This is a problem of ordinary linear optimization. In this book we shall not go into discrete approximation problems and their connection with optimization any further, but rather refer the reader to Collatz and Wetterling [71] and Suchowitzki and Awdejewa [69].

I.2.2 Uniform approximation for the initial boundary value problem (IBVP) for the heat equation

Let B be the rectangle of all points $(x, t) \in \mathbb{R}^2$ with $0 < x < 1, 0 < t < T$. Let $\bar{B}$ indicate the closure of B. Assume that we are given $\Phi \in C[0, 1]$ and $f_0, f_1 \in C[0, T]$ such that

$$\Phi(0) = f_0(0) \qquad \text{and} \qquad \Phi(1) = f_1(0). \tag{I.2.9}$$

We seek a function $u \in C(B)$ with

$$u_{xx}, u_t \in C((0, 1) \times (0, \mathrm{T}])$$

such that

$$u_t(x, t) = u_{xx}(x, t) \qquad \text{for all } (x, t) \in (0, 1) \times (0, T] \tag{I.2.10}$$

$$u(x, 0) = \Phi(x) \qquad \text{for all } x \in [0, 1] \tag{I.2.11}$$

$$\left.\begin{aligned} u(0, t) &= f_0(t) \\ u(1, t) &= f_1(t) \end{aligned}\right\} \qquad \text{for all } t \in [0, T]. \tag{I.2.12}$$

Physically, $u = u(x, t)$ can be interpreted as the temperature distribution in a rod of unit length, where an initial temperature is given by (I.2.11). At the ends $x = 0$ and $x = 1$ of the rod, the temperatures $f_0 = f_0(t)$ and $f_1 = f_1(t)$ for a time period $[0, T]$

are given by (I.2.12). These are boundary conditions. Thus, (I.2.10) is the simplest case of the heat equation.

We define

$$L = \{v \in C(\bar{B}) : v_t, v_{xx} \in C((0, 1) \times (0, T])\}$$

and

$$\Gamma = \{(x, 0) : 0 \leqslant x \leqslant 1\} \cup \{(0, t) : 0 \leqslant t \leqslant T\} \cup \{(1, t) : 0 \leqslant t \leqslant T\}.$$

With that, the so-called maximum–minimum principle holds: for each $u \in L$ satisfying (I.2.10), we have

$$\left\{\begin{matrix}\max \\ \min\end{matrix}\right\}_{(x,t)\in\bar{B}} u(x, t) = \left\{\begin{matrix}\max \\ \min\end{matrix}\right\}_{(x,t)\in\Gamma} u(x, t) \tag{I.2.13}$$

which also implies that

$$\max_{(x,t)\in\bar{B}} |u(x, t)| = \max_{(x,t)\in\Gamma} |u(x, t)|. \tag{I.2.14}$$

(For the proof see Protter and Weinberger [67], page 160.)

Problem I.2.2 Prove the implication (I.2.13) ⇒ (I.2.14) and show that the IBVP (I.2.10)–(I.2.12) has at most one solution.

We consider now the special case $f_0 \equiv 0$ and $f_1 \equiv 0$. Then for each $\Phi \in C[0, 1]$ with $\Phi(0) = \Phi(1) = 0$ there exists a solution $u \in L$ of the IBVP (I.2.10)–(I.2.12). (For the proof see, for example, Petrowski [54].) Therefore, by problem I.2.2, the solution is unique.

In order to obtain an approximation for u, we define functions v_j by

$$v_j(x, t) = \exp[-(j\pi)^2 t] \sin(j\pi x) \qquad \text{for } j = 1, \ldots, n. \tag{I.2.15}$$

Then every linear combination

$$v(x, t) = \sum_{j=1}^{n} a_j v_j(x, t) \tag{I.2.16}$$

belongs to L and satisfies the conditions (I.2.10) and (I.2.12) with $f_0 \equiv f_1 \equiv 0$. The same holds for $v - u$, and (I.2.14) yields

$$\max_{(x,t)\in\bar{B}} |v(x, t) - u(x, t)| = \max_{x\in[0,1]} |v(x, 0) - \Phi(x)|. \tag{I.2.17}$$

Therefore, from the assumption (I.2.16), one obtains a best approximation of u on B if the coefficients a_j are determined so that

$$\max_{x\in[0,1]} \left| \sum_{j=1}^{n} a_j \sin(j\pi x) - \Phi(x) \right|$$

is minimized. This is another problem of uniform approximation in the sense of Section I.2.1.

Problem I.2.3 Treat the case $f_0 \equiv 0, f_1(t) = h(t), \Phi(x) = h(0) \cdot x$ with a suitable analogue of (I.2.16) and reduce the case $f_0 \equiv 0, f_1(t) = h(t), \Phi \in C[0, 1]$ with $\Phi(0) = 0, \Phi(1) = h(0)$ by superposition to the two cases

$$f_0 \equiv 0, f_1(t) = h(t), \Phi(x) = h(0) \cdot x$$

and

$$f_0 \equiv 0, f_1 \equiv 0, \Phi \in C[0, 1] \qquad \text{with } \Phi(0) = \Phi(1) = 0$$

(see Collatz and Krabs [73]).

I.2.3 A linear optimization problem derived from a boundary value problem for the potential equation

Let B be a simply connected bounded domain of the (t, s)-plane with piecewise smooth boundary curve Γ. To be solved is the following boundary value problem (BVP). We seek a $u \in C(\bar{B}) \cap C^2(B)$, where $\bar{B}$ is the closed hull of B, $C(\bar{B})$ is the vector space of continuous real-valued functions on $\bar{B}$, and $C^2(B)$ is the vector space of twice continuously differentiable functions, such that

$$\Delta u = \frac{\partial^2 u}{\partial t^2} + \frac{\partial^2 u}{\partial s^2} = 0 \text{ in } B \tag{I.2.18}$$

and

$$u = f \text{ on } \Gamma \tag{I.2.19}$$

where $f \in C(\Gamma)$ is given and fixed.

This BVP has a unique solution $\hat{u}$ and the following boundary maximum principle holds (see Protter and Weinberger [67], page 53). If a function $\varphi \in C(\bar{B}) \cap C^2(B)$ is given such that

$$\begin{aligned} &\Delta\varphi = 0 \text{ in } B \text{ and } \varphi \geqslant f \text{ or } \varphi \leqslant f \text{ on } \Gamma \text{ respectively} \\ &(\varphi \geqslant f \text{ on } \Gamma \Leftrightarrow \varphi(t, s) \geqslant f(t, s) \text{ for all } (t, s) \in \Gamma) \end{aligned} \tag{I.2.20}$$

then it follows that $\varphi \geqslant \hat{u}$ or, respectively, $\varphi \leqslant \hat{u}$ on $\bar{B} = B \cup \Gamma$. In order to make use of these facts computationally, we imagine functions $\varphi_0, \ldots, \varphi_n \in C(\bar{B}) \cap C^2(B)$ given so that

$$\Delta\varphi_j = 0 \qquad \text{for } j = 0, \ldots, n$$

for example, $\varphi_j = \mathrm{Re}(z^j)$ or $\varphi_j = \mathrm{Im}(z^j)$, where $z = t + \mathrm{i}s$, $\mathrm{i} = \sqrt{-1}$. Then, for every linear combination

$$\varphi(x) = \sum_{j=0}^{n} \varphi_j x_j \qquad \text{for } x \in \mathbb{R}^{n+1}$$

it follows that $\Delta\varphi(x) = 0$. We now seek $\hat{x} \in \mathbb{R}^{n+1}$ such that

$$\varphi(\hat{x}) \geqslant f \qquad \text{and} \qquad \varphi(\hat{x}) \leqslant f \text{ on } \Gamma$$

such that at a given point $(\hat{t}, \hat{s}) \in \bar{B}$ the value $\varphi(\hat{x})(\hat{t}, \hat{s})$ is as small or as large as possible, respectively. For such an $\hat{x} \in \mathbb{R}^{n+1}$ the (unknown) solution $\hat{u}$ of the BVP (I.2.18), (I.2.19) will be approximated according to the boundary maximum principle in $(\hat{t}, \hat{s})$ optimally from above or from below, respectively.

We obtain the linear optimization problem of respectively minimizing or maximizing the linear functional

$$\psi(x) = \sum_{j=0}^{n} \varphi_j(\hat{t}, \hat{s})x_j$$

subject to the respective side conditions

$$\sum_{j=0}^{n} \varphi_j(t, s)x_j \geqslant f(t, s) \qquad \text{(I.2.21}a\text{)}$$

for all $(t, s) \in \Gamma$.

$$\sum_{j=0}^{n} \varphi_j(t, s)x_j \leqslant f(t, s) \qquad \text{(I.2.21}b\text{)}$$

Remark On the basis of the boundary maximum principle (I.2.20), it would also make sense to minimize

$$\left\| \sum_{j=0}^{n} \varphi_j x_j - f \right\|_\infty = \max_{(t,s)\in\Gamma} \left| \sum_{j=0}^{n} \varphi_j(t, s)x_j - f(t, s) \right|$$

subject to the side conditions (I.2.21a), and (I.2.21b), respectively.

This is a problem of one-sided uniform approximation, which is equivalent to the problem of minimizing the number γ in the inequalities

$$\sum_{j=0}^{n} \varphi_j(t, s)x_j - f(t, s) \begin{Bmatrix} \leqslant \gamma \\ \geqslant -\gamma \end{Bmatrix}, (t, s) \in \Gamma$$

respectively, subject to the side conditions (I.2.21a), and (I.2.21b) (Proof = Exercise).

If

$$\hat{\varphi} = \sum_{j=0}^{n} \varphi_j \hat{x}_j$$

is a solution of this problem and again $\hat{u}$ is the solution of the BVP (I.2.18), and (I.2.19), then

$$\hat{\varphi} \geqslant \hat{u} \text{ and } \hat{\varphi} \leqslant \hat{u} \text{ on } \bar{B}$$

respectively (Proof = Exercise). Moreover $\hat{\varphi}$ is obviously optimal in the sense that it best approximates $\hat{u}$ from one side uniformly on $\bar{B}$.

I.2.4 Linear boundary value problems and uniform approximation

I.2.4.1 The general case

Consider now a linear differential equation of the form

$$L[y](t) = y^{(n)}(t) + a_1(t)y^{(n-1)}(t) + \cdots + a_n(t)y(t) = r(t) \qquad \text{(I.2.22)}$$

for all $t \in [a, b]$ with n linear boundary conditions of the form

$$U_\mu[y] = \sum_{k=0}^{n-1} [y^{(k)}(a) + \beta_{\mu k} y^{(k)}(b)] = \gamma_{\mu k} \qquad \text{(I.2.23)}$$

for $\mu = 1, \ldots, n; n \geqslant 2$.

Here $a_1, \ldots, a_n, r \in C[a, b]$ are given functions and $\beta_{\mu k}, \gamma_{\mu k} \in \mathbb{R}$ are given constants. We seek a function $y \in C^n[a, b]$ which satisfies (I.2.22) and (I.2.23).

We assume that the completely homogeneous BVP

$$L[y] = 0 \text{ on } [a, b] \text{ and } U_\mu[y] = 0 \text{ for all } \mu = 1, \ldots, n$$

has only the trivial solution in $C^n[a, b]$. Then the BVP (I.2.22)–(I.2.23) has precisely one solution $\hat{y} \in C^n[a, b]$ which may be represented in the form $\hat{y} = y_1 + y_2$ with $y_1, y_2 \in C^n[a, b]$ such that

$$L[y_1] = 0 \text{ on } [a, b], U_\mu[y_1] = \gamma_\mu \qquad \text{for all } \mu \qquad \text{(I.2.24)}$$

and

$$L[y_2] = r \text{ on } [a, b], U_\mu[y_2] = 0 \qquad \text{for all } \mu. \qquad \text{(I.2.25)}$$

In addition there exists a so-called Green's function $G \in C([a, b] \times [a, b])$ with

$$y_2(t) = \int_a^b G(t, s)r(s)\, ds \qquad \text{(I.2.26)}$$

(see, e.g., Coddington and Levinson [55]).

G and y_2 may be constructed explicitly if a so-called fundamental system of linear independent solutions $z_1, \ldots, z_n \in C^n[a, b]$ of the homogeneous equation $L[y] = 0$ is known. However, as a rule, that is not the case, although the existence of fundamental systems is guaranteed.

In order to obtain an approximate solution for the (unknown) solution $\hat{y}$ of the BVP (I.2.22)–(I.2.23), we choose $(s + 1)$ functions $v_0, \ldots, v_s \in C^n[a, b]$ such that

$$U_\mu[v_0] = \gamma_\mu, U_\mu[v_k] = 0 \qquad \text{for all } k = 1, \ldots, s, \mu = 1, \ldots, n. \qquad \text{(I.2.27)}$$

For each function

$$v(t) = v_0(t) + \sum_{j=1}^{s} v_j(t)a_j \qquad \text{for } a_j \in \mathbb{R} \qquad \text{(I.2.28)}$$

it is true that $U_\mu[v] = \gamma_\mu$ for $\mu = 1, \ldots, n$, and $\hat{y} - v$ is a solution of the semi-homogeneous problem

$$L[y] = r - L[v], U_\mu[y] = 0 \qquad \text{for all } \mu = 1, \ldots, n.$$

From (I.2.26) it follows, therefore, that

$$\hat{y}(t) - v(t) = \int_a^b G(t, s) \{r(s) - L[v](s)\} \, ds.$$

Consequently, there is an error estimate

$$\| \hat{y} - v \|_\infty \leqslant \max_{t \in [a,b]} \int_a^b | G(t, s) | \, ds \, \| r - L[v] \|_\infty . \tag{I.2.29}$$

In order to obtain as sharp an error estimate as possible in (I.2.29), one chooses the coefficients a_j in (I.2.28) so that $\| r - L[v] \|_\infty$ is as small as possible. Setting

$$f = r - L[v_0] \qquad \text{and} \qquad w = \sum_{j=1}^{s} L[v_j] a_j \tag{I.2.30}$$

the problem therefore arises of approximating the function $f \in C[a, b]$ by functions w uniformly, i.e., in the sense of the maximum norm (I.2.1) with $M = [a, b]$, as well as possible. If an upper bound for

$$\max_{t \in [a,b]} \int_a^b | G(t, s) | \, ds$$

is known, then one can use (I.2.29) to estimate the error $\| y - v \|_\infty$.

I.2.4.2 Problems of monotonic type and one-sided approximation

The linear BVP (I.2.22)–(I.2.23) is said to be of monotonic type if the following implications hold:

$$\left.\begin{array}{ll} L[\varphi](t) \geqslant 0 & \text{for } t \in [a, b] \\ U_\mu[\varphi] = 0 & \text{for } \mu = 1, \ldots, n \end{array}\right\} \Rightarrow \varphi(t) \geqslant 0 \qquad \text{for } t \in [a, b]. \tag{I.2.31}$$

Sufficient conditions for problems to be of monotonic type can be found in Collatz [68], and Protter and Weinberger [67]. If monotonicity is present, then the associated completely homogeneous problem has only the trivial solution and, as remarked above, the BVP (I.2.22)–(I.2.23) is uniquely solvable.

Let $\hat{y} \in C^n[a, b]$ be the solution. To obtain an approximate solution, we again choose functions $v_0, \ldots, v_r \in C^n[a, b]$ satisfying (I.2.27) and determine the a_j in (I.2.28) such that

$$L[v](t) - r(t) \begin{cases} \geqslant 0 \\ \leqslant 0 \end{cases} \qquad \text{for all } t \in [a, b].$$

From (I.2.31) it follows that

$$v(t) \geqslant \hat{y}(t) \text{ and } v(t) \leqslant \hat{y}(t) \qquad \text{for all } t \in [a, b]$$

and the form of the error estimate (I.2.29) again suggests choosing the a_j in (I.2.28) so that

$$\| L[v] - r \|_\infty = \max_{t \in [a,b]} \{L[v](t) - r(t)\}$$

and

$$\| L[v] - r \|_\infty = \max_{t \in [a,b]} \{r(t) - L[v](t)\}$$

respectively, are minimized. With f and w given by (I.2.30), there arises a one-sided uniform approximation problem of minimizing

$$\| w - f \|_\infty = \max_{t \in [a,b]} | w(t) - f(t) |$$

subject to the respective side conditions

$$w(t) - f(t) \geqslant 0 \text{ and } f(t) - w(t) \geqslant 0 \qquad \text{for all } t \in [a, b].$$

I.2.5 A linear control–approximation problem

Let $C[0, 1]^n$, $C[0, 1]^r$, respectively, be the vector space of vector-valued functions $x(t)$, $u(t)$, respectively, with values in $\mathbb{R}^n$, $\mathbb{R}^r$, and which are defined and continuous on $[0, 1]$. Further, let U be a (e.g. finite dimensional) subspace of $C[0, 1]^r$, let $A(t)$, $B(t)$, respectively, be continuous $n \times n$, $n \times r$, matrix functions on $[0, 1]$, let $r > 0$ be a given constant, let $x_0 \in \mathbb{R}^n$ and let $\hat{x} \in C[0, 1]^n$. Let $\| \ \|_\infty$ denote the maximum norm in $\mathbb{R}^n$, $\mathbb{R}^r$, respectively. We seek $x \in C[0, 1]^n$ such that $\max_{t \in [0,1]} \| x(t) - \hat{x}(t) \|_\infty$ is minimized under the conditions

$$\dot{x}(t) = \frac{dx}{dt}(t) = A(t)x(t) + B(t)u(t) \qquad \text{for } t \in [0, 1] \tag{I.2.32}$$

$$x(0) = x_0 \tag{I.2.33}$$

$$u \in U \text{ and } \| u(t) \|_\infty \leqslant \gamma \qquad \text{for all } t \in [0, 1]. \tag{I.2.34}$$

Physically this is a question of controlling a motion described by (I.2.32), and (I.2.33) with a fixed initial point $x_0 \in \mathbb{R}^n$ with the help of a control function u satisfying (I.2.34) so that the maximal deviation from a given trajectory $x \in C[0, 1]^n$ is as small as possible. As is well known, for every $u \in C[0, 1]^r$ there exists precisely one solution $x = x_u \in C[0, 1]^n$ of (I.2.32), (I.2.33), which is given by

$$x_u(t) = Y(t)\left(x_0 + \int_0^t Y(s)^{-1} B(s)u(s)\, ds\right). \tag{I.2.35}$$

Here $Y(t)$ is an $n \times n$ matrix function which is nonsingular for each $t \in [a, b]$ and

$$\dot{Y}(t) = A(t)Y(t) \qquad \text{for } t \in [0, 1] \tag{I.2.36}$$

$$Y(0) = I = n \times n \text{ unit matrix.} \tag{I.2.37}$$

$Y(t)$ is determined uniquely by (I.2.36) and (I.2.37); see, for example, Coddington and Levinson [55].

The problem (I.2.32)–(I.2.34) is therefore equivalent to the approximation problem of minimizing the deviation

$$\| x_u - \hat{x} \|_\infty = \max_{t \in [0,1]} \| x_u(t) - \hat{x}(t) \|_\infty$$

subject to the side conditions

$$u \in U, \| u(t) \|_\infty \leqslant \gamma \qquad \text{for all } t \in [0, 1]$$

where x_u is given by (I.2.35). If one defines $u = (u^1, \ldots, u^r)^T$ and $x = (x^1, \ldots, x^n)^T$, then it turns out that this problem is equivalent to the linear optimization problem of minimizing the linear functional $\psi(u, \delta) = \delta$ subject to the (infinitely many) side conditions

$$\begin{aligned} &u \in U, -\gamma \leqslant u^k(t) \leqslant +\gamma && \text{for all } t \in [0, 1], k = 1, \ldots, r \\ &-\delta \leqslant x_u^i(t) - \hat{x}^i(t) \leqslant +\delta && \text{for all } t \in [0, 1], i = 1, \ldots, n. \end{aligned}$$

In general, the function $Y(t)$ determined by (I.2.36)–(I.2.37), and thus also x_u, are not known explicitly so that one has to think of x_u as given only implicitly by (I.2.32)–(I.2.33).

I.3 THE GENERAL LINEAR OPTIMIZATION PROBLEM

I.3.1 Posing the problem; a weak duality theorem and simple consequences

Let E and F be two partially ordered normed vector spaces whose order relations we denote equivalently by '$\geqslant$' or '$\leqslant$'. Let K_E and K_F be the associated cones for the ordering (see Section IV.1.1). Let a continuous linear mapping $A : E \to F$, a continuous linear form $c : E \to \mathbb{R}$ (see Section IV.1.2), and a fixed element $b \in F$ be given.

Problem (P) Minimize $c(x)$ subject to the side conditions

$$\begin{aligned} &A(x) \geqslant b \qquad x \geqslant \Theta_E \\ &(\Leftrightarrow A(x) - b \in K_F, x \in K_E. \end{aligned} \tag{I.3.1}$$

A so-called 'dual' problem can be associated with this problem. For this we consider the mapping $A^* : F^* \to E^*$ which is adjoint to A (see Section IV.1.3) and define the following problem.

Problem (D) Maximize $\Phi_b(y^*) = y^*(b)$ subject to the side conditions

$$\begin{aligned} &A^*(y^*) \leqslant c \qquad y^* \geqslant \Theta_{F^*} \\ &(\Leftrightarrow A^*(y^*)(x) = y^*(A(x)) \leqslant c(x) \text{ for all } x \in K_E \text{ and } y^*(y) \geqslant 0 \text{ for all } y \in K_F). \end{aligned} \tag{I.3.2}$$

We set

$$M = \{x \in E : A(x) \geqslant b, x \geqslant \Theta_E\} \tag{I.3.3}$$

$$N = \{y^* \in F^* : A^*(y^*) \leqslant c, y^* \geqslant \Theta_{F^*} \tag{I.3.4}$$

and define

$$\alpha = \begin{cases} \inf\limits_{x \in M} c(x) & \text{if } M \text{ is non-empty}^1 \\ +\infty & \text{if } M \text{ is empty} \end{cases} \tag{I.3.5}$$

and

$$\beta = \begin{cases} \sup\limits_{y^* \in N} y^*(b) & \text{if } N \text{ is non-empty}^1 \\ -\infty & \text{if } N \text{ is empty.} \end{cases} \tag{I.3.6}$$

Theorem I.3.1 α, β *satisfy* $\beta \leqslant \alpha$ (I.3.7)

Proof If M or N are empty, the inequality (I.3.7) follows immediately from the definitions (I.3.5) and (I.3.6). If M and N are not empty, then for each $x \in M$ and $y^* \in N$ it follows that

$$y^*(b) \leqslant y^*(A(x)) = A^*(y^*)(x) \leqslant c(x).$$

If one keeps $y^* \in N$ fixed and varies x in M, then it follows that

$$y^*(b) \leqslant \alpha = \inf_{x \in M} c(x).$$

From this inequality, which holds for all $y^* \in N$, we conclude the validity of (I.3.7).

We call each $x \in M$ admissible, and each $y^* \in N$ admissible for the dual problem.

If, in addition, $c(x) = \alpha$, and $y^*(b) = \beta$, respectively, then $x \in M$, $y^* \in N$, respectively, are said to be optimal. From Theorem I.3.1 follows directly the next theorem.

Theorem I.3.2 If x, y^* *are admissible, respectively, for the dual problem, and if* $c(x) = y^*(b)$ *holds, then* x *and* y^* *are optimal and* $\beta = \alpha$.

The content of these two theorems is usually referred to as a weak duality theorem. In addition, we have the following Complementary Slackness Theorem.

Theorem I.3.3 Let $\hat{x} \in M$ *and* $y^* \in N$ *be given. The following two statements are equivalent:*

(*a*) $\hat{x}$ *and* $\hat{y}$ *are optimal and* $\beta = \alpha$

(*b*) $y^*(A(\hat{x}) - b) = 0$ *and* $(A^*(\hat{y}^*) - c)(\hat{x}) = 0$. (I.3.8)

Proof Suppose (*a*) is satisfied. Then

$$c(\hat{x}) = \hat{y}^*(b) \leqslant \hat{y}^*(A(\hat{x})) = A^*(\hat{y}^*)(\hat{x}) \leqslant c(\hat{x})$$

[1] The values $\alpha = -\infty$ and $\beta = +\infty$ are allowed.

from which (I.3.8) follows. Now let (b) be satisfied. Then

$$c(\hat{x}) = A^*(\hat{y}^*)(\hat{x}) = \hat{y}^*(A(\hat{x})) = \hat{y}^*(b)$$

and, by Theorem I.3.2, (a) follows.

I.3.2 The semi-infinite case

In numerous applications we have $E = \mathbb{R}^n$ equipped with some norm and partially ordered by means of the cone

$$K_E = K_r^n = \{x = (x_1, \ldots, x_n)^T : x_i \geqslant 0 \text{ for } i = 1, \ldots, r\}$$

where $0 \leqslant r \leqslant n$ and $K_0^n = \mathbb{R}^n$.

If $\{e_1, \ldots, e_n\}$ is any basis of $\mathbb{R}^n$ (for example, $e_j^i = \delta_{ij}$ the Kronecker delta for $i, j = 1, \ldots, n$), then every (continuous) linear mapping $A : \mathbb{R}^n \to F$ can be represented in the form

$$A(x) = \sum_{j=1}^{n} f_j x_j \qquad \text{for } x = (x_1, \ldots, x_n)^T$$

with $f_j = A(e_j) \in F, j = 1, \ldots, n$. By Section IV.1.2, every $c \in E^*$ can be represented in the form

$$c(x) = \sum_{j=1}^{n} c_j x_j \qquad \text{for } x = (x_1, \ldots, x_n)^T$$

with $c_j = c(e_j) \in \mathbb{R}, j = 1, \ldots, n$. With that, problem (P) reads: minimize

$$c(x) = \sum_{j=1}^{n} c_j x_j$$

subject to the side conditions

$$A(x) = \sum_{j=1}^{n} f_j x_j \geqslant b \qquad \text{for } x \in K_r^n. \tag{I.3.9}$$

Optimization problems of this kind are called semi-infinite. The mapping $A^* : F^* \to E^* (= \mathbb{R}^n)$ which is adjoint to A is defined by

$$A^*(y^*)(x) = y^*(A(x)) = \sum_{j=1}^{n} y^*(f_j) x_j \qquad \text{for all } y^* \in F \text{ and } x \in E.$$

The statement $A^*(y^*) \leqslant c$ is equivalent to

$$\sum_{j=1}^{n} \{y^*(f_j) - c_j\} x_j \leqslant 0 \qquad \text{for all } x \in K_r^n \tag{I.3.10}$$

i.e., for all $x \in \mathbb{R}^n$ with $x_i \geqslant 0, i = 1, \ldots, r\ (\leqslant n)$. The statement (I.3.10) is, however, equivalent to

$$y^*(f_j) \leqslant c_j \qquad \text{for } j = 1, \ldots, r,$$
$$y^*(f_j) = c_j \qquad \text{for } j = (r+1), \ldots, n$$

(Proof = Exercise).

For this reason, the dual problem (D) is equivalent to the problem of maximizing maximizing

$$\Phi_b(y^*) = y^*(b)$$

subject to the side conditions

$$\begin{aligned} &y^*(f_j) \leqslant c_j && \text{for } j = 1, \ldots, r \\ &y^*(f_j) = c_j && \text{for } j = (r+1), \ldots, n \\ &y^* \geqslant \Theta_{F^*} (\Leftrightarrow y^*(y) \geqslant 0 && \text{for all } y \in K_F). \end{aligned} \tag{I.3.11}$$

I.3.3 Semi-infinite problems in function spaces

As in Section I.3.2 let $E = \mathbb{R}^n$, equipped with some norm and partially ordered by means of the cone K_r^n, $0 \leqslant r \leqslant n$. Further, let M be a compact metric space and $F = C(M)$ be the vector space of continuous real-valued functions on M, equipped with the maximum norm (I.1.1) and the ordering relation

$$y \geqslant z \Leftrightarrow y(t) \geqslant z(t) \qquad \text{for all } t \in M. \tag{I.3.12}$$

Then the corresponding cone for F is given by

$$K_F = \{y \in C(M) : y(t) \geqslant 0 \qquad \text{for all } t \in M\} \tag{I.3.13}$$

and the associated semi-infinite problem (P) reads: minimize

$$\sum_{j=1}^{n} c_j x_j$$

subject to the side conditions

$$\begin{aligned} &\sum_{j=1}^{n} f_j(t) x_j \geqslant b(t) && \text{for all } t \in M \\ &x_j \geqslant 0 && \text{for } j = 1, \ldots, r. \end{aligned} \tag{I.3.9$'$}$$

If one chooses a finite number of points $t_1, \ldots, t_m \in M$ and defines for every $y \in F = C(M)$

$$y^*(y) = \sum_{i=1}^{m} y_i^* y(t_i) \tag{I.3.14}$$

whereby $y_1^*, \ldots, y_m^* \in \mathbb{R}$ are fixed non-negative numbers, then y^* is a linear functional on F. Because of

$$|y^*(y)| \leqslant \sum_{i=1}^{n} |y_i^*| \, \|y\|_\infty \qquad \text{for all } y \in F$$

from Theorem IV.1.2, y^* is continuous and consequently $y^* \in F$. Furthermore, we obviously have

$$y^*(y) \geqslant 0 \qquad \text{for all } y \in K_F, \text{ i.e. } y^* \geqslant \Theta_{F^*}.$$

For every $y^* \in F^*$ of the form (I.3.14) the side conditions (I.3.11) take the form

$$\sum_{i=1}^{m} y_i^* f_j(t_i) \leqslant c_j \qquad \text{for } j = 1, \ldots, r$$

$$\sum_{i=1}^{m} y_i^* f_j(t_i) = c_j \qquad \text{for } j = (r+1), \ldots, n \tag{I.3.11$'$}$$

$$y_i^* \geqslant 0, t_i \in M \qquad \text{for } i = 1, \ldots, m.$$

If the conditions (I.3.11$'$) are satisfied, then it follows from Theorem I.3.1 that

$$\sum_{i=1}^{m} y_i^* b(t_i) \leqslant \alpha = \inf_{x \in M} c(x).$$

If in addition there exists $x \in \mathbb{R}^n$ satisfying (I.3.9$'$) and

$$\sum_{i=1}^{m} y_i^* b(t_i) = \sum_{j=1}^{n} c_j x_j$$

then, by Theorem I.3.2, x and y defined by (I.3.14) are optimal.

The Complementary Slackness Theorem (Theorem I.3.3) yields the following assertion.

Corollary *Let $x \in \mathbb{R}^n$ with (I.3.9$'$) and $y^* \in F^*$ be given according to (I.3.14) with (I.3.11$'$); then the following two assertions are equivalent:*

(a) x and y^* are optimal and $\beta = \alpha$.
(b) The following two implications hold:

$$y_i^* > 0 \Rightarrow \sum_{j=1}^{n} f_j(t_i) x_j = b(t_i) \qquad (i = 1, \ldots, m) \tag{I.3.8$'a$}$$

$$x_j > 0 \Rightarrow \sum_{i=1}^{m} y_i^* f_j(t_i) = c_j \qquad (j = 1, \ldots, r). \tag{I.3.8$'b$}$$

The proof is a consequence of Theorem I.3.3 and the fact that in this special case the conditions (I.3.8) become

$$\sum_{i=1}^{m} y_i^* \left(\sum_{j=1}^{n} f_j(t_i) x_j - b(t_i) \right) = 0$$

and

$$\sum_{j=1}^{r}\left(\sum_{i=1}^{m} y_i^* f_j(t_i) - c_j\right) x_j = 0$$

which, however, are equivalent to (I.3.8′*a*) and (I.3.8′*b*).

We shall now elaborate the facts derived above by considering the following two examples.

I.3.3.1 Application to a boundary value problem for the potential equation

We consider as in Section I.2.3 the boundary value problem

$$\Delta u = 0 \text{ in } B, u = f \text{ on } \Gamma$$

with

$$B = \{(t, s) : |t| < 1, |s| < 1\}$$

and

$$\Gamma = \{(t, s) : |t| = 1, |s| \leqslant 1 \text{ or } |s| = 1, |t| \leqslant 1\}$$

$$f(t, \pm 1) = \cos(\tfrac{1}{2}\pi t) \qquad |t| \leqslant 1$$

$$f(\pm 1, s) = \cos(\tfrac{1}{2}\pi s) \qquad |s| \leqslant 1.$$

For the application of the boundary maximum principle described in Section I.2.3, we make the assumption

$$\varphi(x, t, s) = x_1 + (t^4 - 6t^2 s^2 + s^4)x_2 \qquad \text{for } x = (x_1, x_2) \tag{I.3.15}$$

which reflects the symmetry of the problem and for which, with an arbitrary choice of $x \in \mathbb{R}^2$, the equation

$$\Delta\varphi(x, t, s) = 0 \qquad \text{for all } (t, s) \in B$$

is valid. In order, say, to achieve $\varphi(x) \geqslant f$ on Γ, it obviously suffices, because of the symmetry of the problem and the assumption (I.3.15), to satisfy the inequality

$$\varphi(x, t, 1) = x_1 + (t^4 - 6t^2 + 1)x_2 \geqslant \cos(\tfrac{1}{2}\pi t) \qquad \text{for all } t \in M = [0, 1]. \tag{I.3.9″}$$

In order to find an optimal upper bound for the unknown solution $\hat{u} \in C(\bar{B}) \cap C^2(B)$ of the boundary value problem at the point $(\hat{t}, \hat{s}) = (0, 0)$ by the boundary maximum principle, according to Section I.2.3 we have to solve the following problem: minimize

$$\varphi(x, 0, 0) = x_1$$

subject to the side conditions (I.3.9″). Therefore we have a problem of the form (I.3.9′) with $r = 0$ and $c_1 = 1, c_2 = 0, n = 2$.

For $x_1 = 0.8, x_2 = 0.2$ the side conditions (I.3.9″) are fulfilled, which implies $\alpha \leqslant 0.8$.

On the other hand, if one chooses $t_1 = 0, t_2 = 1, y_1^* = 0.8, y_2^* = 0.2$, then, for $f_1(t) \equiv 1, f_2(t) = t^4 - 6t^2 + 1, t \in [0, 1]$, one has

$$\begin{aligned} &y_1^* f_1(t_1) + y_2^* f_1(t_2) = 1 = c_1 \\ &y_1^* f_2(t_1) + y_2^* f_2(t_2) = 0 = c_2 \\ &y_1^* \geqslant 0, y_2^* \geqslant 0, t_1, t_2 \in M. \end{aligned} \tag{I.3.11″}$$

Thus, $y^*(y) = y_1^* y(0) + y_2^* y(1), y \in C(M)$, is admissible for the dual problem which implies

$$y_1^* \cos(\tfrac{1}{2}\pi t_1) + y_2^* \cos(\tfrac{1}{2}\pi t_2) = 0.8 + 0.2 \cdot 0 = 0.8 \leqslant \alpha.$$

Hence, $\alpha = 0.8$ is an optimal upper bound of $\hat{u}(0, 0)$ with the assumption (I.3.15) and $x = (0.8, 0.2)$ is optimal.

Problem I.3.1 Show that if t_1 is the zero of the polynomial $t^4 - 6t^2 + 1$ in $[0, 1]$, then (I.3.15) and $\cos(\frac{1}{2}\pi t_1) \approx 0.7957$ yield an optimal lower bound for $\hat{u}(0, 0)$. In summary, therefore, we obtain the inclusion

$$0.7957 \leqslant \hat{u}(0, 0) \leqslant 0.8$$

and $x = (x_1, x_2)$ with

$$x_1 = \cos(\tfrac{1}{2}\pi t_1) \qquad x_2 = \tfrac{1}{8}\pi t_1^{-1}(3 - t_1^2)^{-1} \sin(\tfrac{1}{2}\pi t_1)$$

is optimal.

I.3.3.2 Application to a linear boundary value problem of monotonic type

Next we consider the linear boundary value problem (see Section I.2.4.2):

$$L[y](s) = -y''(s) + (1 + s^2)y(s) = s^2 \qquad \text{for } s \in [-1, +1] \tag{I.3.16}$$

$$y(-1) = y(+1) = 0. \tag{I.3.17}$$

This problem is of monotonic type (see Collatz [68], page 300) and, consequently, is uniquely solvable. For the functions

$$v_j(s) = 1 - s^{2j} \qquad j = 1, \ldots, p$$

the boundary conditions (I.3.17) are satisfied and, therefore, also for every linear combination

$$v(s) = \sum_{j=1}^{p} v_j(s) x_j \qquad \text{for } s \in [-1, +1].$$

For reasons of symmetry, we set $t = s^2$ and define

$$w_j(t) = L[v_j](s) \qquad j = 1, \ldots, p$$

for $t \in [0, 1]$. If one chooses the x_j such that

$$\sum_{j=1}^{p} w_j(t)x_j \geqslant t \qquad \text{for all } t \in [0, 1] \tag{I.3.18}$$

then, according to Section I.2.4.2, $v(s) \geqslant \hat{y}(s)$ for all $s \in [-1, +1]$, and $\hat{y} \in C^2[-1, 1]$ is the solution of the BVP (I.3.16)–(I.3.17). On the basis of the error estimate (I.2.29), it is reasonable to choose the x_j such that (I.3.18) is satisfied and

$$\max_{t \in [0,1]} \left(\sum_{j=1}^{p} w_j(t)x_j - t \right)$$

is minimized.

This problem is equivalent to the problem of minimizing the functional $c(x_1, \ldots, x_{p+1}) = x_{p+1}$ subject to the side conditions

$$\left.\begin{aligned} &\sum_{j=1}^{p} w_j(t)x_j \geqslant t \\ &-\sum_{j=1}^{p} w_j(t)x_j + x_{p+1} \geqslant -t. \end{aligned}\right\} \quad \text{for all } t \in [0, 1] \tag{I.3.9'''}$$

Setting:

$n = (p + 1), E = \mathbb{R}^n, K_E = K_0^n = \mathbb{R}^n, F = C[0, 1] \times C[0, 1]$

$K_F = \{(y_1, y_2)^T \in F : y_1(t) \geqslant 0 \text{ and } y_2(t) \geqslant 0 \text{ for all } t \in [0, 1]$

$f_j(t) = (w_j(t), -w_j(t))^T, j = 1, \ldots, p, f_n(t) = (0, 1)^T$

$b(t) = (t, -t)^T$ for all $t \in [0, 1]$

$c_1 = \cdots = c_p = 0, c_n = 1$

we have, therefore, a problem of the form (I.3.9) with $r = 0$. Now, if one chooses distinct points $t_1^1, \ldots, t_{m_1}^1, t_1^2, \ldots, t_{m_2}^2$ in $[0, 1]$ and corresponding numbers $y_1^1 \geqslant 0, \ldots, y_{m_1}^1 \geqslant 0, y_1^2 \geqslant 0, \ldots, y_{m_2}^2 \geqslant 0$, and defines for every $y = (y_1, y_2)^T \in F$,

$$y^*(y) = \sum_{i=1}^{m_1} y_i^1 y_1(t_i^1) + \sum_{i=1}^{m_2} y_i^2 y_2(t_i^2)$$

then $y^* \in F^*$, if one norms F, say, by $\| (y_1, y_2)^T \| = \| y_1 \|_\infty + \| y_2 \|_\infty$, $\| \; \|_\infty$ = maximum norm in $C[0, 1]$ and also $y^* \geqslant \Theta_{F^*}$. If, in addition, one chooses $t_i^1, t_i^2, y_i^1, y_i^2$ such that

$$\sum_{i=1}^{m_1} w_j(t_i^1)y_i^1 - \sum_{i=1}^{m_2} w_j(t_i^2)y_i^2 = 0 \qquad j = 1, \ldots, p \qquad \sum_{i=1}^{m_2} y_i^2 = 1 \tag{I.3.11'''}$$

then the side conditions (I.3.11) are satisfied and it follows from Theorem I.3.1 that

$$\sum_{i=1}^{m_1} t_i^1 y_i^1 - \sum_{i=1}^{m_2} t_i^2 y_i^2 \leqslant \alpha = p_\infty(f, W) \tag{I.3.19}$$

with

$$W = \left\{ w = \sum_{j=1}^{p} x_j w_j : w(t) \geqslant t \text{ for all } t \in [0, 1] \right\} \qquad f \equiv t$$

and

$$\rho_\infty(f, W) = \inf_{w \in W} \| w - f \|_\infty .$$

Example Let $p = 2$; then one obtains

$$w_1(t) = -t^2 + 3$$
$$w_2(t) = -t^3 - t^2 + 13t + 1.$$

If one chooses $x_1 = -\frac{1}{34}$ and $x_2 = \frac{3}{34}$, then one gets

$$w(t) - t = -\tfrac{1}{34} w_1(t) + \tfrac{3}{34} w_2(t) - t$$
$$= t(\tfrac{5}{34} - \tfrac{1}{17} t - \tfrac{3}{34} t^2) \geqslant 0 \qquad \text{for all } t \in [0, 1]$$

i.e. $\hat{w} \in W$. Further, $\hat{w}(t) - t = 0$ for $t = 0$ and $t = 1$ and also (with $f(t) = t$)

$$\hat{w}(t) - t = \| \hat{w} - f \|_\infty = \tfrac{200}{4131} \geqslant \rho_\infty(f, W) = \alpha \text{ for } t = \tfrac{5}{9}.$$

In order to be able to apply the condition (I.3.19), we choose $t_1^1 = 0$, $t_2^1 = 1$ ($m_1 = 2$) and $t_1^2 = \frac{5}{9}$ ($m_2 = 1$). The conditions (I.3.11″) then read

$$3y_1^1 + 2y_2^1 - \tfrac{218}{81} y_1^2 = 0$$
$$y_1^1 + 12y_2^1 - \tfrac{5644}{729} y_1^2 = 0$$
$$y_1^2 = 1$$

and are satisfied for $y_1^1 = \frac{1}{3}\,\frac{6128}{4131}$ and $y_2^1 = \frac{2495}{4131}$. Then it follows from (I.3.19) that

$$0\, y_1^1 + 1\, y_2^1 - \tfrac{5}{9} y_1^2 = \tfrac{2495}{4131} - \tfrac{5}{9} = \tfrac{200}{4131} \leqslant \alpha = \rho_\infty(f, W).$$

Therefore

$$\rho_\infty(f, W) = \tfrac{200}{4131} \approx 0.0485$$

and

$$\hat{w}(t) = -\tfrac{3}{34} t^3 - \tfrac{1}{17} t^2 + \tfrac{39}{34} t$$

is a minimal solution. Therefore, if one chooses $x_1 = -\frac{1}{34}$ and $x_2 = \frac{3}{34}$, then the conditions (I.3.18) are satisfied uniformly for $p = 2$ in an optimal way and, with the chosen functions v_j, lead to an optimal error estimate (I.2.29).

If one defines

$$\hat{v}(s) = -\tfrac{1}{34}(1 - s^2) + \tfrac{3}{34}(1 - s^4)$$

and

$$\hat{r}(s) = L[\hat{v}](s) = -\hat{v}''(s) + (1 + s^2)\hat{v}(s) \qquad \text{for } s \in [-1, +1]$$

then

$$\hat{v}(-1) = \hat{v}(+1) = 0$$

and

$$\max_{s \in [-1,+1]} |\hat{r}(s) - s^2| = \tfrac{200}{4131} \approx 0.0485.$$

If $\hat{y} \in C^2[-1, +1]$ is the (unknown) solution of the BVP (I.3.16)–(I.3.17), then it follows by Section I.2.4.1 that

$$\hat{y}(t) = \int_{-1}^{+1} G(t, s)[s^2 - (1 + s^2)\hat{y}(s)]\, ds$$

and

$$\hat{v}(t) = \int_{-1}^{+1} G(t, s)[\hat{r}(s) - (1 + s^2)\hat{v}(s)]\, ds \qquad t \in [-1, +1]$$

where $G = G(t, s)$ is the Green's function for the BVP

$$-y''(s) = r(s) \qquad s \in [-1, +1], y(-1) = y(+1) = 0$$

($r \in C[-1, +1]$). In this case, G is given by

$$G(t, s) = \begin{cases} \frac{1}{2}(s + 1)(1 - t) & \text{for } 1 \leqslant s \leqslant t \leqslant +1 \\ \frac{1}{2}(t + 1)(1 - s) & \text{for } -1 \leqslant t \leqslant s \leqslant +1. \end{cases}$$

Therefore, one has

$$\hat{y}(t) - \hat{v}(t) = \int_{-1}^{+1} G(t, s)[s^2 - \hat{r}(s)]\, ds - \int_{-1}^{+1} G(t, s)(1 + s^2)[\hat{y}(s) - \hat{v}(s)]\, ds$$

which implies

$$\|\hat{y} - \hat{v}\|_\infty = \max_{t \in [-1,+1]} \int_{-1}^{+1} G(t, s)\, ds \cdot \max_{t \in [-1,+1]} |t^2 - \hat{r}(t)|$$

$$+ \max_{t \in [-1,+1]} \int_{-1}^{+1} G(t, s)(1 + s^2)\, ds \, \|\hat{y} - \hat{v}\|_\infty.$$

Now

$$\max_{t \in [-1,+1]} \int_{-1}^{+1} G(t, s)\, ds = \tfrac{1}{2} \qquad \text{and} \qquad \max_{t \in [-1,+1]} \int_{-1}^{+1} G(t, s)(1 + s^2)\, ds = \tfrac{7}{12}$$

and thus

$$\|\hat{y} - \hat{v}\|_\infty \leqslant \tfrac{12}{5}\, \tfrac{1}{2}\, \tfrac{200}{4131} = \tfrac{6}{5}\, \tfrac{200}{4131} \leqslant 0.0581.$$

Problem I.3.2a Consider the (one-sided) approximation problem (in the manner of Section I.2.4.2) with respect to $f(t) = t^3$ and

$$V = \{v(t) = x_1 t^2 + x_2 t + x_3 : v(t) \geqslant f(t) \text{ for all } t \in [0, 1]\}.$$

By using the statement (I.3.19) show that $\hat{v}(t) = \frac{3}{2}t^2 - \frac{9}{16}$ is a minimal solution and

$$\rho_\infty(f, V) = \inf_{v \in V} \| v - f \|_\infty = \frac{1}{16}.$$

Hint Choose $t_1^1 = \frac{1}{4}, t_2^1 = 1$ $(m_1 = 2), t_1^2 = 0$ $t_2^2 = \frac{3}{4}$ $(m_2 = 2)$.

Problem I.3.2b Consider the approximation problem (in the manner of Section I.2.1.1) with respect to $f(t) = t^3$ and

$$V = \{v(t) = x_1 t^2 + x_2 t + x_3 : x_1, x_2, x_3 \in \mathbb{R}, t \in [0, 1]\}$$

and show analogously that $\hat{v}(t) = \frac{3}{2}t^2 - \frac{9}{16}t + \frac{1}{32}$ is a minimal solution and

$$\rho_\infty(f, V) = \frac{1}{32}.$$

Problem I.3.2c Consider the uniform approximation problem with respect to $f \in C(M)$, the vector space of continuous real-valued functions defined on a compact metric space M, and V, the n-dimensional linear subspace of $C(M)$ which is spanned by $v_1, \ldots, v_n \in C(M)$. Prove, by analogy to the derivation of the statement (I.3.19), the following assertion: If $t_1, \ldots, t_m \in M$ are given distinct points and $y_1, \ldots, y_m \in R$ are also given numbers such that

$$\sum_{i=1}^{m} v_j(t_i) y_i = 0 \qquad \text{for } j = 1, \ldots, n$$

$$\sum_{i=1}^{m} |y_i| \leqslant 1$$

then it follows that

$$\left| \sum_{i=1}^{m} y_i f(t_i) \right| \leqslant \rho_\infty(f, V). \tag{I.3.20}$$

Remark This statement can also be proven directly.

Problem I.3.2d Apply (I.3.20) to problem I.3.2*b*.

I.3.4 Counter-examples to the validity of general existence and duality statements

In order to guarantee the solvability of problems (P) and (D) with the side conditions (I.3.1) and (I.3.2), in general, it is not sufficient to assume that the sets M, N, respectively, defined by (I.3.3), (I.3.4), of the consistent elements for the dual are not empty as the following example of a semi-infinite problem shows.

I.3.4.1 Unsolvability of a semi-infinite problem

Let $E = \mathbb{R}^2$ and let it be equipped with some norm and the trivial partial ordering with the positive cone $K_0^2 = \mathbb{R}^2$. Further, let $B = [0, 1]$ and $F = C(B)$, equipped with the maximum norm (I.2.1) with B in place of M and the partial ordering

(I.3.12). Finally, let $b(t) = t$ for $t \in [0, 1]$, and $A : E \to F$ be defined by $A(x)(t) = t^2 x_1 + x_2$, for $t \in [0, 1]$, $x = (x_1, x_2)^T$.

As Problem (P) we consider the problem of minimizing the continuous linear functional $c(x) = 0 \cdot x_1 + x_2$ subject to the side conditions

$$x_1, x_2 \in \mathbb{R}, t^2 x_1 + x_2 \geqslant t \qquad \text{for all } t \in [0, 1].$$

One can easily see that the set M of consistent elements is non-empty and that $\alpha = 0$ (with α as in (I.3.5)) (Proof = Exercise). There is, however, no $\hat{x} \in M$ with $c(\hat{x}) = 0$, i.e. the problem is not solvable (see Figure I.3.1).

According to Section I.3.2, the dual problem is equivalent to the problem of minimizing the linear functional $y^*(b)$ subject to the side conditions

$$\begin{aligned} y^*(f_1) &= 0 \\ y^*(f_2) &= 1 \end{aligned} \qquad y^* \geqslant \Theta_{F^*}.$$

Here $f_1(t) = t^2$ and $f_2(t) = 1$ for all $t \in [0, 1]$.

If one defines

$$y^*(y) = y(0) \qquad \text{for all } y \in F = C[0, 1]$$

then $y^* \in F^*$, $y^* \geqslant \Theta_{F^*}$, $y^*(f_1) = 0$ and $y^*(f_2) = 1$. Thus y^* is consistent for the dual problem. Furthermore, $y^*(b) = 0 = \alpha$, from which it follows by Theorem I.3.1 that $y^*(b) = \beta = \alpha = 0$. The dual problem is, therefore, solvable and the extreme values of both problems coincide.

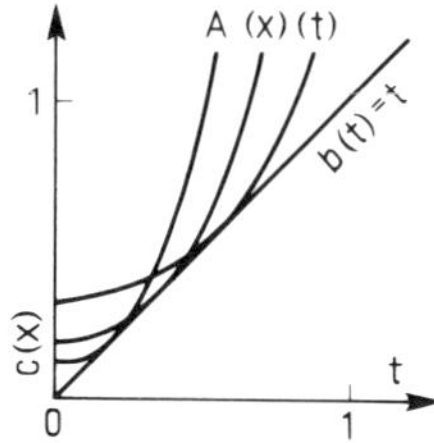

Figure I.3.1 Unsolvability of a semi-infinite problem

I.3.4.2 Occurrence of a duality gap

It is also possible that both problems are solvable and the extrema do not coincide. In this case one speaks of a duality gap, and we now give an example of such a case. Let E be the set of all infinite real sequences $x = \{x_n\}_{n=0,1,2,\ldots}$ in which only a finite number of terms x_n are non-zero. If one defines addition and scalar multiplication in E componentwise, then E becomes a linear vector space over $\mathbb{R}$. Further, let E be equipped with the norm

$$\| x \| = | x_0 | + \sum_{n=1}^{\infty} n \, | x_n |$$

and partially ordered by the relation $x \geqslant \hat{x} \Leftrightarrow x_n \geqslant \hat{x}_n$ for all n. The positive cone of E is then

$$K_E = \{x = \{x_n\}_{n=0,1,2,\ldots} \in E : x_n \geqslant 0 \text{ for all } n\}.$$

Let $F = \mathbb{R}^2$ be equipped with some norm and partially ordered by means of the positive cone $K_F = \{\Theta_2\}$. From this it follows that $K_{F^*} = F^*$ and for every $y^* \geqslant \Theta_{F^*}$ (i.e. $y^* \in K_{F^*}$) there is, by Section IV. 1.2 (Example 3) a unique $\hat{y} \in F$ with

$$y^*(y) = \langle \hat{y}, y \rangle = \hat{y}_1 y_1 + \hat{y}_2 y_2 \qquad \text{for all } y \in F.$$

Let $A : E \to F$ be defined by

$$A(x) = \begin{pmatrix} x_0 + \sum_{n=1}^{\infty} n x_n \\ \sum_{n=1}^{\infty} x_n \end{pmatrix} \qquad \text{for } x = \{x_n\}_{n=0,\,1,\,2,\dots} \in E.$$

A is linear and continuous (Proof = Exercise).

Further, let $b = (1, 0)^T$ and $c(x) = x_0, x \in E$. Problem (P) then reads: minimize $c(x) = x_0$ subject to the side conditions

$$\left.\begin{aligned} x_0 + \sum_{n=1}^{\infty} n x_n &= 1 \\ \sum_{n=1}^{\infty} x_n &= 0 \end{aligned}\right\} \quad \text{for } x = \{x_n\}_{n=0,\,1,\,2,\dots} \in E,\ x_n \geqslant 0 \text{ for all } n.$$

Obviously $x \in E$ is admissible if and only if

$$x_0 = 1 \text{ and } x_n = 0 \qquad \text{for all } n \geqslant 1.$$

Consequently $\alpha = 1$ and $c(x)$ equals α for this single $x \in M$, i.e. Problem (P) is solvable.

The adjoint mapping $A^* : F^* \to E$ is given by

$$\begin{aligned} A^*(y^*)(x) = y^*(A(x)) &= \hat{y}_1 \left\{ x_0 + \sum_{n=1}^{\infty} n x_n \right\} + \hat{y}_2 \left\{ \sum_{n=1}^{\infty} x_n \right\} \\ &= \hat{y}_1 x_0 + \sum_{n=1}^{\infty} (\hat{y}_1 n + \hat{y}_2) x_n. \end{aligned}$$

The statement

$$A^*(y^*) \leqslant c \quad y^* \geqslant \Theta_{F^*}$$

therefore is equivalent to

$$A^*(y^*)(x) = \hat{y}_1 x_0 + \sum_{n=1}^{\infty} (\hat{y}_1 n + \hat{y}_2) x_n \leqslant c(x) = x_0$$

$$\text{for all } \{x_n\}_{n=0,\,1,\,2,\dots} \in E \text{ with } x_n \geqslant 0 \text{ for all } n.$$

This statement in turn is equivalent to

$$\hat{y}_1 \leqslant 1 \quad \text{and} \quad \hat{y}_1 n + \hat{y}_2 \leqslant 0 \quad \text{for all } n \geqslant 1$$

which is equivalent to

$$\hat{y}_1 \leqslant 0 \quad \text{and} \quad \hat{y}_1 + \hat{y}_2 \leqslant 0.$$

The dual problem (D) is therefore equivalent to the problem of maximizing the linear functional $y^*(b) = \langle \hat{y}, b \rangle = \hat{y}_1$ subject to the side conditions

$$\hat{y}_1 \leqslant 0 \quad \text{and} \quad \hat{y}_1 + \hat{y}_2 \leqslant 0.$$

Obviously, $\beta = 0$ and $\hat{y}_1 = \hat{y}_2 = 0$ is a solution of problem (D). Thus, both problems are solvable but

$$\beta = 0 < 1 = \alpha.$$

As a further example of this state of affairs, we consider the following problem: let $E = L_2\,[0, 1] \times \mathbb{R}$. Here $L_2\,[0, 1]$ is the Hilbert space of equivalence classes of measurable and Lebesgue square-integral functions on $[0, 1]$ with the usual norm

$$\| f \|_2 = \left(\int_0^1 | f(t) |^2 \, dt \right)^{1/2}$$

and the partial ordering

$$f \geqslant g \Leftrightarrow f(t) \geqslant g(t) \quad \text{for almost all } t \in [0, 1].$$

Let E be normed by

$$\| (f, r) \| = \| f \|_2 + | r |$$

and partially ordered by

$$(f, r) \geqslant (g, s) \Leftrightarrow f \geqslant g \quad \text{and} \quad r \geqslant s.$$

Further, let $F = L_2\,[0, 1]$ be equipped with the above norm and partial ordering. For every $y^* \in F$ there exists, by Theorem IV.1.3, a unique $y \in F$ with

$$y^*(f) = \int_0^1 y(t) f(t) \, dt \quad \text{for all } f \in F.$$

Let $A : E \to F$ be defined by

$$A(f, r)(t) = \int_t^1 f(s) \, ds + r \quad \text{for } t \in [0, 1].$$

Then A is linear and continuous. Finally, let $b \equiv 1$ and

$$c(f, r) = \int_0^1 t f(t) \, dt + 2r.$$

Then c is a continuous linear functional, and for problem (P) we obtain the problem of minimizing

$$\int_0^1 t f(t) \, dt + 2r$$

subject to the side conditions

$$\int_t^1 f(s) \, ds + r \geqslant 1 \quad \text{for almost all } t \in [0, 1]$$

$$f(t) \geqslant 0 \qquad \text{for almost all } t \in [0, 1] \text{ and } r \geqslant 0.$$

Problem I.3.3a Show that $\alpha = 2$ and $(0, 1)$ is optimal.

Problem I.3.3b Show that the dual problem is equivalent to the problem: maximize the linear functional

$$\int_0^1 g(t)\,dt$$

subject to the side conditions

$$\int_0^t g(s)\,ds \leqslant t$$

$$g(t) \geqslant 0 \quad \text{for almost all } t \in [0, 1]$$

and

$$\int_0^1 g(t)\,dt \leqslant 2.$$

Problem I.3.3c Show that $\beta = 1$ and $g(t) = 1$ for all $t \in [0, 1]$ is optimal. Again both problems are solvable but

$$\beta = 1 < 2 = \alpha.$$

I.3.5 Bibliographical remarks

The last example in Section I.3.4.2, showing the appearance of a duality gap, is problably the first one which appears in the literature about infinite linear optimization problems. It is taken from a paper of Kretschmer [61] and is presented in detail in the book of Göpfert [73] on mathematical optimization in general vector spaces.

The first example in Section I.3.4.2 for the same type of situation was taken from a paper of van Slyke and Wets [68]. It is due to Gale and is a simplified variant of two examples for duality gaps to be found in Duffin and Karlovitz [65]. In one of these two examples, the problem (P) is solvable, but the dual problem (D) is not even consistent so that its extreme value is, by definition, $-\infty$.

It is characteristic of the examples for duality gaps mentioned until now that at least one of the two vector spaces is of infinite dimension. However, the same phenomenon can also occur if both spaces are of finite dimension, e.g. $E = \mathbb{R}^n$, $F = \mathbb{R}^n$. In such a case, one has to assume that at least one of the positive cones in E or F is infinitely generated, since for a finitely generated positive cone no duality gap can occur, by the fundamental theorem of ordinary linear programming.

Ky Fan [69], [70] gave a solvable problem (P) such that the dual problem is also solvable; however, the extremal values do not coincide. This problem is also described in the appendix of the book Göpfert [73] cited above.

The consistent but unsolvable semi-infinite optimization problem given in Section I.3.4.1, for which the dual problem is solvable and the extreme values coincide, is to be found in Krabs [68] and is due to Wetterling.

I.4 EXISTENCE AND DUALITY THEOREMS

I.4.1 Double dualization of an optimization problem

As in Section I.3.1 let E and F be two partially ordered linear vector spaces, $A : E \to F$ be a continuous linear mapping, $c : E \to \mathbb{R}$ be a continuous linear functional, and $b \in F$. The general linear optimization problem then consists of minimizing the linear functional $c(x)$ subject to the side conditions

$$A(x) \geqslant b \quad x \geqslant \Theta_E. \tag{I.4.1}$$

The associated dual problem (D) is then: maximize the linear functional

$$\Phi_b(y^*) = y^*(b)$$

subject to the side conditions

$$A^*(y^*) \leqslant c \quad y^* \geqslant \Theta_{F^*}. \tag{I.4.2}$$

Here $A^* : F^* \to E^*$ is the adjoint mapping of the topological space F^* dual to F into the topological space E^* dual to E. By Section IV.1.3, A^* is linear and continuous if one norms E^* and F^* in the natural way. The symbols '$\leqslant$', and '$\geqslant$' in (I.4.2) refer to the partial ordering induced in E^*, F^* respectively (see Section IV.(1.14). Obviously, the dual problem is equivalent to the problem of minimizing the linear form

$$-\Phi_b(y^*) = -y^*(b)$$

subject to the side conditions

$$-A^*(y^*) \geqslant -c \quad y^* \geqslant \Theta_{F^*}.$$

This problem again has the form of the original problem and therefore can be dualized. The problem arising in this way is obviously equivalent to the problem (P*) of minimizing the linear functional $c^{**}(x^{**})$ subject to the side conditions

$$A^{**}(x^{**}) \geqslant \Phi_b \quad x^{**} \geqslant \Theta_{E^{**}}. \tag{I.4.3}$$

Here $A^{**} = E^{**} \to F^{**}$ is the mapping which is the adjoint to $A^* : E^* \to E^*$, $c^{**} : E^{**} \to \mathbb{R}$ is defined by

$$c^{**}(x^{**}) = x^{**}(c) \quad \text{for all } x^{**} \in E^{**}$$

and the symbol '$\geqslant$' denotes the ordering induced in F^{**}, E^{**} respectively (see Section IV.2.2).

If, for an $x \in E$ the side conditions (I.4.1) are satisfied, then we define $x^{**} \in E^{**}$ by $x^{**} = \Phi_x$, that is, by

$$x^{**}(x^*) = x^*(x) \quad \text{for all } x^* \in E^*.$$

Then, since $x \geqslant \Theta_E$, it follows that

$$x^{**}(x^*) \geqslant 0 \text{ for all } x^* \geqslant \Theta_{E^*} \Rightarrow x^{**} \geqslant \Theta_{E^{**}}.$$

Furthermore

$$A^{**}(x^{**})(y^*) = x^{**}(A^*(y^*)) = \Phi_x(A^*(y^*))$$
$$= A^*(y^*)(x) = y^*(A(x)) \qquad \text{for all } y^* \in F$$

and, accordingly, since $A(x) \geqslant b$, one has

$$A^{**}(x^{**})(y^*) \geqslant y^*(b) = \Phi_b(y^*) \qquad \text{for all } y^* \geqslant \Theta_F$$

whence $A^{**}(x^{**}) \geqslant \Phi_b$. Finally, one obtains $c^{**}(x^{**}) = x^{**}(c) = c(x)$.

Result If $x \in E$ satisfies the side conditions (I.4.1), then $x^{**} = \Phi_x \in E^{**}$ satisfies the side conditions (I.4.3) and $c^{**}(x^{**}) = c(x)$.

The question arises whether for a given $x^{**} \in E^{**}$ with (I.4.3) there also exists an $x \in E$ satisfying (I.4.1) and $c^{**}(x^{**}) = c(x)$.

An answer to this question is contained in the following theorem.

Theorem 4.1 *Let E be reflexive and let the positive cones*

$$K_E = \{x \in E : x \geqslant \Theta_E\} \text{ and } K_F = \{y \in F : y \geqslant \Theta_F\}$$

in E, F respectively, be closed. If for an $x^{**} \in E^{**}$, the side conditions (I.4.3) are satisfied, then there is an $x \in E$ satisfying (I.4.1) and $c^{**}(x^{**}) = c(x)$. In other words: the problems (P) and (P) are equivalent and have the same extrema, i.e. by dualizing the dual (double dualization) of problem (P), one obtains again problem (P).*

Proof Let $x^{**} \in E^{**}$ satisfying (I.4.3) be given. Since E is reflexive, there is an $x \in E$ with $x^{**} = \Phi_x$ (see Section IV.1.2). It follows by (I.4.3) that

$$x^{**}(x^*) = x^*(x) \geqslant 0 \qquad \text{for all } x^* \geqslant \Theta_E.$$

Since K_E is closed, it follows that $x \in K_E$, i.e. $x \geqslant \Theta_E$ by Theorem IV.2.3. Moreover, by (I.4.3), one has

$$(A^{**}(x^{**}) - \Phi_b)(y^*) = y^*(A(x) - b) \geqslant 0 \qquad \text{for all } y^* \geqslant \Theta_F.$$

Since K_F is closed, it follows that $A(x) - b \geqslant \Theta_F$, i.e. $A(x) \geqslant b$ by Theorem IV.2.3. In summary, therefore, $x \in E$ satisfies the side conditions (I.4.1), and the fact that $c^{**}(x^{**}) = c(x)$ has been shown above.

I.4.2 Subconsistency and normality of an optimization problem

We introduce the cartesian product $\mathbb{R} \times F$ normed by $\|(\lambda, y)\| = |\lambda| + \|y\|$, $\lambda \in \mathbb{R}, y \in F$, and define in $\mathbb{R} \times F$ a convex cone (with $(0, \Theta_F)$ as the vertex) by

$$K(A, c) = \{(c(x) + r, A(x) - y) : r \geqslant 0, x \in K_E, y \in K_F\}. \tag{I.4.4}$$

Further, we define

$$L_b = \{(\alpha, b) : \alpha \in \mathbb{R}\} = \mathbb{R} \times \{b\}. \tag{I.4.5}$$

Obviously the set

$$M = \{x \in E : A(x) \geqslant b, x \geqslant \Theta_E\}$$

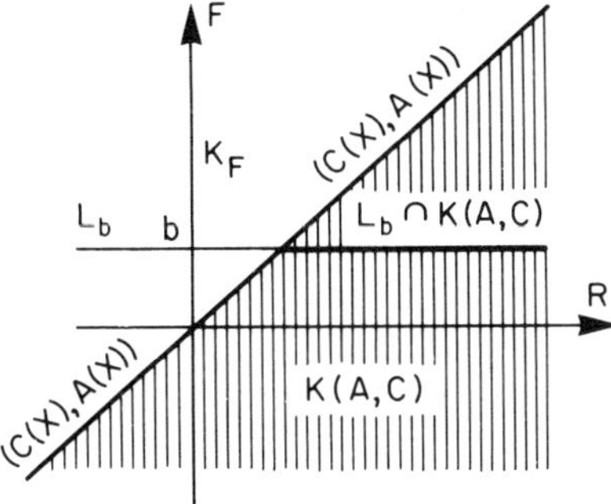

Figure I.4.1 The set $L_b \cap K(A, c)$

of the consistent elements of problem (P) is non-empty if and only if $L_b \cap K(A, c)$ is non-empty, and problem (P) is equivalent to the problem of finding an element $(\hat{\alpha}, \hat{z}) \in L_b \cap K(A, c)$ such that

$$\hat{\alpha} \leqslant \alpha \qquad \text{for all } (\alpha, z) \in L_b \cap K(A, c).$$

Stated geometrically, we seek the point of the set $L_b \cap K(A, c)$ in $\mathbb{R} \times F$ (see Figure I.4.1) lying farthest to the left. Therefore, we can define the extremal value of problem (P) as

$$v(A, c, b) = \begin{cases} \inf\limits_{(\alpha, b) \in K(A, c)} \alpha & \text{if } L_b \cap K(A, c) \neq \phi \\ +\infty & \text{if } L_b \cap K(A, c) = \phi. \end{cases} \tag{I.4.6}$$

Definition The problem (P) is called subconsistent if the intersection $L_b \cap \overline{K(A, c)}$ is not empty. Here $\overline{K(A, c)}$ is the closed hull of $K(A, c)$ in $\mathbb{R} \times F$.

We define the subvalue of the problem (P) by

$$v_s(A, c, b) = \begin{cases} \inf\limits_{(\alpha, b) \in \overline{K(A, c)}} \alpha & \text{if } L_b \cap \overline{K(A, c)} \neq \phi \\ +\infty & \text{if } L_b \cap \overline{K(A, c)} = \phi. \end{cases} \tag{I.4.7}$$

An immediate consequence of these definitions is the following lemma.

Lemma I.4.2 If the problem (P) is consistent, i.e. $L_b \cap K(A, c) \neq \phi$, then (P) is also subconsistent and the following inequalities hold:

$$v_s(a, c, b) \leqslant v(A, c, b) < +\infty. \tag{I.4.8}$$

Definition The problem (P) is called normal if

$$\overline{L_b \cap K(A, c)} = L_b \cap \overline{K(A, c)}. \tag{I.4.9}$$

Problem I.4.1 Show that if $K(A, c)$ is closed in $\mathbb{R} \times F$, then problem (P) is normal.

Lemma I.4.3 Let the problem (P) be normal. Then the following two assertions hold:
(a) (P) is consistent if and only if (P) is subconsistent.
(b) $v(a, c, b) = v_s(A, c, b)$.

Proof The proof of (a) follows directly from (I.4.9). The proof of (b) follows from

$$v(A, c, b) = \inf_{(\alpha, z) \in \overline{L_b \cap K(A, c)}} \alpha = \inf_{(\alpha, b) \in \overline{K(A, c)}} \alpha$$

in the case that $L_b \cap K(A, c) \neq \phi$ and is otherwise a direct consequence of the definitions (I.4.6) and (I.4.7).

If the problem (P) is not normal, then in general $v_s(A, c, b) < v(A, c, b)$ as the first example in Section I.3.4.2 shows. Here one has

$$K(A, c) = \left\{ \begin{pmatrix} x_0 + r \\ x_0 + \sum_{n=1}^{\infty} n x_n \\ \sum_{n=1}^{\infty} x_n \end{pmatrix} : \begin{array}{l} x_n \geqslant 0 \text{ for all } n = 0, 1, 2, \ldots, \\ x_n = 0 \text{ for almost all } n \\ \text{and } r \geqslant 0 \end{array} \right\}$$

$$L_b = \left\{ \begin{pmatrix} \alpha \\ 1 \\ 0 \end{pmatrix} : \alpha \in \mathbb{R} \right\}$$

and

$$L_b \cap K(A, c) = \left\{ \begin{pmatrix} 1 + r \\ 1 \\ 0 \end{pmatrix} : r \geqslant 0 \right\} = \overline{L_b \cap K(A, c)}.$$

One is easily convinced that

$$L_b \cap \overline{K(A, c)} = \left\{ \begin{pmatrix} r \\ 1 \\ 0 \end{pmatrix} : r \geqslant 0 \right\}$$

(Proof = Exercise), i.e. that

$$\overline{L_b \cap K(A, c)} \neq L_b \cap \overline{K(A, c)}.$$

Further (see Figure I.4.2) one has

$$v_s(A, c, b) = 0 < v(A, c, b) = 1.$$

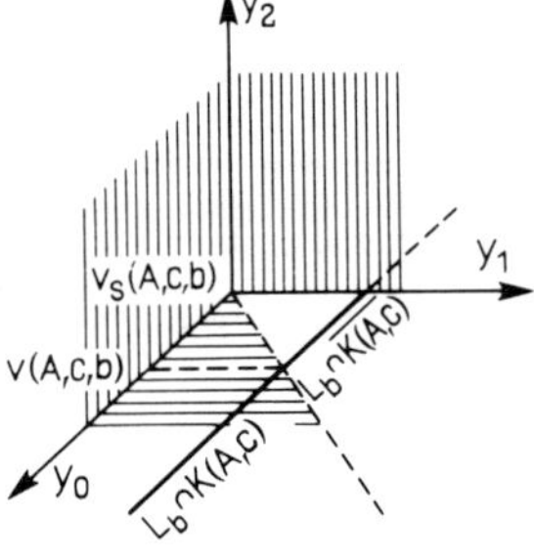

Figure I.4.2 Example of a non-normal problem

$$K(A, c) = \{(y_0, y_1, y_2) : y_0 \geqslant 0, y_1 \geqslant y_2 > 0\} \cup \{(x_0 + r, x_0, 0) : x_0, r \geqslant 0\}$$

$$\overline{K(A, c)} = \{(y_0, y_1, y_2) : y_0 \geqslant 0, y_1 \geqslant y_2 \geqslant 0\}$$

We call the dual problem consistent if

$$N = \{y^* \in F^* : A^*(y^*) \leqslant c, y^* \geqslant \Theta_{F^*}\}$$

is not empty and denote the associated extremal value by

$$v^*(A^*, b, c).$$

Lemma I.4.4 *If the problem (P) is subconsistent and the dual problem (D) is consistent, then one has*

$$-\infty < v^*(A^*, b, c) \leqslant v_s(A, c, b) < +\infty. \tag{I.4.10}$$

Proof Let $(\alpha, b) \in \overline{K(A, c)}$ and $y^* \in N$. Then there are sequences $\{r_n\}, r_n \geqslant 0$, $\{x_n\}, x_n \in K_E$ and $\{y_n\}, y_n \in K_F$ with

$$\alpha = \lim_{n\to\infty} \{c(x_n) + r_n\} \qquad b = \lim_{n\to\infty} \{A(x_n) - y_n\}.$$

Furthermore, for all n, the following inequalities hold

$$c(x_n) + r_n \geqslant c(x_n) \geqslant y^*(A(x_n)) \geqslant y^*(A(x_n) - y_n).$$

By the continuity of y^* it follows that $\alpha \geqslant y^*(b)$, which implies (I.4.10).

We shall 'sharpen' this result (see Theorem I.4.7). By combining Lemmas I.4.2 and I.4.4 we obtain the resulting lemma below.

Lemma I.4.5 *If the problems (P) and (D) are both consistent, then (P) is also subconsistent and*

$$-\infty < v^*(A^*, b, c) \leqslant v_s(A, c, b) \leqslant v(A, c, b) < +\infty.$$

I.4.3 General existence and duality theorems

First we prove the following theorem.

Theorem I.4.6 *Let N be non-empty, i.e. the dual problem (D) is consistent. Then for $(\alpha, b) \in \mathbb{R} \times F$ the assertion*

$$(\alpha, b) \in \overline{K(A, c)} \Leftrightarrow \alpha \geqslant v^*(A^*, b, c) \tag{I.4.11}$$

with

$$v^*(A^*, b, c) = \sup_{y^* \in N} y^*(b)$$

holds.

Proof

1. Let $(\alpha, b) \in \overline{K(A, c)}$.

For every $y^* \in N$ it follows then, as in the proof of Lemma I.4.4, that $\alpha \geqslant y^*(b)$, which yields $\alpha \geqslant v^*(A^*, b, c)$.

Remark For N empty the implication '⇒' is trivially true, since then $v^*(A^*, b, c) = -\infty$.

2. Let $(\alpha, b) \notin \overline{K(A, c)}$.
Since $\overline{K(A, c)}$ is a convex closed cone (Proof = Exercise), Theorem IV.2.3 and the representation formula (IV.1.18) (for $E = \mathbb{R}$) yield the existence of $(\lambda, y^*) \in \mathbb{R} \times F$ with

$$\lambda\alpha + y^*(b) < 0 \leqslant \lambda\beta + y^*(z) \qquad \text{for all } (\beta, z) \in \overline{K(A, c)}. \tag{I.4.12}$$

Therefore, in particular, one has

$$\lambda\{c(x) + r\} + y^*(A(x) - y) \geqslant 0 \qquad \text{for all } x \geqslant \Theta_E, r \geqslant 0, y \geqslant \Theta_F.$$

If one chooses, for example, $x = \Theta_E$ and $r = 0$, then it follows that $(-y^*)(y) \geqslant 0$ for all $y \geqslant \Theta_F$ implies $-y^* \geqslant \Theta_F$. If one chooses $x = \Theta_E$ and $y = \Theta_F$, then it follows that $\lambda r \geqslant 0$ for all $r \geqslant 0$ implies $\lambda \geqslant 0$.

We distinguish two cases: $(\alpha)\lambda = 0$, and $(\beta)\lambda > 0$. $(\alpha)\lambda = 0$. Then (for $y = \Theta_F) y^*(A(x)) \geqslant 0$ for all $x \geqslant \Theta_E$, and consequently $A^*(-y^*) \leqslant \Theta_E$. Since N is not empty, there exists $\hat{y}^* \geqslant \Theta_F$ with $A^*(\hat{y}^*) \leqslant c$. If one defines for every $\rho \geqslant 0$

$$y_\rho^* = \hat{y}^* - \rho y^*$$

then $y_\rho^* \geqslant \Theta_{F^*}$ and $A^*(y_\rho^*) \leqslant c$ implies $y_\rho^* \in N$ and $y_\rho^*(b) = \hat{y}^*(b) - \rho y^*(b)$ with $y^*(b) < 0$ (by (I.4.12)). Therefore, $v^*(A^*, b, c) \geqslant y_\rho^*(b)$ for all $\rho \geqslant 0$ and since $\lim_{\rho\to\infty} y_\rho^*(b) = +\infty$, then necessarily $v^*(A^*, b, c) = +\infty$, which implies $\alpha < v^*(A^*, b, c)$. (β) $\lambda > 0$. If one sets $\hat{y}^* = -y^*/\lambda$, then $\hat{y}^* \geqslant \Theta_{F^*}$ and $c(x) - \hat{y}^*(A(x)) \geqslant 0$ for all $x \geqslant \Theta_E$, so it follows that $A^*(\hat{y}^*) \leqslant c$, i.e. $\hat{y}^* \in N$, and from (I.4.12) it follows that $\alpha < \hat{y}^*(b) \leqslant v^*(A^*, b, c)$.

The decisive theorem of the theory is now given.

Theorem I.4.7a *The problem (P) is subconsistent and its subvalue $v_s(A, c, b)$ (see (I.4.7)) is finite if and only if the dual problem (D) is consistent and the dual extreme value $v^*(A^*, b, c)$ is finite. In both cases, one has*

$$-\infty < v_s(A, c, b) = v^*(A^*, b, c) < +\infty.$$

Theorem I.4.7b *If the problem (P) is not subconsistent and the dual problem (D) is consistent then it follows that*

$$v_s(A, c, b) = v^*(A^*, b, c) = +\infty.$$

Theorem I.4.7c *If the problem (P) is subconsistent and the dual problem (D) is not consistent then it follows that*

$$v_s(A, c, b) = v^*(A^*, b, c) = -\infty.$$

Proof The assertions (b) and (c) are immediate consequences of the assertion (a).

Proof of (a) is now given.

(α) Suppose the problem (D) is consistent and $v^*(A^*, b, c) < +\infty$. Then, according to Theorem I.4.6, $(v^*(A^*, b, c), b) \in \overline{K(A, c)}$, i.e. the problem (P) is subconsistent and therefore we obtain $v_s(A, c, b) \leqslant v^*(A^*, b, c)$. The equality of both values follows from Lemma I.4.4.

(β) Suppose the problem (P) is subconsistent and $v_s(A, c, b) > -\infty$. If one chooses $\beta < v_s(A, c, b)$ arbitrary, then $(\beta, b) \notin \overline{K(A, c)}$. By Theorem IV.2.3, and (IV.1.18) we obtain the existence of $(\lambda, y^*) \in \mathbb{R} \times F$ with

$$\lambda\beta + y^*(b) < 0 \leqslant \lambda\alpha + y^*(z) \qquad \text{for all } (\alpha, z) \in \overline{K(A, c)}. \tag{I.4.13}$$

Since $(v_s(A, c, b), b) \in \overline{K(A, c)}$, this yields

$$\lambda\{v_s(A, c, b) - \beta\} > 0 \Rightarrow \lambda > 0.$$

Furthermore, we have

$$\lambda\{c(x) + r\} + y^*(A(x) - y) \geqslant 0 \qquad \text{for all } x \geqslant \Theta_E, r \geqslant 0, y \geqslant \Theta_F.$$

From this we are led to conclude, as in the proof of Theorem I.4.6 (part 2, case (β)), that $\hat{y}^* = (-1/\lambda)y^* \in N$, i.e. the problem (D) is consistent. From (I.4.13) it follows further that

$$\beta < \hat{y}^*(b) \leqslant v^*(A^*, b, c).$$

Since $\beta < v_s(A, c, b)$ was chosen arbitrarily, then necessarily one has

$$v_s(A, c, b) \leqslant v^*(A^*, b, c).$$

Equality follows again from Lemma I.4.4.

Corollary *If the problem (P) is normal, then in Theorem I.4.1 one can replace the word subconsistent by consistent and the word subvalue of problem (P) by the word value in all assertions according to Lemma I.4.3.*

Existence Theorem *Suppose the convex cone $K(A, c)$ in (I.4.4) is closed. Then for every $b \in F$ the following assertions are valid:*
(a) Problem (P) is consistent and its value is finite if and only if the dual problem (D) is consistent and its value is finite. In both cases the problem (P) is solvable, i.e. the infimum is assumed and

$$-\infty < v(A, c, b) = v^*(A^*, b, c) < +\infty.$$

(b) If problem (P) is not consistent and the dual problem (D) is consistent, then one has

$$v(A, c, b) = v^*(A^*, b, c) = +\infty.$$

(c) If the problem (P) is consistent and the dual problem (D) is not consistent, then one has

$$v(A, c, b) = v^*(A^*, b, c) = -\infty.$$

Proof Since $K(A, c)$ is closed, problem (P) is normal and all assertions except the solvability of (P) in (*a*) are guaranteed by the corollary to Theorem I.4.7. Because of the fact that $K(A, c)$ is closed, however, we have

$$v(A, c, b) = \min_{(\alpha, b) \in K(A, c)} \alpha = \min_{x \in M} c(x)$$

when problem (P) is consistent and $v(A, c, b) > -\infty$.

For variable $b \in F$, the closedness of $K(A, c)$ is also characteristic for the solvability of the problem under the assumption that the dual problem (D) is consistent and its value is finite. In fact the following theorem is true.

Theorem I.4.8 *Suppose N is non-empty. Then the convex cone $K(A, c)$ defined by (I.4.4) is closed if and only if, for every $b \in F$ with $v^*(A, b, c) < \infty$, there exists an $x_b \in M$ such that $c(x_b) = v^*(A^*, b, c) = v(A, c, b)$.*

Proof

1. If $K(A, c)$ is closed, then for every $b \in F$ satisfying $v^*(A^*, b, c) < \infty$, the solvability of problem (P) and the equality of the extremal values $v^*(A, b, c)$ and $v(A, c, b)$ follow from the existence theorem.

2. For the proof in the other direction, we choose $(\alpha, b) \in \overline{K(A, c)}$. By Theorem I.4.6, one has $v^*(A^*, b, c) \leqslant \alpha < +\infty$. Therefore there is an $x_b \in M$ with $c(x_b) = v(A, c, b) = v^*(A^*, b, c)$.

Thus, $\alpha = c(x_b) + r$ with $x_b \geqslant \Theta_E$, $r \geqslant 0$ and $b = A(x_b) - y$ with $y \geqslant \Theta_F$, i.e. $(\alpha, b) \in K(A, c)$.

We shall elaborate this theorem with the example of Section I.3.4.1. Here $F = C[0, 1]$ and

$$K(A, c) = \left\{ \begin{array}{ll} (x_2 + r, f_1 x_1 + f_2 x_2 - y) : (x_1, x_2)^{\mathrm{T}} \in \mathbb{R}^2, r \geqslant 0, \\ y \geqslant \Theta_F (\Leftrightarrow y(t) \geqslant 0 & \text{for all } t \in [0, 1]) \end{array} \right\}$$

with $f_1(t) = t^2$ and $f_2(t) = 1$ for all $t \in [0, 1]$. Since the associated problem (P) for $b(t) = t$ for all $t \in [0, 1]$ does not have a solution, by Theorem I.4.7, $K(A, c)$ cannot be closed. In fact $(0, b) \notin K(A, c)$. Nevertheless, if one defines for every natural number n

$$x_1^n = \frac{(n-1)^2}{4n} \qquad x_2^n = \frac{1}{n} \qquad \text{and} \qquad r_n = 0$$

as well as

$$y^n(t) = t^2 x_1^n + x_2^n - \left(1 - \frac{1}{n}\right) t \qquad \text{for all } t \in [0, 1]$$

then

$$y^n(t) = \frac{1}{n} \left[\tfrac{1}{2}(n-1)t - 1\right]^2 \geqslant 0 \qquad \text{for all } t \in [0, 1] \text{ and all } n$$

that is

$$(x_2^n, f_1 x_1^n + f_2 x_2^n - y^n) \in K(A, c) \qquad \text{for all } n.$$

Furthermore, it follows from $x_2^n = 1/n$ and

$$f_1(t) x_1^n + f_2(t) x_2^n - y^n(t) = \left(1 - \frac{1}{n}\right) t$$

that

$$\lim_{n\to\infty} x_2^n = 0, \lim_{n\to\infty} f_1 x_1^n + f_2 x_2^n - y^n = b \Rightarrow (0, b) \in \overline{K(A, c)}.$$

One is easily convinced that

$$\overline{L_b \cap K(A, c)} = L_b \cap \overline{K(A, c)} = \{(r, b) : r \geqslant 0\}$$

i.e. in this case the problem (P) is normal. Hence, on the basis of the corollary to Theorem I.4.7, we also conclude that $v(A, c, b) = v^*(A^*, b, c) = 0$.

Theorem I.4.9 *Suppose N is non-empty. Then Problem (P) is normal if and only if the following implication is true:*

$$v^*(A^*, b, c) < +\infty \Rightarrow M \neq \phi \qquad \text{and} \qquad v(A, c, b) = v^*(A^*, b, c).$$

Problem I.4.2 Prove Theorem I.4.9.

I.4.4 Existence theorems for the dual problem

I.4.4.1 Dualization of the existence theorem in Section I.4.3

Making use of Theorem I.4.1 and the existence theorem above one comes easily to existence assertions for the dual problem as well. For this purpose we formulate this as a problem of minimizing the linear form $-\Phi_b(y^*) = -y^*(b)$ subject to the side conditions

$$-A^*(y^*) \geqslant -c \qquad y^* \geqslant \Theta_{F^*}.$$

Further we define, by analogy to $K(A, c)$ in (I.4.4), the convex cone

$$K(A^*, b) = \{(-y^*(b) + s, -A^*(y^*) - x^*) : y^* \geqslant \Theta_{F^*}, s \geqslant 0, x^* \geqslant \Theta_{E^*}\} \tag{I.4.14}$$

in $\mathbb{R} \times E^*$, equipped with the norm

$$\| (\alpha, x^*) \| = |\alpha| + \| x^* \| \qquad \text{with } \| x^* \| = \sup_{\|x\| \leqslant 1} |x^*(x)|.$$

From Theorem I.4.1 and the above existence theorem one obtains the following theorem.

Theorem I.4.10 *Suppose E is reflexive, the positive cones*

$$K_E = \{x \in E : x \geqslant \Theta_E\} \qquad \textit{and} \qquad K_F = \{y \in F : y \geqslant \Theta_F\}$$

of E and F respectively, are closed, and the convex cone $K(A^*, b)$ defined by (I.4.14) is also closed in $\mathbb{R} \times E^*$ (equipped with the above norm). Then the following assertions hold:

(a) The dual problem (D) is consistent and its value is finite if and only if the problem (P) is consistent and its value is finite. In both cases the dual problem (D)

is solvable and

$$-\infty < v^*(A^*, b, c) = v(A, c, b) < +\infty.$$

(b) and (c) are the same assertions as in the above existence theorem.
(Proof = Exercise.)

I.4.4.2 Application to the semi-infinite problem

Consider the semi-infinite problem (P) of minimizing the linear form

$$c(x) = \sum_{j=1}^{n} c_j x_j$$

subject to the side conditions (I.3.9)

$$A(x) = \sum_{j=1}^{n} f_j x_j \geqslant b \qquad \text{for } x \in K_r^n. \tag{I.4.15}$$

Here $E = \mathbb{R}^n$ is reflexive and $K_E = K_r^n$ (see (IV.1.7′)) is closed. The dual problem according to Section I.3.2 is equivalent to the problem of minimizing the linear form $-y^*(b)$ subject to the side conditions (I.3.11)

$$\begin{aligned} -y^*(f_j) &\geqslant -c_j \qquad j = 1, \ldots, r \\ -y^*(f_j) &= -c_j \qquad j = (r+1), \ldots, n \\ y^* &\geqslant \Theta_{F^*}. \end{aligned}$$

$K(A^*, b) \subseteq \mathbb{R}^{n+1}$ is given by

$$\begin{aligned} K(A^*, b) = \{(-y^*(b) + s_0, -y^*(f_1) - s_1, \ldots, -y^*(f_r) - s_r, \\ -y^*(f_{r+1}), \ldots, -y^*(f_n)) : s_0 \geqslant 0, s_1 \geqslant 0, \ldots, s_r \geqslant 0, \\ y^* \geqslant \Theta_{F^*}\}. \end{aligned}$$

If one defines $f_0 = b$, then one can also represent $K(A^*, b)$ in the form

$$K(A^*, b) = K^* - K \tag{I.4.16}$$

with

$$K^* = \{(-y^*(f_0), \ldots, -y^*(f_n)) : y^* \geqslant \Theta_{F^*}\} \tag{I.4.17}$$

and

$$K = \{x \in \mathbb{R}^{n+1} : x_0 \leqslant 0, x_1 \geqslant 0, \ldots, x_r \geqslant 0, x_{r+1} = \cdots = x_n = 0\}. \tag{I.4.18}$$

K is obviously a closed convex cone in $\mathbb{R}^{n+1}$.

Lemma I.4.11 *Let the interior* $\mathring{K}_F$ *of the positive cone* K_F *of* F *be non-empty and suppose there is an* $\hat{x} \in K_r^n$ *with*

$$\hat{f} = \sum_{j=1}^{n} f_j \hat{x}_j - b \in \mathring{K}_F. \tag{I.4.19}$$

Then according to (I.4.17) for K^* and (I.4.18) for K it follows that

$$K^* \cap K = \{\Theta_{n+1}\}.$$

Proof Suppose $k \in K^* \cap K$. Then k has the form

$$k = (-y^*(f_0), \ldots, -y^*(f_n)) \qquad \text{for a } y^* \geqslant \Theta_{F^*}$$

and

$$-y^*(f_0) = s_0 \leqslant 0$$
$$-y^*(f_j) = s_j \geqslant 0 \qquad \text{for } j = 1, \ldots, r$$
$$-y^*(f_j) = 0 \qquad \text{for } j = (r+1), \ldots, n.$$

It follows that

$$y^*(\hat{f}) = s_0 - \sum_{j=1}^{r} \hat{x}_j s_j \leqslant 0$$

with $\hat{f}$ defined according to (I.4.19). From Theorem IV.1.5, therefore we obtain $y^* = \Theta_{F^*}$, which implies $k = \Theta_{n+1}$.

Lemma I.4.12. Suppose the interior $\mathring{K}_F$ of the positive cone K_F of F is non-empty and there is an $\hat{x} \in K_r^n$ satisfying (I.4.19). Suppose further that the cone K^* in $\mathbb{R}^{n+1}$ given by (I.4.17) is closed. Then the cone $K(A^*, b)$ defined by (I.4.16) is a closed cone in $\mathbb{R}^{n+1}$.

Proof The assertion is an immediate consequence of Lemma I.4.11 and Theorem IV.2.2.

Theorem I.4.10 and Lemma I.4.12 yield an existence theorem for the dual problem of the semi-infinite problem subject to the side conditions (I.3.9) = (I.4.15), in fact one can state this as follows.

Theorem I.4.13 Suppose the assumptions of Lemma I.4.12 are satisfied and suppose further that the positive cone K_F of F is closed. If the value of problem (P) subject to the side conditions (I.4.15) is finite, then the dual problem subject to the side conditions (I.3.11) is solvable and the values of both problems coincide.

I.4.4.3 A direct existence theorem for the dual problem

Theorem I.4.13 is a special case of the following general existence for the dual problem subject to the side conditions (I.4.2).

Theorem I.4.14 Suppose that the interior $\mathring{K}_F$ of the positive cone K_F is non-empty and suppose there is an $\hat{x} \in K_E$ with $A(\hat{x}) - b \in \mathring{K}_F$. If then the value of the problem (P) subject to the side conditions (I.4.1) is finite, then the dual problem (D) subject to the side conditions (I.4.2) is solvable and the values of both problems coincide.

Proof We consider the cone $K(A, c)$ after the manner of (I.4.4), which one can also represent as follows:

$$K(A, c) = \bigcup_{x \in K_E} \{(c(x) + r, A(x) - y) : r \geqslant 0, y \in K_F\}.$$

This cone contains the open convex cone

$$K_0 = \bigcup_{x \in K_E} \{(c(x) + r, A(x) - y) : r > 0, y \in \mathring{K}_F\}$$

and $K(A, c) \subseteq \overline{K}_0$.

In fact if $(c(x) + r, A(x) - y) \in K(A, c)$ is given, then we choose $r_0 > 0$ and $y_0 \in \mathring{K}_F$ arbitrarily. Then, for all $\lambda \in [0, 1]$, one has

$$r_\lambda = \lambda r + (1 - \lambda) r_0 > 0 \qquad \text{and} \qquad y_\lambda = \lambda y + (1 - \lambda) y_0 \in \mathring{K}_F$$

by Theorem IV.3.3, and hence

$$(c(x) + r_\lambda, A(x) - y_\lambda) \in K_0$$

as well as

$$(c(x) + r, A(x) - y) = \lim_{\lambda \to 1} (c(x) + r_\lambda, A(x) - y_\lambda).$$

Obviously $(v(P), b) \notin K_0$, where $v(P) = v(A, c, b)$; for otherwise there would be an $x \in K_E$ with $A(x) - b \in \mathring{K}_F$ and $c(x) < v(P)$, a contradiction to the definition of $v(P)$. By applying the separation theorem 1 in Section IV.3.2. (for $A = \{(v(P), b)\}$ and $B = \mathring{B} = K_0$) and the representation formula (IV.1.18), we conclude therefore the existence of $(\lambda, y^*) \in \mathbb{R} \times F^*$ and $\rho \in \mathbb{R}$ with

$$\lambda v(P) + y^*(b) \leqslant \rho < \lambda\alpha + y^*(z) \qquad \text{for all } (\alpha, z) \in K_0$$

which, because of $K(A, c) \subseteq \overline{K}_0$, implies

$$\lambda v(P) + y^*(b) \leqslant \rho \leqslant \lambda\{c(x) + r\} + y^*(A(x) - y) \tag{I.4.20}$$

$$\text{for all } x \in K_E, r \geqslant 0 \text{ and } y \in K_F.$$

Further, since $\lambda \geqslant 0$, also $-y^* = \hat{y}^* \in K_{F^*}$. If λ were 0, then for $x = \hat{x}$ and $y = \Theta_F$ in (I.4.20) we would have $\hat{y}^*(A(\hat{x}) - b) \leqslant 0$ and thus $\hat{y}^* = \Theta_{F^*}$ by Theorem IV.1.5, from which would follow $y^* = \Theta_{F^*}$ and that together with $\lambda = 0$ is impossible.

Therefore we can assume without loss of generality that $\lambda = 1$. If one chooses $r = 0$ and $y = \Theta_F$ in (I.4.20), then it follows that

$$\rho \leqslant c(x) - \hat{y}^*(A(x)) = (c - A^*(\hat{y}^*))(x) \qquad \text{for all } x \in K_E$$

which has the consequence $A^*(\hat{y}^*) \leqslant c$.

Now $\hat{y}^*$ is consistent in the sense of the dual which, by Theorem I.3.1, implies $\hat{y}^*(b) \leqslant v(P)$. If one chooses $x = \Theta_E$, $y = \Theta_F$, and $r = 0$ in (I.4.20), in then it follows that

$$v(P) + y^*(b) \leqslant 0 \Rightarrow v(P) \leqslant -y^*(b) = \hat{y}^*(b)$$

and hence $\hat{y}^*(b) = v^*(A^*, b, c)$, i.e. $\hat{y}^*$ is a solution of problem (D).

I.4.5 Bibliographical remarks

The first contributions to the theory of infinite, linear optimization stem from Bratton [55], Duffin [56], Hurwicz [58], and Kretschmer [61]. The paper of Bratton appears only in the form of a Cowles Commission Discussion Paper and is therefore not generally accessible. Duffin and Kretschmer, from whom also the second example in Section I.3.4.2 for a duality gap stems, take as their basis a symmetric formulation for the original problem, by starting from two dual pairs (U, U^*), (V, V^*) of locally convex topological vector spaces U and V and the associated dual spaces U^* and V^* and consider a linear mapping A of U^* into V. U and V are again partially ordered by convex cones, whereby partial orderings are induced also in U^* and V^*. For given $b \in V$ and $a \in U$, the problem of minimizing the linear form $(a, x) = x(a)$ subject to the side conditions $x \geqslant \Theta_{U^*}, Ax \geqslant b$ is considered. To the mapping A, an adjoint mapping $A^* : V^* \to U$ is associated via the identity

$$v^*(Au^*) = (Au^*, v^*) = (A^*v^*, u^*) = u^*(A^*v^*)$$

for all $u^* \in U^*$ and $v^* \in V^*$. Taking $a^* = -b, b^* = -a$ the dual problem is considered to be that of minimizing the linear form $(a^*, y) = y(a^*)$ subject to the side conditions $y \geqslant \Theta_{V^*} A^*y \geqslant b^*$. This formulation has the advantage of complete symmetry so that the starting problem can also be considered as the dual problem for the dual problem. For applications, this formulation is not always advantageous, because one must choose the domain of definition E for the original problem immediately as the dual space U^* of a topological vector space U.

Duffin [56] also introduces the concept of subconsistency and proves as the first main result of his paper the analogue to Theorem I.4.7 in his terminology. The existence of a point $x \geqslant \Theta_{U^*}$ with $(Ax - b)$ in the interior of the positive cone of V, Duffin calls superconsistency of the problem (P) and proves the following analogue to Theorem I.4.14: if the dual problem is superconsistent and has a finite extremal value, then the problem is solvable and the sum of the extremal values is equal to zero (which corresponds to the equality of the extremal values in our formulation; see Theorem I.4.14).

For the case that U and V are Hilbert spaces, Göpfert [73] also takes up the Duffin theory in his book. He denotes subconsistency as asymptotic consistency.

Hurwicz [58] in his long paper pursues among other topics a generalization of a lemma of Farkas [02] which is basic to the theory of ordinary linear optimization. He begins for this with a topological vector space X, a partially ordered locally convex topological vector space Y, and a continuous linear mapping $T : X \to Y$. Let $T^* : Y^* \to X^*$ be the adjoint mapping of the dual space Y^* of Y, equipped with the induced partial ordering, into the dual space X^* of X. Then for the sets

$$Z_T = \{x^* \in X^* : x^* = T^*(y^*), y^* \geqslant \Theta_{Y^*}\}$$

and

$$V_T = \{x^* \in X^* : T(x) \geqslant \Theta_Y, x \in X \Rightarrow x^*(x) \geqslant 0\}$$

we get the relation $Z_T \subseteq V_T$. Hurwicz gives a necessary and sufficient condition for the equality $Z_T = V_T$, which in the case $X = \mathbb{R}^n$, $Y = \mathbb{R}^m$, equipped with the componentwise ordering, is precisely the statement of the classical Farkas Lemma. Similarly to the case of the Farkas Lemma, one can use its generalization $Z_T = V_T$ to prove general existence and duality theorems.

A somewhat different generalization is given by Ben-Israel and Charnes [68] who, with Kortanek [69], make a consequence of this the basis of a general linear optimization theory. This consequence in the finite dimensional case is the following. Let C be a closed convex cone in $\mathbb{R}$, A be an $m \times n$ matrix, and $b \in \mathbb{R}^m$. Then the following two statements are equivalent:
(a) there exists a sequence $\{x_k\}$, $x_k \in C$ with $\lim_{k\to\infty} Ax_k = b$
(b) if $A^{\mathrm{T}}y \in C^*$, $y \in \mathbb{R}^m$, then $b^T y \geqslant 0$, whereby $C^* = \{y \in \mathbb{R}^n : x \in C \Rightarrow y^{\mathrm{T}}x \geqslant 0\}$ is the cone adjoint to C.

For $C = \mathbb{R}^n$, the equivalence of (a) and (b) is again a restatement of the Farkas Lemma. Making use of this equivalence, Ben-Israel, Charnes and Kortanek in [69] give a complete classification of all relationships which can exist using the concepts of boundedness, unboundedness, asymptotic boundedness, asymptotic unboundedness, improper asymptotic consistency and strict and asymptotic inconsistency between a linear optimization problem and its dual. In this, all phenomena of infinite optimization already occur in finite dimensional spaces, if one uses arbitrary convex cones as positive cones.

The classification just referred to was then taken up by Kallina and Williams [71] in a large survey article and further expanded upon. They call an optimization problem normal if its (extremal) value is finite and coincides with its sub-value (see here Lemma I.4.3). They relate this concept of normality, which is equivalent with the non-appearance of duality gaps, to the concept of stability of convex optimization problems introduced by Rockafellar [67].

The classification scheme given by Ben-Israel, Charnes and Kortanek [69] was also demonstrated again by Gustafson, Kortanek and Rom [70] by way of example on certain moment problems, whereby it is here a matter of dealing with semi-infinite linear optimization problems in function spaces. Parallel to the developments discussed so far, ran the investigation of semi-infinite optimization problems which, going back to a paper of Haar [24], was begun by Charnes, Cooper and Kortanek [63]. Their results, which already contained a correction of the original paper of Haar, were then expanded upon and corrected in the article of Duffin and Karlovitz [65] and were then further developed by themselves in [65]. At the centre of all these investigations also stands a generalization of the classical Farkas Lemma to infinite systems of linear inequalities which had first been given incorrectly by Haar and then in corrected form by Duffin and Karlovitz [65]. A special setting, which often occurs in applications, is found in Krabs [68]. There are also existence and duality assertions to be found, which are based on the closedness of certain subcones of the cone $K(A, c)$ (I.4.4) and $K(A^*, b)$ (I.4.14) (see the existence theorem in Section I.4.3 and Theorem I.4.10). The closedness of these cones were later characterized by Krabs [71], whereby somewhat more complicated statements than Theorem I.4.8 came out. The existence and duality assertions mentioned with respect to $K(A, c)$ and $K(A^*, b)$ are also to be found in Dieter [66]. There it is also shown that for the example of Kretschmer of Section I.3.4.2 (see also Section I.3.5) for a duality gap, the cones are not closed. Based on these investigations, Göpfert [73] shows that in this example the duality gap does not appear if one considers the subvalue instead of the (extremal) value. A presentation of the results of Dieter can also be found in Sander [73].

Vershik [70] points out that for duality statements to hold the proper choice of the spaces E and F and the ordering cones in the problem (P) of Section I.3.1 is important. He also indicates how the problem (P) and its dual problem (D) can be approximated in a natural way by problems of finite dimensions.

The investigations of infinite linear optimization problems referred to until now all take place in locally convex topological vector spaces and use the separation theorems for convex sets which are valid in these spaces. Now such separation theorems can also be proven if one leaves out the topology and instead uses suitable algebraic concepts. Such separation theorems have been given by Klee [69] and have been used by Lempio [71a] for the construction of a general nonlinear optimization theory. They were also used especially for linear problems by Lempio [71b] in order to obtain a general maximum principle and duality assertions for linear optimization problems in infinite dimensional vector spaces. In closing, we wish to mention another interesting class of linear optimization problems which occur in the so-called bottle-neck process and have been presented in detail by Göpfert [73]. We refer also to the special literature given there, which is connected with the names of Tyndall, Levinson and Grinold. Recently these investigations were taken up again by Schechter [72] and treated in the realm of a duality theory (Schechter [73]) in which one no longer operates explicitly with linear mappings but only with convex cones and their adjoints.

I.5 APPLICATIONS

I.5.1 Semi-infinite problems in function spaces

As in Section I.3.3 we consider the problem (P): minimize

$$\sum_{j=1}^{n} c_j x_j$$

subject to the side conditions

$$\sum_{j=1}^{n} f_j(t) x_j \geqslant b(t) \qquad \text{for all } t \in M \tag{I.5.1}$$

$$x_j \geqslant 0 \qquad \text{for } j = 1, \ldots, r (0 \leqslant r \leqslant n). \tag{I.5.2}$$

Here $f_1, \ldots, f_n, b \in C(M)$ are given functions and M is a compact metric space. $C(M)$ is equipped with the maximum norm and partially ordered in the natural way.

In problem (P) nothing is changed if instead of $C(M)$ we consider the finite dimensional subspace V of $C(M)$ spanned by the functions $f_1, \ldots, f_n, b$. V is normed in the same way and partially ordered as $C(M)$.

According to Section I.3.2 the problem dual to (P) is equivalent to problem (D): maximize the linear functional

$$\Phi_b(y^*) = y^*(b)$$

subject to the side conditions

$$\begin{aligned} &y^*(f_j) \leqslant c_j && \text{for } j = 1, \ldots, r \\ &y^*(f_j) = c_j && \text{for } j = (r+1), \ldots, n \\ &y^* \geqslant \Theta_{V^*} \end{aligned}$$

Assumption There exists an $\hat{x} \in \mathbb{R}^n$ with (I.5.2) and

$$\sum_{j=1}^{n} f_j(t)\hat{x}_j > b(t) \qquad \text{for all } t \in M. \tag{I.5.3}$$

By (I.5.3) the condition (IV. 2.16) is implied for V. On the basis of the corollary to Theorem IV.2.6, the problem (D) is therefore equivalent to the problem: maximize

$$y^*(b) = \sum_{i=1}^{m} y_i^* b(t_i)$$

subject to the side conditions

$$\begin{aligned} &\sum_{i=1}^{m} y_i^* f_j(t_i) \leqslant c_j && \text{for } j = 1, \ldots, r \\ &\sum_{i=1}^{m} y_i^* f_j(t_i) = c_j && \text{for } j = (r+1), \ldots, n \\ &y_i^* \geqslant 0 && \text{for } i = 1, \ldots, m, \{t_1, \ldots, t_m\} \subseteq M \end{aligned} \tag{I.5.4}$$

where n runs through the natural numbers. For the case $r = 0$, we shall show (see Theorem I.5.2) that one can restrict oneself to the case $m \leqslant n$.

By applying Theorem I.4.14 one obtains immediately the following theorem.

Theorem I.5.1 *If there is an $\hat{x} \in \mathbb{R}^n$ with $\hat{x}_j \geqslant 0$ for $j = 1, \ldots, r$ and (I.5.3) is satisfied and if the infimum α of the semi-infinite problem (P) is finite, then the dual problem (D) satisfying the side conditions (I.5.4) is solvable, and for every optimal solution $(y_i^*, t_i)_{i=1,\ldots,m}$ of (I.5.4) we have*

$$\sum_{i=1}^{m} y_i^* b(t_i) = \alpha.$$

Under the hypothesis (I.5.3) the method in Section I.3.3 therefore furnishes optimal lower bounds which coincide with α by means of the solution of (I.3.11′).

The example in Section I.3.4.1 shows that problem (P) is not necessarily solvable, even if the assumption (I.5.3) is satisfied and the infimum is finite. If the problem

(P) is solvable, then the corollary in Section I.3.3 to the complementary slackness theorem (Theorem I.3.3) subject to the assumption (I.5.3) may be sharpened: let $x \in \mathbb{R}^n$ be given satisfying (I.5.1), (I.5.2) and $(y_i^*, t_i)_{i=1,\ldots,m}$ with (I.5.4). Then the following two statements are equivalent:

(*a*) x and (y_i^*, t_i) are optimal,

(*b*) the following implications are valid:

$$y_i^* > 0 \text{ for some } i = 1, \ldots, m \Rightarrow \sum_{j=1}^{n} f_j(t_i)x_j = b(t_i) \tag{I.5.5a}$$

$$x_j > 0 \text{ for some } j = 1, \ldots, r \Rightarrow \sum_{i=1}^{m} y_i^* f_j(t_i) = c_j. \tag{I.5.5b}$$

In statement (*a*) one does not have to demand the equality of the extremal values in addition to the other assumptions if the assumption (I.5.3) is fulfilled.

We consider now the case $r = 0$. Then the inequalities in (I.5.4) disappear. Furthermore we have another theorem.

Theorem I.5.2 *If $r = 0$ and if there is a solution $(y_i^*, t_i)_{i=1,\ldots,m}$ of (I.5.4), then there is a subset $\{t_{i_1}, \ldots, t_{i_p}\}$ of $\{t_1, \ldots, t_m\}$ with $p \leqslant n$ and numbers $\hat{y}_{i_1}^* \geqslant 0, \ldots, \hat{y}_{i_r}^* \geqslant 0$ such that the vectors $(f_1(t_{i_k}), \ldots, f_n(t_{i_k}))^{\mathrm{T}}, k = 1, \ldots, r$, in $\mathbb{R}^n$ are linearly independent and*

$$\sum_{k=1}^{p} f_j(t_{i_k})\hat{y}_{i_k}^* = c_j \qquad \text{for } j = 1, \ldots, n. \tag{I.5.4'}$$

Proof If the vectors $(f_1(t_i), \ldots, f_n(t_i))^{\mathrm{T}}$ for $i = 1, \ldots, m$ are linearly independent, then necessarily $m \leqslant n$ so that we can choose $p = m$ and $\hat{y}_i^* = y_i^*$, $i = 1, \ldots, m$. If the vectors $(f_1(t_i), \ldots, f_n(t_i))^{\mathrm{T}}$ for $i = 1, \ldots, m$ linearly dependent, then there are numbers $z_1, \ldots, z_m \in \mathbb{R}$ not all zero such that

$$\sum_{i=1}^{m} f_j(t_i)z_i = 0 \qquad \text{for } j = 1, \ldots, n.$$

We can assume without loss of generality that at least one $z_i > 0$. If one defines

$$\tau = \min_{z_i > 0} \frac{y_i^*}{z_i} = \frac{y_{i_0}^*}{z_{i_0}} \quad \text{and} \quad \hat{y}_i^* = y_i^* - \tau z_i \qquad \text{for } i = 1, \ldots, m$$

then $\hat{y}_i^* \geqslant 0$ for all i and $\hat{y}_{i_0}^* = 0$. Furthermore

$$\sum_{\substack{i=1\\ i \neq i_0}}^{m} f_j(t_i)\hat{y}_i^* = \sum_{i=1}^{m} f_j(t_i)y_i^* - \tau \sum_{i=1}^{m} f_j(t_i)z_i = c_j$$

for $j = 1, \ldots, n$. If the remaining vectors $(f_1(t_i), \ldots, f_n(t_i))^{\mathrm{T}}$ for $i = 1, \ldots, m$, $i \neq i_0$, are linearly independent, then necessarily $p = (m - 1) \leqslant n$ and (I.5.4′) is satisfied. In the other case one continues the above construction and arrives after finitely many steps at a representation (I.5.4′) for which $p \leqslant n$ and all vectors

$(f_1(t_{i_k}), \ldots, f_n(t_{i_k}))^{\mathrm{T}}, k = 1, \ldots, p$, are linearly independent, as well as $y_{i_k}^* \geqslant 0$ for $k = 1, \ldots, p$.

As a simple consequence of Theorem I.5.2 we also have our next theorem.

Theorem I.5.3 *Let $r = 0$ and suppose there is an $\hat{x} \in \mathbb{R}^n$ satisfying (I.5.3). If the problem (P) is solvable, then there is an optimal solution $(y_i^*, t_i)_{i=1,\ldots,m}$ of (I.5.4) with $m \leqslant n$ such that the vectors $(f_1(t_i), \ldots, f_n(t_i))^{\mathrm{T}}$ for $i = 1, \ldots, m$ are linearly independent and (I.5.5a) is satisfied for every solution $x \in \mathbb{R}^n$ of problem (P).*

Proof Let $x \in \mathbb{R}^n$ and $(y_i^*, t_i)_{i=1,\ldots,m}$ be solutions of problems (P) and (D) respectively, and $y_i^* > 0$ for all i, which, by (I.5.5*a*) implies

$$\sum_{j=1}^{n} f_j(t_i)x_j = b(t_i) \qquad \text{for all } i.$$

By Theorem I.5.2 there is a subset $\{t_{i_1}, \ldots, t_{i_p}\}$ of $\{t_1, \ldots, t_m\}$ with $p \leqslant n$ and numbers $\hat{y}_{i_1}^* \geqslant 0, \ldots, \hat{y}_{i_p}^* \geqslant 0$ such that the vectors $(t_1(t_{i_k}), \ldots, f_n(t_{i_k}))^{\mathrm{T}}$ for $k = 1, \ldots, p$ are linearly independent and (I.5.4′) is valid. Obviously we have further

$$\hat{y}_{i_k}^* > 0 \text{ for some } k = 1, \ldots, p \Rightarrow \sum_{j=1}^{n} f_j(t_{i_k})x_j = b(t_{i_k}). \tag{I.5.5a′}$$

From this and the equivalence of the two statements (*a*) and (*b*) above, the optimality of $(y_{i_k}^*, t_{i_k})_{n=1,\ldots,p}$ follows.

Because of this equivalence, the implication (I.5.5*a*′) holds for every solution of the problem (P).

I.5.2 Uniform approximation of functions

We take up the problem in Section I.2.1.1 again. Suppose therefore that M is a compact metric space and $C(M)$ is the vector space of continuous real-valued functions on M equipped with the maximum norm (I.2.1). Let a fixed function $f \in C(M)$ be given and suppose there is an n-dimensional linear subspace V of $C(M)$ spanned by the linearly independent functions $v_1, \ldots, v_n \in C(M)$. We again seek a function $\hat{v} \in V$ satisfying (I.2.2), i.e. a function $\hat{v} \in V$ which best approximates the function f among all elements of V in the uniform sense. The existence of $\hat{v}$ is guaranteed by Theorem II.5.2. In the following the question is one of giving necessary and sufficient conditions for $\hat{v}$. For this purpose we start from problem (P): minimize the linear functional $c(x, \gamma) = \gamma$ with $x \in \mathbb{R}^n, \gamma \in \mathbb{R}$ subject to the side conditions (I.2.5)

$$\left.\begin{aligned} &\sum_{j=1}^{n} v_j(t)x_j + \gamma \geqslant f(t) \\ &\sum_{j=1}^{n} -v_j(t)x_j + \gamma \geqslant -f(t) \end{aligned}\right\} \quad \text{for all } t \in M. \tag{I.5.6}$$

Problem I.5.1a Show that if $(\hat{x}, \hat{\gamma})$ is a solution of problem (P), then

$$\hat{v} = \sum_{j=1}^{n} v_j \hat{x}_j \tag{I.5.7}$$

also is a solution of the approximation problem and

$$\hat{\gamma} = \| \hat{v} - f \|_\infty = \rho_\infty(f, V) \tag{I.5.8}$$

with $\rho_\infty(f, V)$ given by (I.2.3).

Problem I.5.1b If $\hat{v}$ is a best approximation of f of the form (I.5.7), then show that $(\hat{x}, \hat{\gamma})$ with $\hat{\gamma}$ given by (I.5.8) is a solution of problem (D).

Let Y be the finite dimensional linear subspace of $C(M)$ spanned by $v_1, \ldots, v_n$, $e = 1$ and f, and $F = Y \times Y$ equipped with the norm

$$\| (y_1, y_2) \| = \| y_1 \|_\infty + \| y_2 \|_\infty \qquad \text{for } y_1, y_2 \in Y$$

and partially ordered by

$$(y_1, y_2) \geqslant \Theta_F \Leftrightarrow \begin{Bmatrix} y_1 \geqslant \Theta_Y \\ y_2 \geqslant \Theta_Y \end{Bmatrix} \Leftrightarrow \begin{Bmatrix} y_1(t) \geqslant 0 \\ y_2(t) \geqslant 0 \end{Bmatrix} \text{ for all } t \in M.$$

If one defines further $E = \mathbb{R}^{n+1}$ with the positive cone $K_E = \mathbb{R}^n \times \mathbb{R}^+$, $\mathbb{R}^+ = \{r \in \mathbb{R} : r \geqslant 0\}$, then

$$A(x, \gamma) = \begin{pmatrix} \sum_{j=1}^{n} v_j x_j + \gamma e \\ \\ \sum_{j=1}^{n} -y_j x_j + \gamma e \end{pmatrix} \qquad \text{for } x \in \mathbb{R}^n, \gamma \in \mathbb{R} \tag{I.5.9}$$

is a continuous linear mapping of E into F. Setting $b = (f, -f)$ the side conditions (I.5.6) can be written in the form

$$A(x, \gamma) \geqslant b \qquad \text{for } (x, \gamma) \in K_E$$

(Proof = Exercise). the problem (P) is a problem of semi-infinite optimization in the sense of Section I.3.2. In this case the cone $K(A, c)$ defined by (I.4.4) is

$$K(A, c) = \Big\{ (\gamma + r, \sum_{j=1}^{n} v_j x_j + \gamma e - y_1, \sum_{j=1}^{n} -v_j x_j + \gamma e - y_2) : x \in \mathbb{R}^n,$$

$$\gamma \geqslant 0, r \geqslant 0, (y_1, y_2) \geqslant \Theta_F \Big\} \quad (\subseteq \mathbb{R} \times Y \times Y).$$

Assertion $K(A, c)$ *is closed.*

Proof Let $\overline{K(A,c)}$ be the closure of $K(A,c)$ and $(\alpha, z_1, z_2) \in \overline{K(A,c)}$. Then there are sequences $(x^k, \gamma^k) \in \mathbb{R}^{n+1}, r^k \geq 0, \gamma^k \geq 0$ and $(y_1^k, y_2^k) \geq \Theta_F$ with

$$\alpha = \lim_{k\to\infty} (\gamma^k + r^k) \qquad z_1 = \lim_{k\to\infty} z_1^k \qquad z_2 = \lim_{k\to\infty} z_2^k,$$

$$z_1^k = \sum_{j=1}^{n} v_j x_j^k + \gamma^k e - y_1^k \qquad z_2^k = \sum_{j=1}^{n} -v_j x_j^k + \gamma^k e - y_2^k.$$

From $\alpha = \lim_{k\to\infty} (\gamma^k + r^k)$ follows the existence of a number $\beta > 0$ with $0 \leq r^k$, $\gamma^k \leq \beta$ for all k. Since $z_1 = \lim_{k\to\infty} z_1^k$ and $z_2 = \lim_{k\to\infty} z_2^k$, there are constants β_1, β_2 with $\| z_1^k \|_\infty \leq \beta$, and $\| z_2^k \|_\infty \leq \beta_2$ for all k. Since

$$z_1^k \leq \sum_{j=1}^{n} v_j x_j^k + \gamma^k e \Rightarrow z_1^k - \gamma^k e \leq \sum_{j=1}^{n} v_j x_j^k$$

$$z_2^k \leq \sum_{j=1}^{n} -v_j x_j^k + \gamma^k e \Rightarrow \sum_{j=1}^{n} v_j x_j^k \leq -z_2^k + \gamma^k e$$

we have

$$\left\| \sum_{j=1}^{n} v_j x_j^k \right\|_\infty \leq \max(\beta + \beta_1, \beta + \beta_2) = \delta.$$

Since the set

$$\left\{ \sum_{j=1}^{n} v_j x_j : \left\| \sum_{j=1}^{n} v_j x_j \right\|_\infty \leq \delta \right\}$$

is compact in Y, for suitable subsequences

$$\lim_{i\to\infty} \sum_{j=1}^{n} v_j x_j^{k_i} = \sum_{j=1}^{n} v_j x_j \qquad \text{(since } V \text{ is closed)}$$

$$\lim_{i\to\infty} \gamma^{k_i} = \gamma \geq 0 \qquad \lim_{i\to\infty} r^{k_i} = r = \alpha - \gamma \geq 0$$

$$y_1 = \lim_{i\to\infty} y_1^{k_i} = \sum_{j=1}^{n} v_j x_j + \gamma e - z_1 \geq \Theta_Y$$

$$y_2 = \lim_{i\to\infty} y_2^{k_i} = \sum_{j=1}^{n} -v_j x_j + \gamma e - z_2 \geq \Theta_Y$$

(Details = Exercise), from which it follows that $(\alpha, z_1, z_2) \in K(A,c)$. Since the extremal value of problem (P) is bounded below by zero, we obtain the solvability of problem (P) from the existence theorem in Section I.4.3 and, together with problem I.5.1, also the solvability of the posed approximation problem.

We now turn to the dual problem (D). The mapping $A^* : F^* \to E^*$ (by Section

IV.1.3) which is adjoint to A (by (I.5.9)) is defined by

$$A^*(y^*)(x,\gamma) = y^*(A(x,\gamma))$$

$$= y_1^*\left(\sum_{j=1}^n v_j x_j + \gamma e\right) + y_2^*\left(\sum_{j=1}^n -v_j x_j + \gamma e\right) \qquad \text{(see (IV.1.18))}$$

$$= \sum_{j=1}^n [y_1^*(v_j) - y_2^*(v_j)]x_j + \gamma(y_1^*(e) + y_2^*(e))$$

for all $y_1^*, y_2^* \in Y^*$, $x \in \mathbb{R}^n$ and $\gamma \in \mathbb{R}$. Since $c(x,\gamma) = \gamma$, the inequality $A^*(y^*) \leqslant c$ is, by definition (IV.1.14), equivalent to

$$\sum_{j=1}^n [y_1^*(v_j) - y_2^*(v_j)]x_j + \gamma(y_1^*(e) + y_2^*(e) - 1) \leqslant 0$$

$$\text{for all } x \in \mathbb{R}^n \text{ and } \gamma \geqslant 0. \qquad \text{(I.5.10)}$$

Problem I.5.2a Show that for each $y^* \in F^*$, i.e. $y^*(y_1, y_2) = y_1^*(y_1) + y_2^*(y_2)$, $y_1, y_2 \in Y$ with uniquely determined $y_1^*, y_2^* \in Y^*$ (see (IV.1.18)), there follows

$$y^* \geqslant \Theta_{F^*} \Leftrightarrow \{y_1^* \geqslant \Theta_{Y^*} \text{ and } y_2^* \geqslant \Theta_{Y^*}\}.$$

Problem I.5.2b Show that the assertion (I.5.10) is equivalent to

$$y_1^*(v_j) - y_2^*(v_j) = 0 \qquad \text{for } j = 1, \ldots, n \qquad \text{(I.5.11)}$$

$$y_1^*(e) + y_2^*(e) \leqslant 1. \qquad \text{(I.5.12)}$$

With that, the problem dual to problem (P) is equivalent to problem (D): maximize the continuous linear form $y_1^*(f) - y_2^*(f)$ subject to the side conditions (I.5.11), (I.5.12) and $y_1^* \geqslant \Theta_{Y^*}$, $y_2^* \geqslant \Theta_{Y^*}$.

In order to show the solvability of problem (D), we note first that the positive cone K_F of F has a non-empty interior, which is given by

$$\overset{\circ}{K}_F = \{(y_1, y_2) \in F : y_1(t) > 0 \text{ and } y_2(t) > 0 \text{ for all } t \in M\}$$

(Proof = Exercise)

If one now chooses $\hat{x} = \Theta_n$ = zero vector of $\mathbb{R}^n$ and $\gamma > \|f\|_\infty$, then the side conditions (I.5.6) are satisfied with strict inequality, i.e. $A(\hat{x},\hat{\gamma}) - b \in \overset{\circ}{K}_F$ holds with $A(\hat{x},\hat{\gamma})$ from (I.5.9) and $b = (f, -f)$. From Theorem I.4.14 follow immediately the solvability of the problem (D) and the fact that its extremal value coincides with that of problem (P).

Since $e \in Y (e = 1)$ the condition (IV.2.16) is satisfied for y (in place of V). Hence, by the corollary to Theorem IV.2.6, each $y^* \geqslant \Theta_{Y^*}$ can be represented in the form

$$y^*(y) = \sum_{i \in I} \lambda_i y(t_i)$$

with $\lambda_i \geqslant 0$ for all $i \in I$ and distinct points $t_i \in M$, $i \in I$, whereby I is a suitable finite index set. The problem (D) is therefore equivalent to: problem (D*):

maximize

$$\sum_{i\in I_1} \lambda_i^1 f(t_i^1) - \sum_{i\in I_2} \lambda_i^2 f(t_i^2)$$

subject to the side conditions

$$\sum_{i\in I_1} \lambda_i^1 v_j(t_i^1) - \sum_{i\in I_2} \lambda_i^2 y_j(t_i^2) = 0 \qquad \text{for } j = 1, \ldots, n \tag{I.5.11*}$$

$$\sum_{i\in I_1} \lambda_i^1 + \sum_{i\in I_2} \lambda_i^2 \leqslant 1 \tag{I.5.12*}$$

$$\lambda_i^1 \geqslant 0, i \in I_1 \qquad \text{and} \qquad \lambda_i^2 \geqslant 0, i \in I_2 \tag{I.5.13}$$

as well as

$$\{t_i^1 : i \in I_1\} \cup \{t_i^2 : i \in I_2\} \subseteq M.$$

Here I_1 and I_2 run through all finite index sets. The complementary slackness theorem of Section I.3.1 (Theorem I.3.3) can now be sharpened to give the theorem below.

Theorem I.5.4 *Suppose* $(x, \gamma) \in \mathbb{R}^n \times \mathbb{R}^+$ *is consistent for the problem (P), i.e.* (x, γ) *satisfies the side conditions (I.5.6) and* $(\lambda_i^1, t_i^1)_{i\in I_1}, (\lambda_i^2, t_i^2)_{i\in I_2}$ *are consistent for the problem (D*), i.e. they satisfy the conditions (I.5.11*), (I.5.12*), and (I.5.13). Then the following two statements are equivalent:*
(α) (x, γ) *and* $(\lambda_i^k, t_i^k)_{i\in I_k}, k = 1,2,$ *are optimal,*
(β) *the following implications hold:*

$$\lambda_i^1 > 0 \Rightarrow \sum_{j=1}^{n} v_j(t_i^1)x_j - f(t_i^1) = -\gamma \tag{I.5.14a}$$

$$\lambda_i^2 > 0 \Rightarrow \sum_{j=1}^{n} v_j(t_i^2)x_j - f(t_i^2) = \gamma \tag{I.5.14b}$$

$$\gamma > 0 \Rightarrow \sum_{i\in I_1} \lambda_i^1 + \sum_{i\in I_2} \lambda_i^2 = 1. \tag{I.5.15}$$

Proof
1. Suppose that (α) holds. Then the extremal values of problems (P) and (D*) coincide. Therefore one has

$$\gamma = \sum_{i\in I_1} \lambda_i^1 f(t_i^1) - \sum_{i\in I_2} \lambda_i^2 f(t_i^2). \tag{I.5.16}$$

From (I.5.11*) we get

$$\sum_{i\in I_1} \lambda_i^1 \sum_{j=1}^{n} v_j(t_i^1)x_j - \sum_{i\in I_2} \lambda_i^2 \sum_{j=1}^{n} v_j(t_i^2)x_j = 0 \tag{I.5.17}$$

and (I.5.12*), (I.5.16), and (I.5.17) yield

$$\sum_{i\in I_1} \lambda_i^1 \left\{ \sum_{j=1}^{n} v_j(t_i^1)x_j - f(t_i^1) + \gamma \right\} + \sum_{i\in I_2} \lambda_i^2 \left\{ - \sum_{j=1}^{n} v_j(t_i^2)x_j + f(t_i^2) + \gamma \right\}$$

$$= - \sum_{i\in I_1} \lambda_i^1 f(t_i^1) + \sum_{i\in I_2} \lambda_i^2 f(t_i^2) + \gamma \left(\sum_{i\in I_1} \lambda_i^1 + \sum_{i\in I_2} \lambda_i^2 \right) \leqslant 0 \tag{I.5.18}$$

which is possible only if the implications (I.5.14*a*), (I.5.14*b*) and (I.5.15) hold.
2. Suppose (β) is true. Then there is equality in (I.5.18) which implies (I.5.16). From that and Theorem I.3.2, we have the optimality of (x, γ) and $(\lambda_i^k, t_i^k)_{i \in I_k}$, $k = 1,2$.

Now let (x, γ) be a solution of problem (P) with $\gamma > 0$. Then it follows from (I.5.15) that

$$\sum_{i \in I_1} \lambda_i^1 + \sum_{i \in I_2} \lambda_i^2 = 1$$

for every solution $\{(\lambda_i^1, t_i^1)_{i \in I_1}, (\lambda_i^2, t_i^2)_{i \in I_2}\}$ of problem (D*). So we can assume without loss of generality that equality in (I.5.12*) holds.

By an appropriate application of Theorem I.5.2 one obtains the further theorem now given.

Theorem I.5.5 *If* $(\lambda_i^k, t_i^k)_{i \in I_k}$, $i = 1,2$ *is a solution to (I.5.11*) and (I.5.12*) with equality rather than inequality, then there are subsets* $\tilde{I}_k$, $k = 1,2$ *of* I_k *such that* $\tilde{I}_1 \cup \tilde{I}_2$ *consists of at most* $(n + 1)$ *indices and there are numbers* $\tilde{\lambda}_i^k \geqslant 0, i \in \tilde{I}_k, k = 1,2$ *such that*

$$\sum_{i \in I_1} \tilde{\lambda}_i^1 v_j(t_i^1) - \sum_{i \in \tilde{I}_2} \tilde{\lambda}_i^2 v_j(t_i^2) = 0 \qquad \text{for } j = 1, \ldots, n$$

$$\sum_{i \in \tilde{I}_1} \tilde{\lambda}_i^1 + \sum_{i \in \tilde{I}_2} \tilde{\lambda}_i^2 = 1.$$

By analogy with Theorem I.5.3 one finally obtains, with the aid of Theorem I.5.4, the existence of a solution $(\lambda_i^k, t_i^k)_{i \in I_k}$, $k = 1,2$ of the problem (D*) such that $I_1 \cup I_2$ consists of at most $(n + 1)$ points.

Theorem I.5.6 *Suppose M consists of at least* $(n + 1)$ *points. A function* $\hat{v} \in V$ *is a best approximation of f in V, i.e.*

$$\| f - \hat{v} \|_\infty = \rho_\infty(f, V) = \inf_{v \in V} \| f - v \|_\infty \tag{I.5.19}$$

holds if and only if there are $s \leqslant (n + 1)$ *distinct points* $\hat{t}_1, \ldots, \hat{t}_s$ *in*

$$E_{\hat{v}} = \{t \in M : | f(t) - \hat{v}(t) | = \| f - \hat{v} \|_\infty\} \tag{I.5.20}$$

and numbers $\hat{y}_1, \ldots, \hat{y}_s \in \mathbb{R}$ *such that*

$$\sum_{i=1}^{s} | \hat{y}_i | = 1 \tag{I.5.21}$$

$$\sum_{i=1}^{s} \hat{y}_i v_j(\hat{t}_i) = 0 \qquad \text{for } j = 1, \ldots, n \tag{I.5.22}$$

$$\left(\Leftrightarrow \sum_{i=1}^{s} \hat{y}_i v(\hat{t}_i) = 0 \qquad \text{for all } v \in V \right)$$

$$\hat{y}_i \neq 0 \Rightarrow f(\hat{t}_i) - \hat{v}(\hat{t}_i) = \| f - \hat{v} \|_\infty \operatorname{sgn} \hat{y}_i. \tag{I.5.23}$$

Proof

1. If for a given $\hat{v} \in V$ there are distinct points $\hat{t}_1, \ldots, \hat{t}_s \in E_{\hat{v}}, s \leqslant (n+1)$, and numbers $\hat{y}_i, i = 1, \ldots, s$, satisfying (I.5.21), (I.5.22), and (I.5.23), then it follows for every $v \in V$ that

$$\| f - \hat{v} \|_\infty = \sum_{i=1}^{s} \hat{y}_i(f(\hat{t}_i) - \hat{v}(\hat{t}_i))$$

$$= \sum_{i=1}^{s} \hat{y}_i f(\hat{t}_i)$$

$$= \sum_{i=1}^{s} \hat{y}_i(f(\hat{t}_i) - v(\hat{t}_i)) \leqslant \| f - v \|_\infty,$$

i.e. (I.5.19) is satisfied.

2. Suppose that (I.5.19) is satisfied for $\hat{v} \in V$. Further let $(\lambda_i^k, t_i^k)_{i \in I_k}, k = 1,2$ be optimal for the problem (D*) (which is solvable on the basis of the above considerations). By Theorem I.5.4 and problem I.5.1, the implications (I.5.14*a*), (I.5.14*b*), and (I.5.15) hold, whereby

$$x = \hat{x} \qquad \gamma = \| f - \hat{v} \|_\infty \qquad \text{and} \qquad \hat{v} = \sum_{j=1}^{n} \hat{x}_j v_j.$$

Now there are two possible cases.

(α) $\gamma = 0$. If one then chooses $s = (n+1)$ points $\hat{t}_i \in M, i = 1, \ldots, s$, arbitrarily, then (I.5.21), (I.5.22), and (I.5.23) can be satisfied.

(β) $\gamma > 0$. Let $E_k^+ = \{t_i^k : i \in I_k, \lambda_i^k > 0\}, k = 1,2$. Because of (I.5.14*a*), and (I.5.14*b*) $E_1^+ \cap E_2^+ = \phi$, and because of (I.5.15) both sets E_k^+ are not simultaneously empty, and $E^+ = E_1^+ \cup E_2^+$ is a non-empty subset of $E_{\hat{v}}$ by (I.5.20). If one defines $I_k^+ = \{i \in I_k : \lambda_i^k > 0\}, k = 1,2$ and $\hat{y}_i^1 = \lambda_i^1$ for $i \in I_1^+$ as well as $\hat{y}_i^2 = -\lambda_i^2$ for $i \in I_2^+$, then it follows that $\hat{y}_i^k \neq 0$ for all $i \in I_k^+, k = 1,2$ and further from (I.5.12*) one has

$$\sum_{i \in I_1^+} | \hat{y}_i^1 | + \sum_{i \in I_2^+} | \hat{y}_i^2 | = 1. \tag{I.5.21$'$}$$

Furthermore

$$\sum_{i \in I_1^+} \hat{y}_i^1 v_j(t_i^1) + \sum_{i \in I_2^+} \hat{y}_i^2 v_j(t_i^2) = 0 \qquad \text{for } j = 1, \ldots, n \tag{I.5.22$'$}$$

by (I.5.11*), and finally

$$f(t_i^k) - \hat{v}(t_i^k) = \| f - \hat{v} \|_\infty \operatorname{sgn} \hat{y}_i^k \qquad \text{for } i \in I_k^+, k = 1,2 \tag{I.5.23$'$}$$

by (I.5.14*a*), and (I.5.14*b*). With that, (I.5.12), (I.5.22), and (I.5.23) are satisfied and s is then the number of points $\hat{t}_i \in E^+$. On the basis of the remarks at the end of Theorem I.5.5 we can regard the $(\lambda_i^k, t_i^k)_{i \in I_k}, k = 1,2$ to be chosen so that $I_1 \cup I_2$ consists of at most $(n+1)$ points, which implies that $s \leqslant (n+1)$.

Problem I.5.3a Show that, in the inequality (I.3.20), the equality sign is attainable if one chooses suitable t_i and y_i.

Problem I.5.3b Show that, under the assumption (I.5.22), the statement (I.5.23) is equivalent to

$$\sum_{i=1}^{s} \hat{y}_i f(\hat{t}_i) = \| f - \hat{v} \|_\infty \sum_{i=1}^{s} | \hat{y}_i |.$$

Problem I.5.3c Let $M = [0, 1], f(t) = t^2$ and $V = \{x \cdot t : x \in \mathbb{R}, t \in M\}$. Show, by using Theorem I.5.6, that $\hat{v}(t) = 2\{\sqrt{2} - 1\}t$ is a best approximation of f in V and that $\rho_\infty(f, V) = 3 - 2\sqrt{2}$.
Hint Choose $\hat{t}_1 = 1$ and $\hat{t}_2 = \sqrt{2} - 1$ (see Figure I.5.1).

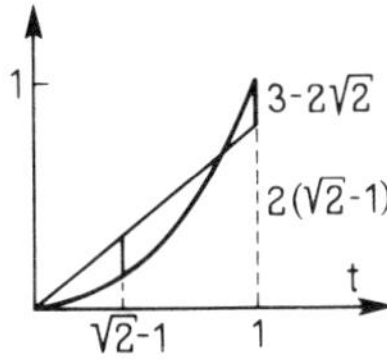

Figure I.5.1 Best approximation of $f(t) = t^2$ in $\{x \cdot t : x \in \mathbb{R}, t \in [0, 1]\}$

I.5.3 One-sided uniform approximation

The problem of one-sided uniform approximation has arisen in connection with boundary value problems of monotonic type already in Sections I.2.4.2 and I.3.3.2 and will be systematically investigated here. Since the considerations are partly analogous to those of Section I.5.2, occasionally details in the proofs will be omitted.

As in Section I.5.2 let V be an n-dimensional subspace of $C(M)$ (spanned by the linearly independent functions $v_1, \ldots, v_n \in C(M)$) and let a fixed element $f \in C(M)$ be given. Then we define the subsets of V

$$V^+ = \{v \in V : v(t) \geqslant f(t) \text{ for all } t \in M\} \tag{I.5.24$^+$}$$

and

$$V^- = \{v \in V : v(t) \leqslant f(t) \text{ for all } t \in M\} \tag{I.5.24$^-$}$$

respectively, and consider the (one-sided) approximation problem. We seek a $\hat{v} \in V^+$, and V^-, respectively, such that

$$\| \hat{v} - f \|_\infty = \rho_\infty(f, V^+) = \inf_{v \in V} \| v - f \|_\infty \tag{I.5.25$^+$}$$

and

$$\| \hat{v} - f \|_\infty = \rho_\infty(f, V^-) = \inf_{v \in V} \| v - f \|_\infty. \tag{I.5.25$^-$}$$

Here $\| \ \|_\infty$ again denotes the maximum norm in $C(M)$.

Problem I.5.4 Show that V^+, and V^-, respectively, are closed convex subsets of V and therefore also of each subspace of $C(M)$ which contains V (why?).

Now let W be the finite dimensional linear subspace of $C(M)$ which is spanned by V and f. Then W is reflexive (see Section IV.2.2) and by problem I.5.4 V^+, V^-, are convex closed subsets of W. Moreover the functional $\phi(w) = \| w - f \|_\infty$ is continuous and convex on W (Proof = Exercise). By applying the existence theorem in Section II.2.1, one then obtains another theorem.

Theorem I.5.7 *If the closed, convex subsets V^+, V^-, of V defined by (I.5.24$^+$), (I.5.24$^-$), respectively, are not empty, then the one-sided approximation problem is solvable. i.e. there is a $\hat{v} \in V^+$, V^-, with (I.5.25$^+$), (I.5.25$^-$), respectively.*

In addition we have the following theorem.

Theorem I.5.8 *If V contains the constant functions, then the one-sided approximation problem is solvable and for the minimal deviations (I.5.25$^+$), (I.5.25$^-$), respectively, we have*

$$\rho_\infty(f, V^+) = \rho_\infty(f, V^-) = 2\rho_\infty(f, V) \tag{I.5.26}$$

$$\rho_\infty(f, V) = \inf_{v \in V} \| v - f \|_\infty . \tag{I.5.27}$$

Furthermore, $\hat{v} \in V$ is a minimal solution with respect to f in V if and only if $\hat{v}^+ = \hat{v} + \rho_\infty(f, V)$, $\hat{v}^- = \hat{v} - \rho_\infty(f, V)$, respectively, is a minimal solution of f in V^+, V^-, respectively (see Figure I.5.2).

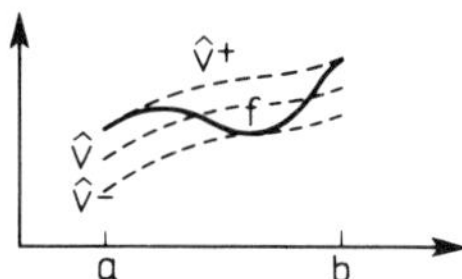

Figure I.5.2 Relation between one-sided and symmetric uniform approximation

Proof Since V contains the constant functions, V^+ and V^- are certainly not empty (why?), so that the solvability of the one-sided approximation problem is assured by Theorem I.5.7. Now let $\hat{v} \in V$ be a minimal solution with respect to f in V. Then

$$v^+ = \hat{v} + \rho_\infty(f, V) \in V^+ \qquad \hat{v}^- = \hat{v} - \rho_\infty(f, V) \in V^-$$

respectively, and

$$\| \hat{v}^+ - f \|_\infty = \| \hat{v} - f \|_\infty + \rho_\infty(f, V) = 2\rho_\infty(f, V)$$

$$\| \hat{v}^- - f \|_\infty = \| f - \hat{v}^- \|_\infty = \| f - \hat{v} \|_\infty + \rho_\infty(f, V) = 2\rho_\infty(f, V)$$

which implies

$$\rho_\infty(f, V^+) \quad \text{and} \quad \rho_\infty(f, V^-) \leqslant 2\rho_\infty(f, V)$$

respectively. If one had $\rho_\infty(f, V^+) < 2\rho_\infty(f, V)$, then for every $v^+ \in V^+$ with $\| v^+ - f \|_\infty = \rho_\infty(f, V^+)$ and $v = v^+ - \frac{1}{2}\rho_\infty(f, V^+) \in V$

$$\| v - f \|_\infty = \tfrac{1}{2}\rho_\infty(f, V^+) < \rho_\infty(f, V)$$

which is a contradiction to (I.5.27). Consequently $\rho_\infty(f, V^+) = 2\rho_\infty(f, V)$ and $\hat{v}^+ = \hat{v} + \rho_\infty(f, V)$ is a minimal solution of f in V^+. It is easy to see that all minimal solutions of f in V^+ are obtained in this way.

The conclusion for V^- is analogous.

For the rest of the section we consider the one-sided approximation problem only for V^- according to (I.5.24$^-$). Again, necessary and sufficient conditions for minimal solutions will be given. For this purpose we begin with problem (P$^-$): minimize the linear functional $c(x, \gamma) = \gamma$ with $x \in \mathbb{R}^n$, $\gamma \in \mathbb{R}$ subject to the side conditions

$$\left.\begin{aligned} &\sum_{j=1}^{n} v_j(t)x_j + \gamma \geqslant f(t) \\ &\sum_{j=1}^{n} -v_j(t)x_j \geqslant -f(t) \end{aligned}\right\} \quad \text{for all } t \in M. \tag{I.5.28}$$

Problem I.5.5a Show that if $(\hat{x}, \hat{\gamma})$ is a solution of problem (P$^-$), then $\hat{v}$ satisfying (I.5.7) is a solution of the one-sided approximation problem (I.5.25$^-$) and

$$\hat{\gamma} = \| v - f \|_\infty = \rho_\infty(f, V^-) \tag{I.5.29}$$

Problem I.5.5b Show that if $\hat{v} \in V^-$ is a best approximation of f of the form (I.5.7) in V^-, then $(\hat{x}, \hat{\gamma})$ with $\hat{\gamma}$ given by (I.5.29) is a solution of the problem (P$^-$).

Now let Y be the linear subspace of $C(M)$ defined as in Section I.5.2 and $F = Y \times Y$. If one again defines $E = \mathbb{R}^{n+1}$ with the positive cone $K_E = \mathbb{R}^n \times \mathbb{R}^+$ then

$$A(x, \gamma) = \begin{pmatrix} \sum_{j=1}^{n} v_j x_j + \gamma e \\ \\ \sum_{j=1}^{n} -v_j x_j \end{pmatrix} \quad \text{for } x \in \mathbb{R}^n, \gamma \in \mathbb{R} \tag{I.5.30}$$

is a continuous linear mapping of E into F. Setting $b = (f, -f)$, the side conditions (I.5.28) can be written in the form

$$A(x, \gamma) \geqslant b \qquad (x, \gamma) \in K_E.$$

The problem (P$^-$) is also a problem of semi-infinite optimization in the sense of Section I.3.2, and the cone $K(A, c)$ defined by (I.4.4) is

$$K(A, c) = \left\{ \left(\gamma + r, \sum_{j=1}^{n} v_j x_j + \gamma e - y_1, \sum_{j=1}^{n} -v_j x_j - y_2 \right) : x \in \mathbb{R}^n, \gamma \geqslant 0, \right.$$
$$\left. r \geqslant 0, (y_1, y_2) \geqslant \Theta_F \right\}.$$

Again it can be shown that $K(A, c)$ is closed (Proof = Exercise; see the corresponding proof in Section I.5.2), so that the solvability of problem (P^-) follows from the existence theorem in Section I.4.3 in the event that the side conditions (I.5.28) can be satisfied. That is precisely the case if V^- (see (I.5.24$^-$)) is not empty so that, by means of problem I.5.5, the solvability of the one-sided approximation problem again follows.

In analogy to Section I.5.2 one shows further that the problem which is dual to problem (P^-) is equivalent to problem (D^-): maximize

$$\sum_{i\in I_1} \lambda_i^1 f(t_i^1) - \sum_{i\in I_2} \lambda_i^2 f(t_i^2)$$

subject to the side conditions

$$\sum_{i\in I_1} \lambda_i^1 v_j(t_i^1) - \sum_{i\in I_2} \lambda_i^2 v_j(t_i^2) = 0 \qquad \text{for } j = 1, \ldots, n \tag{I.5.11**}$$

$$\sum_{i\in I_1} \lambda_i^1 \leqslant 1 \tag{I.5.12**}$$

$$\lambda_i^1 \geqslant 0, i \in I_1 \qquad \text{and} \qquad \lambda_i^2 \geqslant 0, i \in I_2 \tag{I.5.13}$$

as well as

$$\{t_i^1 : i \in I_1\} \cup \{t_i^2 : i \in I_2\} \subseteq M.$$

Here I_1 and I_2 run through all finite index sets. We now make the following assumption.

Assumption Suppose there is an $\hat{x} \in \mathbb{R}^n$ with

$$f(t) - \sum_{j=1}^{n} v_j(t)\hat{x}_j > 0 \qquad \text{for all } t \in M. \tag{I.5.31}$$

This assumption can be satisfied, for example, if V contains the constant functions. From this it follows that the side conditions (I.5.28) can be satisfied with strict inequality, if one chooses

$$\gamma = \hat{\gamma} > \left\| f - \sum_{j=1}^{n} v_j \hat{x}_j \right\|_\infty .$$

From this we conclude further that $A(\hat{x}, \hat{\gamma}) - b \in \overset{\circ}{K}_F$ with $b = (f, -f)$ and $\overset{\circ}{K}_F$ is the interior of K_F (see Section I.5.2). From Theorem I.4.14 therefore again follows the solvability of problem (D^-) and the fact that its extremal value coincides with that of problem (P^-) and consequently with the minimal deviation (I.5.25$^-$) of f with respect to V^-. Under the assumption (I.5.31) one can also prove an equilibrium theorem corresponding to Theorem I.5.4, which will not be carried out further here.

The following analogue of Theorem I.5.6 is decisive.

Theorem I.5.9 Under the assumption (I.5.31), $\hat{v}^- \in V^-$ is a best approximation of f in V^-, i.e. (I.5.25$^-$) holds with $\hat{v} = \hat{v}^-$, if and only if there are $s \leqslant (n + 1)$ distinct *points*

$$t_i^1 \in \{t \in M : f(t) - \hat{v}^-(t) = \| f - \hat{v}^- \|_\infty\} \qquad \text{for } i = 1, \ldots, s_1$$
$$t_i^2 \in \{t \in M : f(t) - \hat{v}^-(t) = 0\} \qquad \text{for } i = 1, \ldots, s_2$$

$(s_1 + s_2) = s$, *and numbers* $\lambda_i^1 \geqslant 0, i = 1, \ldots, s_1, \lambda_i^2 \geqslant 0, i = 1, \ldots, s_2$, *satisfying*

$$\sum_{i=1}^{s_1} \lambda_i^1 \leqslant 1 \tag{I.5.32}$$

$$\sum_{i=1}^{s_1} \lambda_i^1 v_j(t_i^1) - \sum_{i=1}^{s_2} \lambda_i^2 v_j(t_i^2) = 0 \qquad \text{for } j = 1, \ldots, n \tag{I.5.33}$$

$$\sum_{i=1}^{s_1} \lambda_i^1 f(t_i^1) - \sum_{i=1}^{s_2} \lambda_i^2 f(t_i^2) = \| f - \hat{v}^- \|_\infty. \tag{I.5.34}$$

If $\rho_\infty(f, V) > 0$, *then equality holds in (I.5.32).*

The proof follows along the same lines as that of Theorem I.5.6 and will therefore be omitted.

I.5.4 Application to a boundary value problem for the potential equation

We again take up the boundary value problem of Section I.3.3.1, which is to be solved approximately by means of functions of the form (I.3.15). On the basis of the remark at the end of Section I.2.3, the boundary maximum principle (I.2.20) leads in this case to the following one-sided approximation problem: find $\hat{v}^-$ in

$$V^- = \{v(t) = x_1 + (t^4 - 6t^2 + 1)x_2 : x_1, x_2 \in \mathbb{R}\}$$

and

$$x_1 + (t^4 - 6t^2 + 1)x_2 \leqslant \cos(\tfrac{1}{2}\pi t) \text{ for all } t \in [0, 1]$$

with

$$\max_{t \in [0,1]} \{\cos(\tfrac{1}{2}\pi t) - \hat{v}^-(t)\} \leqslant \max_{t \in [0,1]} \{\cos(\tfrac{1}{2}\pi t) - v(t)\} \qquad \text{for all } v \in V^-.$$

In this case, therefore, $v_1 = 1$, $v_2(t) = t^4 - 6t^2 + 1$ and $f(t) = \cos(\frac{1}{2}\pi t)$. Since V contains the constant functions, the problem is solvable by Theorem I.5.8.

In order to obtain a solution, Theorem I.5.9 will be applied. For this purpose we first seek a

$$v(t) = x_1 + (t^4 - 6t^2 + 1)x_2$$

such that

$$f(0) - v(0) = \eta \tag{I.5.35}$$

$$f(1) - v(1) = \eta \tag{I.5.36}$$

$$f(t_1) - v(t_1) = 0 \qquad \text{and} \qquad f'(t_1) - v'(t_1) = 0 \tag{I.5.37}$$

hold with suitable $\eta > 0$ and $t_1 \in (0, 1)$. Then requirements (I.5.35), (I.5.36) yield

$$x_1 = 0.8 - \eta \qquad x_2 = 0.2$$

and, from (I.5.37), we obtain the equations

$$0.8 - \eta + (t_1^4 - 6t_1^2 + 1)\, 0.2 = \cos(\tfrac{1}{2}\pi t_1) \tag{I.5.38}$$

$$0.8t_1(t_1^2 - 3) = -\tfrac{1}{2}\pi \sin(\tfrac{1}{2}\pi t_1). \tag{I.5.39}$$

From (I.5.39) we obtain the approximation $t_1 \approx 0.64$. Then (I.5.38) yields $\eta \approx 0.00624$ and hence $x_1 \approx 0.79376$. It is not difficult to verify

$$\hat{v}^-(t) = 0.79376 + 0.2\,(t^4 - 6t^2 + 1) \tag{I.5.40}$$

belongs to V^- and $\eta = \| f - \hat{v}^- \|_\infty = 6.24 \times 10^{-3}$.

In order now to show that this $\hat{v}^- \in V^-$ also solves the one-sided approximation problem, we choose $s_1 = 2, t_1^1 = 0, t_2^1 = 1, s_2 = 1, t_1^2 = 0.64$ and solve (I.5.33) with $\lambda_1^2 = 1$, i.e.

$$\lambda_1^1 + \lambda_2^1 - 1 = 0 (\Rightarrow \text{(I.5.32)}),$$

$$\lambda_1^1 - 4\lambda_2^1 + 1.2898 = 0$$

which yields $\lambda_1^1 = 0.54204$ and $\lambda_2^1 = 0.45796$. From this we have

$$\begin{aligned}\lambda_1^1 f(t_1^1) + \lambda_2^1 f(t_2^1) - \lambda_1^2 f(t_1^2) &\approx 0.54204 + 0 - 0.5358\\ &= 0.00624 = \| f - \hat{v}^- \|_\infty.\end{aligned}$$

By Theorem I.5.9, $\hat{v}^-$ in (I.5.40) is therefore, except possibly for round-off errors, a solution of the one-sided approximation problem and $\rho_\infty(f, V^-) = 6.24 \times 10^{-3}$. On the basis of the remark at the end of Section I.2.3, one obtains, therefore, for $x_1 \approx 0.79376, x_2 = 0.2$ in (I.3.15), an approximate solution of the given boundary value problem which, among all functions of the form (I.3.15), approximates best in the uniform sense the unknown solution $\hat{u}$ of the boundary value problem from below on B. In fact, one has

$$0 \leqslant \hat{u}(t, s) - 0.79376 - 0.2(t^4 - 6t^2s^2 + s^4) \leqslant 6.24 \times 10^{-3} \qquad \text{for all } (t, s) \in \bar{B}.$$

From Theorem I.5.8 one has further that

$$\hat{v}(t) = \hat{v}^-(t) + \tfrac{1}{2}\rho_\infty(f, V^-) = 0.79688 + 0.2(t^4 - 6t^2 + 1)$$

is a best approximation of $f(t) = \cos(\frac{1}{2}\pi t)$ in

$$V = \{v(t) = x_1 + x_2(t^4 - 6t^2 + 1) : x_1, x_2 \in \mathbb{R}\}$$

and $\rho_\infty(f, V) = 3.12 \times 10^{-3}$ holds.

Problem I.5.6 Show, with the help of the maximum principle (I.2.20), that for

$$\varphi(t, s) = 0.79688 + 0.2(t^4 - 6t^2s^2 + s^4)$$

one has

$$\max_{(t,s)\in B} | \hat{u}(t,s) - \varphi(t,s) | \leqslant 3.12 \times 10^{-3}$$

whereby again $\hat{u}$ is the solution of the boundary value problem.

I.5.5 A control–approximation problem in heat conduction

In Section I.1.2 we considered a control–approximation problem in connection with the heating of a metal plate in a stove and transformed this into a problem of infinite linear optimization. In the previous notation this reads: minimize γ subject to the side conditions (I.1.10)

$$\left.\begin{aligned} B(u)(x) + \gamma &\geqslant \hat{y}(x) \\ -B(u)(x) + \gamma &\geqslant -\hat{y}(x) \end{aligned}\right\} \quad \text{for } x \in [-1, +1] \tag{I.5.41}$$

and (I.1.4)

$$\left.\begin{aligned} u(t) &\geqslant -1 \\ -u(t) &\geqslant -1 \end{aligned}\right\} \quad \text{for } t \in [0, T]. \tag{I.5.42}$$

Here $B : C[0, T] \to C[-1, 1]$ is the linear mapping defined by (I.1.8) and $\hat{y} \in C[-1, +1]$ is given by (I.1.9). For each $u \in C[0, T]$, $B(u)$ is the unique solution of the initial boundary value problem (I.1.2), (I.1.3), (I.1.5) with $y_0 = 0$. If one equips both $C[0, T]$ and $C[-1, +1]$ with the maximum norms, denoted in both cases by $\|\ \|_\infty$, then an *a priori* estimate found in Friedman [64] implies

$$\| B(u) - B(\hat{u}) \|_\infty \leqslant K \| u - \hat{u} \|_\infty \tag{I.5.43}$$

for any two functions $u, \hat{u} \in C[0, T]$.
Now let $E \in \mathbb{R} \times C[0, T]$, equipped with the norm

$$\| (\gamma, u)^{\mathrm{T}} \| = | \gamma | + \| u \|_\infty$$

and with the trivial partial ordering where $K_E = E$ is the positive cone. Further let $F = (C[-1, +1])^2 \times (C[0, T])^2$, equipped with the norm

$$\| (y_1, y_2, u_1, u_2)^{\mathrm{T}} \| = \| y_1 \|_\infty + \| y_2 \|_\infty + \| u_1 \|_\infty + \| u_2 \|_\infty$$

and the partial ordering

$$(y_1, y_2, u_1, u_2)^{\mathrm{T}} \geqslant (\hat{y}_1, \hat{y}_2, \hat{u}_1, \hat{u}_2)^{\mathrm{T}} \Leftrightarrow \begin{cases} y_i(x) \geqslant \hat{y}_i(x) & \text{for all } x \in [-1, +1] \\ u_i(t) \geqslant \hat{u}_i(t) & \text{for all } t \in [0, T] \\ i = 1,2. & \end{cases}$$

Also define a mapping $A : E \to F$ by

$$A(\gamma, u) = \begin{pmatrix} \gamma e_1 + B(u) \\ \gamma e_1 - B(u) \\ u \\ -u \end{pmatrix} \qquad (e_1 = 1 \text{ on } C[-1, +1]) \tag{I.5.44}$$

and a functional $c : E \to \mathbb{R}$ by

$$c(\gamma, u) = \gamma. \tag{I.5.45}$$

Then A and c are linear and continuous (Proof = Exercise). Setting $b = (\hat{y}, -\hat{y}, -e_2, -e_2)^{\mathrm{T}}$, $e_2 = 1$ on $C[0, T]$, the side conditions (I.5.41), (I.5.42) go over into

$$A(\gamma, u) \geqslant b \qquad \text{for } (\gamma, u) \in E. \tag{I.5.46}$$

So we obtain the problem of minimizing the continuous linear functional c (I.5.45) subject to the side conditions (I.5.46). That is a linear optimization problem in the sense of Section I.3.1. The solvability of this problem is not assured. However, with the aid of Theorem I.4.14 the solvability of the associated dual problem and the coincidence of the extremal values can be proven. For this purpose we remark first that the interior $\overset{\circ}{K}_F$ of the positive cone K_F consists of all elements $(y_1, y_2, u_1, u_2) \in (C[-1, +1])^2 \times (C[0, T])^2$ with $y_i(x) > 0$ for all $x \in [-1, 1]$, and $u_i(t) > 0$ for all $t \in [0, T]$ for $i = 1,2$ (Proof = Exercise). If one chooses $\hat{u} \equiv 0$ and $\hat{\gamma} > \| \hat{y} \|_\infty$, then the inequalities (I.5.41), (I.5.42) are satisfied strictly for all $x \in [-1, +1]$ and all $t \in [0, T]$, i.e. $A(\hat{\gamma}, \hat{u}) - b \in \overset{\circ}{K}_F$. Furthermore, the extremal value of $c(\gamma, u) = \gamma$ for (γ, u) satisfying (I.5.46) is bounded from below by zero. By Theorem I.4.14, therefore, the associated dual problem is solvable and its extremal value equals that of c subject to the side conditions (I.5.46). By Section I.3.1, the dual problem is one of maximizing the continuous linear functional $y^*(b)$ subject to the side conditions

$$y^* \geqslant \Theta_{F^*}, y^*(A(\gamma, u)) \leqslant \gamma \qquad \text{for all } (\gamma, u) \in E. \tag{I.5.47}$$

By (IV.1.18), suitably applied, for each $y^* \in F^*$ there are uniquely determined elements $y_i^* \in C[-1, +1]$ and $z_i^* \in C[0, T]^*$, $i = 1,2$, such that

$$y^*(y_1, y_2, u_1, u_2) = y_1^*(y_1) + y_2^*(y_2) + z_1^*(i_1) + z_2^*(u_2)$$
$$\text{for all } (y_1, y_2, u_1, u_2)^{\mathrm{T}} \in F.$$

Moreover,

$$y^* \geqslant \Theta_{F^*} \Leftrightarrow \{y_i^* \geqslant \Theta_{C[-1,+1]^*} \text{ and } z_i^* \geqslant \Theta_{C[0,T]^*} \text{ for } i = 1,2\}$$

(see problem I.5.2), whereby $y_i^* \geqslant \Theta_{C[-1,+1]^*}$, $z_i^* \geqslant \Theta_{C[0,T]^*}$, respectively, are defined in accordance with (IV.1.14).

Now the side conditions (I.5.47) become

$$y_i^* \geqslant \Theta_{C[-1,+1]^*} \text{ and } z_i^* \geqslant \Theta_{C[0,T]^*} \text{ for } i = 1,2$$
$$y_1^*(\gamma e_1 + B(u)) + y_2^*(\gamma e_1 - B(u)) + z_1^*(u) - z_2^*(u) \leqslant \gamma \qquad \text{for all } (\gamma, u) \in E$$

which is equivalent to

$$\begin{aligned} &y_i^* \geqslant \Theta_{C[-1,+1]^*} \text{ and } z_i^* \geqslant \Theta_{C[0,T]^*} && \text{for } i = 1,2 \\ &y_1^*(e_1) + y_2^*(e_1) = 1 && \\ &y_1^*(B(u)) - y_2^*(B(u)) = z_2^*(u) - z_1^*(u) && \text{for all } u \in C[0,T]. \end{aligned} \tag{I.5.48}$$

The dual problem is therefore one of maximizing the continuous linear functional

$$y^*(b) = y_1^*(\hat{y}) - y_2^*(\hat{y}) - z_1^*(e_2) - z_2^*(e_2) \tag{I.5.49}$$

subject to the side conditions (I.5.48).

If one defines

$$U = \{u \in C[0,T] : \| u \|_\infty \leqslant 1\} \tag{I.5.50}$$

then the infimum of c subject to the side conditions (I.5.41), (I.5.42) equals the minimal deviation

$$\rho_\infty(\hat{y}, B(u)) = \inf_{u \in U} \| \hat{y} - B(u) \|_\infty \tag{I.5.51}$$

of the function $\hat{y} \in C[-1,+1]$ from the (convex) set $B(U) = \{B(u) : u \in V\}$ in $C[-1,+1]$. So the control–approximation problem in Section I.1.2 consists of determining a function $\hat{u} \in V$ with $\| \hat{y} - B(\hat{u}) \|_\infty = \rho_\infty(\hat{y}, B(U))$. As already remarked above, the existence of $\hat{u}$ is not guaranteed. However, $\rho_\infty(\hat{y}, B(U))$ can be calculated, in fact by the following theorem.

Theorem I.5.10 *The minimal deviation defined by (I.5.51) is*

$$\rho_\infty(\hat{y}, B(U)) = \max_{\substack{y^* \in C[-1,+1] \\ \|y^*\| \leqslant 1}} \left\{ y^*(\hat{y}) - \sup_{\substack{u \in C[0,T] \\ \|u\|_\infty \leqslant 1}} y^*(B(u)) \right\}. \tag{I.5.52}$$

Proof Let $y^* \in C[-1,+1]$ with $\| y^* \| \leqslant 1$ be given. Then for every $u \in C[0,T]$ with $\| u \|_\infty \leqslant 1$ we have

$$y^*(\hat{y}) - \sup_{\substack{u \in C[0,T] \\ \|u\|_\infty \leqslant 1}} y^*(B(u)) \leqslant y^*(\hat{y}) - y^*(B(u)) \leqslant \| \hat{y} - B(u) \|_\infty$$

whence

$$y^*(\hat{y}) - \sup_{\substack{u \in C[0,T] \\ \|u\|_\infty \leqslant 1}} y^*(B(u)) \leqslant \rho_\infty(\hat{y}, B(U)) \tag{I.5.53}$$

follows. Hence for the proof of (I.5.52), we seek $y^* \in C[-1,+1]^*$ with $\| y^* \| \leqslant 1$, for which equality holds in (I.5.53). For this purpose we start from a solution of the dual problem, i.e. from a quadruple $(y_1^*, y_2^*, z_1^*, z_2^*) \in F^*$, for which the value (I.5.49) is maximized. Since this maximal value equals the infimum of γ under the side conditions (I.5.41), (I.5.42), we obtain from the above considerations

$$y_1^*(\hat{y}) - y_2^*(\hat{y}) - z_1^*(e_2) - z_2^*(e_2) = \rho_\infty(\hat{y}, B(U)). \tag{I.5.54}$$

If one sets $y^* = y_1^* - y_2^*$ and $z^* = z_1^* - z_2^*$, then

$$\| y^* \| \leqslant \| y_1^* \| + \| y_2^* \| = y_1^*(e_1) + y_2^*(e_1) = 1$$

$$\| z^* \| = \sup_{\substack{u \in C[0,T] \\ \|u\|_\infty \leqslant 1}} z^*(u) \leqslant \| z_1^* \| + \| z_2^* \| = z_1^*(e_2) + z_2^*(e_2)$$

by Section IV.2.3.

We now make the assumption that $\| z^* \| < z_1^*(e_1) + z_2^*(e_2)$. By Theorem IV.2.7, there are two linear forms $z_+^*, z_-^* \geqslant \Theta_{C[0,T]^*}$ with

$$z_1^*(u) - z_2^*(u) = z^*(u) = z_+^*(u) - z_-^*(u)$$

and

$$\| z^* \| = z_+^*(e_2) + z_-^*(e_2).$$

From this it follows that

$$y_1^*(B(u)) - y_2^*(B(u)) = z_-^*(u) - z_+^*(u) \qquad \text{for all } u \in C[0, T]$$

i.e. the quadruple $(y_1^*, y_2^*, z_+^*, z_-^*)$ also satisfies (I.5.48), and

$$y_1^*(\hat{y}) - y_2^*(\hat{y}) - z_+^*(e_2) - z_-^*(e_2) > y_1^*(\hat{y}) - y_2^*(\hat{y}) - z_1^*(e_2) - z_2^*(e_2)$$

a contradiction to the optimality of $(y_1^*, y_2^*, z_1^*, z_2^*)$. Consequently $\| z^* \| = z_1^*(e_1) + z_2^*(e_2)$ and, because of the fact that

$$-z^*(u) = y^*(B(u)) \qquad \text{for all } u \in U$$

(see (I.5.48)) by (I.5.54), the statement (I.5.53) is satisfied with the equality sign.

Problem I.5.7 Show that

$$\rho_\infty(\hat{y}, B(U)) = \max_{\substack{y^* \in C[-1,+1]^* \\ \|y^*\| \leqslant 1}} \{ | y^*(\hat{y}) | - \| B^*(y^*) \| \} \tag{I.5.55}$$

where $B^* : C[-1, +1]^* \to C[0, 1]^*$ is the linear mapping adjoint to B (see Section IV.1.3).

According to (I.5.53), in order to determine a lower bound of $\rho_\infty(\hat{y}, B(u))$, one chooses $y^* \in C[-1, +1]^*$ with $\| y^* \| \leqslant 1$ and then calculates the left-hand side of the inequality (I.5.53). In that procedure, generally, the calculation of the supremum will cause difficulties. They can be mitigated, however, by choosing for y^* a positive (continuous) linear functional on $C[-1, +1]$ (see Section IV.2.3). On the basis of a general monotonicity theorem (see, for example, Collatz [68]) the following implication holds:

$$\{u_1, u_2 \in C[0, T], u_1(t) \geqslant u_2(t) \text{ for all } t \in [0, T]\}$$
$$\Rightarrow B(u_1)(x) \geqslant B(u_2)(x) \text{ for all } x \in [-1, +1].$$

If $y^* \in C[-1, +1]$ is positive, then it follows immediately that

$$\sup_{\substack{u \in C[0,T] \\ \|u\|_\infty \leqslant 1}} y^*(B(u)) = y^*(B(e_2)).$$

If, further, $\| y^* \| = y^*(e_1) \leqslant 1$, then one concludes from (I.5.53) that

$$y^*(\hat{y}) - y^*(B(e_2)) \leqslant \rho_\infty(\hat{y}, B(u)). \tag{I.5.56}$$

We shall elaborate this with a simple example. Let $y_0 = 0, y_T = e$. Then we choose

$$y^*(y) = y(0) \qquad \text{for all } y \in C[-1, +1] \tag{I.5.57}$$

and since $\hat{y} = e_1 = 1$ we obtain

$$1 - B(e_2)(0) \leqslant \rho_\infty(\hat{y}, B(U)). \tag{I.5.58}$$

Further, one can show that $u = e_2$ is a solution of the control–approximation problem and

$$\| \hat{y} - y(T, \cdot, e_2) \| = \max_{-1 \leqslant x \leqslant 1} | 1 - B(e_2)(x) | = 1 - B(e_2)(0)$$

so that equality holds in (I.5.58) (see Krabs [75]).

Generally, in the evaluation of (I.5.53) one will begin with the explicit representations (I.1.8), (I.1.9), of $B(u)$, and $\hat{y}$ respectively. Since by problem I.1.1, the series in (I.1.8), (I.1.9), respectively, converge uniformly, i.e. in the sense of the maximum norm in $C[-1, +1]$, it follows for every $y^* \in C[-1, +1]^*$ that

$$y^*(B(u)) = \sum_{k=1}^{\infty} \mu_k^2 B_k y^*(\cos \mu_k \cdot) \int_0^T \exp[-\mu_k^2(T - t)] u(t) \, dt \qquad \text{for } u \in C[0, T] \tag{I.5.59}$$

and

$$y^*(\hat{y}) = y^*(y_T) - y_0 \sum_{k=1}^{\infty} B_k y^*(\cos \mu_k \cdot) \exp(-\mu_k^2 T) \tag{I.5.60}$$

Further, one will choose not an arbitrary continuous linear functional but one whose values can be calculated easily. Here there are two possibilities:

(*a*) One takes finitely many distinct points $x_1, \ldots, x_m$ in $[-1, +1]$ and numbers $y_1, \ldots, y_m \in \mathbb{R}$ with

$$\sum_{i=1}^{m} | y_i | \leqslant 1$$

and defines

$$y^*(g) = \sum_{i=1}^{m} y_i g(x_i) \qquad \text{for } g \in C[-1, +1]. \tag{I.5.61}$$

Then $y^* \in C[-1, +1]$ and

$$\| y^* \| = \sum_{i=1}^{m} | y_i | \leqslant 1$$

(Proof = Exercise, see also Sections IV.2.3 and IV.2.4).

If, further, all $y_i \geqslant 0$, then y^* is positive and from (I.5.56), (I.5.59), and (I.5.60) it follows that

$$\sum_{i=1}^{m} y_i y_T(x_i) - y_0 \sum_{k=1}^{\infty} B_k \sum_{i=1}^{m} y_i \cos(\mu_k x_i) \exp(-\mu_k^2 T)$$

$$- \sum_{k=1}^{\infty} B_k \sum_{i=1}^{m} y_i \cos(\mu_k x_i)[1 - \exp(-\mu_k^2 T)] \leqslant \rho_\infty(\hat{y}, B(U)). \qquad \text{(I.5.62)}$$

(*b*) One takes a function $y \in C[-1, +1]$ with

$$\int_{-1}^{+1} | y(x) | \, dx \leqslant 1$$

and defines

$$y^*(g) \int_{-1}^{+1} y(x) g(x) \, dx \qquad \text{for } g \in C[-1, +1]. \qquad \text{(I.5.63)}$$

Then $y^* \in C[-1, +1]^*$ and

$$\| y^* \| \leqslant \int_{-1}^{+1} | y(x) | \, dx \leqslant 1$$

(Proof = Exercise).

If, further, $y \geqslant 0$, then y^* is positive, and from (I.5.56), (I.5.59), and (I.5.60) it follows that

$$\int_{-1}^{+1} y(x) y_T(x) \, dx - y_0 \sum_{k=1}^{\infty} B_k \int_{-1}^{+1} y(x) \cos(\mu_k x) \, dx \exp(-\mu_k^2 T)$$

$$- \sum_{k=1}^{\infty} B_k \int_{-1}^{+1} y(x) \cos(\mu_k x) \, dx \, [1 - \exp(-\mu_k^2 T)] \leqslant \rho_\infty(\hat{y}, B(U)). \qquad \text{(I.5.64)}$$

It can even be shown that the minimal deviation $\rho_\infty(\hat{y}, B(U))$ can be approximated arbitrarily closely with the expression $| y^*(\hat{y}) | - \| B^*(y^*) \|$, if y^* is chosen suitably according to (I.5.61), and (I.5.63) respectively (see Krabs [75]).

In particular, for $y \equiv \frac{1}{2}$, we get from (I.5.64) the estimate

$$\tfrac{1}{2} \int_{-1}^{+1} y_T(x) \, dx - y_0 \sum_{k=1}^{\infty} B_k \sin \mu_k \exp(-\mu_k^2 T)$$

$$- \sum_{k=1}^{\infty} B_k \, [(\sin \mu_k)/\mu_k] \, [1 - \exp(-\mu_k^2 T)] \leqslant \rho_\infty(\hat{y}, B(U)).$$

From this one obtains for the above example, $y_0 = 0, y_T = e_1 \equiv 1$, by making use of

$$(\sin \mu_k)/\mu_k = b \, (\cos \mu_k)/\mu_k^2$$

(see Section I.1.2), the inequality

$$1 - b \sum_{k=1}^{\infty} B_k \, [(\cos \mu_k)/\mu_k^2] \, [1 - \exp(-\mu_k^2 T)] \leqslant \rho_\infty(\hat{y}, B(U)).$$

On the basis of the above considerations (for purposes of comparison) the equalities

$$\rho_\infty(\hat{y}, B(U)) = 1 - B(e_2)(0) = 1 - \sum_{k=1}^{\infty} B_k [1 - \exp(-\mu_k^2 T)]$$

are valid.

Problem I.5.8a Show that for all $x \in [-1, +1]$

$$\sum_{k=1}^{\infty} B_k \cos(\mu_k x) = 1.$$

Problem I.5.8b Making use of (*a*) show that

$$\sum_{k=1}^{\infty} B_k \, [(\cos \mu_k)/\mu_k^2] = 1/b.$$

In conclusion, we remark that lower bounds can also be obtained for the minimal deviation $\rho_\infty(\hat{y}, B(U))$ without using the explicit representations (I.1.8), and (I.1.9), for $B(u)$, and $\hat{y}$, and these bounds come arbitrarily close to the minimal deviation. This will not be carried out in detail here; we refer the reader to Glashoff and Krabs [75].

Similar methods for obtaining lower bounds for the extremal value of a control problem with a partial differential equation were given by Yavin [71, 73].

In general, the solvability of the problem treated above for continuous control functions is not guaranteed. It is guaranteed, however, if one replaces the vector space of continuous functions by the vector space of functions bounded and measurable on $[0, T]$, whereby the minimal deviation does not change (see Krabs [76]). The existence of optimal measurable control functions was first proven by Yegorov [63] and later by Weck [75], who showed, moreover, that in the case of positive minimal deviation for every optimal control function u the 'bang bang principle' necessarily holds: $|\hat{u}| = 1$ almost everywhere, whence the uniqueness follows immediately. This 'bang bang principle' was refined further by Glashoff and Krabs [75]. Methods for the approximate solution of the problem were given by Arndt [75], Glashoff and Gustafson [76], and Krabs and Weck [74].

SOLUTIONS AND HINTS TO THE PROBLEMS

Problem I.1.1 The relations

$$(k-1)\pi < \mu_k < k\pi \qquad \text{for } k = 1,2,\ldots$$

can be derived from

$$\tan \mu_k = b/\mu_k \qquad \text{for } k = 1,2, \ldots$$

by drawing the graphs of $\tan \mu$ and b/μ and considering the points where they intersect.

The uniform convergence of the two series that define $y(t, x, u)$ is a straightforward consequence of

$$|B_k| = \left| \frac{2 \sin \mu_k}{\mu_k + \frac{1}{2} \sin 2\mu_k} \right| = \frac{2b \mid \cos \mu_k \mid}{\mid \mu_k^2 + \frac{1}{2}\mu_k \sin 2\mu_k \mid} \leqslant \frac{4b}{\mu_k^2} \leqslant \frac{4b}{\pi^2} \frac{1}{(k-1)^2}$$

for $k \geqslant 2$, if the Weierstrass criterion is applied.

Problem I.2.1 The assertion that V_0 is non-empty follows immediately from the well known fact that there is exactly one polynomial v of degree $(r-1)(\leqslant(n-1))$ which interpolates f at $t_1, \ldots, t_r \in M = [a, b]$.

Problem I.2.2 Put $v = u - \max_{\bar{B}} |u|$. Then $v \in L$ and $v \leqslant 0$ on $\bar{B}$. If there exists a point $(x^*, t^*) \in \bar{B}$ such that $v(x^*, t^*) = 0$, then by (I.2.13)

$$\max_{\bar{B}} v = 0 = \max_{\Gamma} v = \max_{\Gamma} u - \max_{\bar{B}} |u|$$

which implies (I.2.14).

If $v < 0$ on $\bar{B}$, then

$$\max_{\bar{B}} |u| = \max_{\bar{B}}(-u) = \max_{\bar{B}}(-v) - \max_{\bar{B}} |u| = \max_{\Gamma}(-v) - \max_{\bar{B}} |u|$$

$$= \max_{\Gamma}(-u) \leqslant \max_{\Gamma} |u| \leqslant \max_{\bar{B}} |u|$$

which also implies (I.2.14).

If $u_1, u_2 \in L$ are two solutions of (I.2.10)–(I.2.12) then $u_1 - u_2 \in L$ and by (I.2.14)

$$\max_{\bar{B}} |u_1 - u_2| = \max_{\Gamma} |u_1 - u_2| = 0.$$

Problem I.2.3 First of all we mention that there exists exactly one solution $u \in L$ of (I.2.10)–(I.2.12) in the case $f_0 \equiv 0, f_1 = h \in C[0, T], \Phi \in C[0, 1]$ with $\Phi(0) = 0, \Phi(1) = h(0)$ (see, for instance, Friedman [64]). If $\Phi(x) = h(0)x$, then we put

$$v(x, t) = \sum_{j=1}^{n} a_j \exp[-(\tfrac{1}{2}j\pi)^2 t] \sin \tfrac{1}{2}j\pi x \qquad \text{for } (x, t) \in \bar{B},$$

with variable coefficients $a_1, \ldots, a_n \in \mathbb{R}$. Obviously $v \in L$ and, if $u \in L$ is the solution of (I.2.10)–(I.2.12), then by (I.2.14)

$$\max_{\bar{B}} | v - u | = \max_{\Gamma} | v - u |$$

$$= \max \left\{ \max_{x \in [0,1]} \left| \sum_{j=1}^{n} a_j \sin \tfrac{1}{2} j\pi x - h(0)x \right|, \right.$$

$$\left. \max_{t \in [0,T]} \left| \sum_{j=1}^{n} a_j \sin \tfrac{1}{2} j\pi \exp[-(\tfrac{1}{2} j\pi)^2 t] - h(t) \right| \right\}.$$

Minimizing $\max_{\bar{B}} | v - u |$ by a suitable choice of $a_1, \ldots, a_n \in \mathbb{R}$ amounts to a problem of simultaneous uniform approximation. If $\Phi \in C[0, 1]$, $\Phi(0) = 0$, $\Phi(1) = h(0)$, then we put

$$v(x, t) \sum_{j=1}^{m} a_j \exp[-(j\pi)^2 t] \sin j\pi x + \sum_{j=1}^{n} b_j \exp[-(\tfrac{1}{2} j\pi)^2 t] \sin \tfrac{1}{2} j\pi x$$

$(x, t) \in \bar{B}$, with variable coefficients $a_1, \ldots, a_m, b_1, \ldots, b_n \in \mathbb{R}$. Again $v \in L$ and, if $u \in L$ is the solution of (I.2.10)–(I.2.12), then by (I.2.14)

$$\max_{\bar{B}} | v - u | = \max_{\Gamma} | v - u |$$

$$= \max \left\{ \max_{x \in [0,1]} \left| \sum_{j=1}^{m} a_j \sin j\pi x + \sum_{j=1}^{n} b_j \sin \tfrac{1}{2} j\pi x - \Phi(x) \right|, \right.$$

$$\left. \max_{t \in [0,T]} \left| \sum_{j=1}^{n} b_j \sin \tfrac{1}{2} j\pi \exp[-(\tfrac{1}{2} j\pi)^2 t] - h(t) \right| \right\}$$

$$\leqslant \max_{x \in [0,1]} \left| \sum_{j=1}^{m} a_j \sin j\pi x - [\Phi(x) - h(0)x] \right|$$

$$+ \max \left\{ \max_{x \in [0,1]} \left| \sum_{j=1}^{n} b_j \sin \tfrac{1}{2} j\pi x - h(0)x \right|, \right.$$

$$\left. \max_{t \in [0,T]} \left| \sum_{j=1}^{n} b_j \sin \tfrac{1}{2} j\pi \exp[-(\tfrac{1}{2} j\pi)^2 t] - h(t) \right| \right\}.$$

Problem I.3.1 We have to solve the problem of maximizing $\varphi(x, 0, 0) = x_1$ subject to

$$\varphi(x, t, 1) = x_1 + (t^4 - 6t^2 + 1)x_2 \leqslant \cos(\tfrac{1}{2}\pi t) \qquad \text{for all } t \in [0,1]$$

which is equivalent to minimizing $-x_1$ subject to

$$-x_1 - (t^4 - 6t^2 + 1)x_2 \geqslant -\cos(\tfrac{1}{2}\pi t) \qquad \text{for all } t \in [0, 1]. \tag{1}$$

If we put $y_1^* = 1$, then we have (with the notation in Section I.3.3.1)

$$y_1^* f_1(t_1) = 1 \qquad y_1^* f_2(t_1) = 0$$

hence

$$-y_1^* \cos(\tfrac{1}{2}\pi t_1) = -\cos(\tfrac{1}{2}\pi t_1) \leqslant \inf\{(-x_1) \text{ subject to } (1)\} = \alpha.$$

From

$$-\varphi(x, t_1, 1) = -x_1 = -\cos(\tfrac{1}{2}\pi t_1)$$

and

$$\varphi'(x, t_1, 1) = -4(t_1^3 - 3t_1)x_2 = \tfrac{1}{2}\pi \sin \tfrac{1}{2}\pi t_1$$

we obtain

$$x_1 = \cos(\tfrac{1}{2}\pi t_1) \qquad \text{and} \qquad x_2 = \tfrac{1}{8}\pi t_1^{-1}(3 - t_1^2)^{-1} \sin \tfrac{1}{2}\pi t_1$$

and one can verify that (1) is satisfied. This implies

$$-x_1 = -\cos(\tfrac{1}{2}\pi t_1) = -\max\{\varphi(x, 0, 0) \mid \varphi(x, t, 1) \leqslant \cos(\tfrac{1}{2}\pi t) \text{ for all } t \in [0, 1]\}.$$

Problem I.3.2a In order to show that $\hat{v} \in V$ we put

$$\hat{v}(t) - f(t) = \tfrac{1}{16} - \varphi(t)$$

where

$$\varphi(t) = t(t - \tfrac{3}{4})^2 \qquad \text{for } t \in [0, 1].$$

Then

$$\varphi'(t) = 3(t - \tfrac{1}{4})(t - \tfrac{3}{4}) \qquad \text{and} \qquad \varphi''(t) = 3(2t - 1).$$

Hence

$$\varphi''(t)\begin{cases} \leqslant 0 & \text{for } t \in [0, \tfrac{1}{2}] \\ \geqslant 0 & \text{for } t \in [\tfrac{1}{2}, 1] \end{cases}$$

and φ is concave in $[0, \frac{1}{2}]$ and convex in $[\frac{1}{2}, 1]$, respectively. This implies

$$\min_{t\in[0,\frac{1}{2}]} \varphi(t) = \min\{\varphi(0), \varphi(\tfrac{1}{2})\} = 0 \qquad \max_{t\in[0,\frac{1}{2}]} \varphi(t) = \varphi(\tfrac{1}{4}) = \tfrac{1}{16}$$

$$\min_{t\in[\frac{1}{2},1]} \varphi(t) = \varphi(\tfrac{3}{4}) = 0 \qquad \max_{t\in[\frac{1}{2},1]} \varphi(t) = \max_{t\in[\frac{1}{2},1]} \{\varphi(\tfrac{1}{2}), \varphi(1)\} = \tfrac{1}{16}$$

whence $\hat{v} \in V$ and $\| \hat{v} - f \|_\infty = \frac{1}{16}$.

If we choose $t_1^1 = \frac{1}{4}, t_2^1 = 1$ $(m_1 = 2)$, $t_1^2 = 0, t_2^2 = \frac{3}{4}$ $(m_2 = 2)$, then (I.3.11‴) reads (with $w_1(t) = 1, w_2(t) = t, w_3(t) = t^2$)

$$y_1^1 + y_2^1 - y_1^2 - y_2^2 = 0$$
$$\tfrac{1}{4}y_1^1 + y_2^1 - 0y_1^2 - \tfrac{3}{4}y_2^2 = 0$$
$$\tfrac{1}{16}y_1^1 + y_2^1 - 0y_1^2 - \tfrac{9}{16}y_2^2 = 0$$
$$y_1^2 + y_2^2 = 1$$

and yields the solution $y_1^1 = \frac{2}{3}, y_1^2 = \frac{1}{3}, y_2^1 = \frac{1}{3}, y_2^2 = \frac{2}{3}$. By (I.3.19) (with t^3 instead of t) we therefore conclude

$$\frac{1}{64}\frac{2}{3} + 1\frac{1}{3} - 0\frac{1}{3} - \frac{27}{64}\frac{2}{3} = \frac{1 + 32 - 27}{96} = \frac{1}{16} \leqslant \rho_\infty(f, V).$$

This implies $\rho_\infty(f, V) = \frac{1}{16}$ because of $\rho_\infty(f, V) \leqslant \| \hat{v} - f \|_\infty = \frac{1}{16}$.

Problem I.3.2b On defining φ as in (*a*) we obtain

$$\hat{v}(t) - f(t) = \tfrac{1}{32} - \varphi(t) \qquad \text{for } t \in [0, 1]$$

whence

$$\| \hat{v} - f \|_\infty = \tfrac{1}{32}.$$

By Section I.2.1.1 the approximation problem is equivalent with minimizing γ subject to

$$\begin{aligned} x_1 + x_2 t + x_3 t^2 + \gamma &\geqslant t^3 \\ -x_1 - x_2 t - x_3 t^2 + \gamma &\geqslant -t^3 \qquad \text{for } t \in [0, 1]. \end{aligned} \tag{2}$$

This is a problem of the form (I.3.9) with $r = 1$ because we can assume $\gamma \geqslant 0$. In order to obtain a feasible solution of the dual problem, one can proceed as follows. Choose $t_1^1, \ldots, t_{m_1}^1, t_1^2, \ldots, t_{m_2}^2 \in [0, 1]$ and numbers $y_1^1 \geqslant 0, \ldots, y_{m_1}^1 \geqslant 0$, $y_1^2 \geqslant 0, \ldots, y_{m_1}^2 \geqslant 0$ such that

$$\begin{aligned} &\sum_{i=1}^{m_1} y_i^1 + \sum_{i=1}^{m_2} y_i^2 \leqslant 1 \\ &\sum_{i=1}^{m_1} y_i^1 - \sum_{i=1}^{m_2} y_i^2 = 0 \\ &\sum_{i=1}^{m_1} y_i^1 t_i^1 - \sum_{i=1}^{m_2} y_i^2 t_i^2 = 0 \\ &\sum_{i=1}^{m_1} y_i^1 (t_i^1)^2 + \sum_{i=1}^{m_2} y_i^2 (t_i^2)^2 = 0 \end{aligned} \tag{3}$$

and define

$$y^*(y) = \sum_{i=1}^{m_1} y_i^1 y^1(t_i^1) + \sum_{i=1}^{m_2} y_i^2(t_i^2) \qquad \text{for } (y^1, y^2) \in C[0, 1]^2. \tag{4}$$

Then y^* satisfies (I.3.11) and by Theorem I.3.1 we obtain

$$\sum_{i=1}^{m_1} y_i^1 (t_i^1)^3 - \sum_{i=1}^{m_2} y_i^2 (t_i^2)^3 \leqslant \inf\{\gamma \text{ subject to (2)}\} = \rho_\infty(f, V).$$

If we choose $m_1 = m_2 = 2$, $t_1^1 = \frac{1}{4}$, $t_2^1 = 1$, $t_2^2 = 1$, $t_1^2 = 0$, $t_2^2 = \frac{3}{4}$ then (3) yields as solution $y_1^1 = \frac{1}{3}$, $y_2^1 = \frac{1}{6}$, $y_1^2 = \frac{1}{6}$, $y_2^2 = \frac{1}{3}$. Hence

$$y_1^1(t_1^1)^3 + y_2^1(t_2^1)^3 - y_1^2(t_1^2)^3 - y_2^2(t_2^2)^3 = \tfrac{1}{32} \leqslant \rho_\infty(f, V).$$

Equality holds because of $\rho_\infty(f, V) \leqslant \| \hat{v} - f \| = \frac{1}{32}$.

Remark The assertion $\rho_\infty(f, V) = \frac{1}{32}$ is also an immediate consequence of (*a*) and Theorem I.5.8.

Problem I.3.2c By Section I.2.1.1 the approximation problem is equivalent to minimizing γ subject to (I.2.5) which again is a problem of the form (I.3.9) with $r = 1$. Let $y_1, \ldots, y_m \in \mathbb{R}$ be given such that

$$\sum_{i=1}^{m} v_j(t_i)y_i = 0 \qquad \text{for } j = 1, \ldots, n, \ \sum_{i=1}^{m} |y_i| \leqslant 1. \tag{5}$$

Then we define $m_1 = m_2 = m$, $t_i^1 = t_i^2 = t_i$ and

$$y_i^1 = \max(0, y_i), y_i^2 = \max(0, -y_i) \text{ for } i = 1, \ldots, m$$

and obtain

$$\sum_{i=1}^{m_1} y_i^1 + \sum_{i=1}^{m_2} y_i^2 \leqslant 1$$

$$\sum_{i=1}^{m_1} y_i^1 v_j(t_i^1) - \sum_{i=1}^{m_2} y_i^2 v_j(t_i^2) = 0 \qquad \text{for } j = 1, \ldots, n$$

because of

$$y_i = y_i^1 - y_i^2, |y_i| = y_i^1 + y_i^2, i = 1, \ldots, m.$$

Again y^* defined by (4) satisfies (I.3.11) and Theorem I.3.1 yields

$$\sum_{i=1}^{m} y_i f(t_i) = \sum_{i=1}^{m_1} y_i^1 f(t_i^1) - \sum_{i=1}^{m_2} y_i^2 f(t_i^2) \leqslant \inf\{\gamma \text{ subject to (I.2.5)}\} = \rho_\infty(f, V).$$

Since the assumption (5) also holds, if we replace each y_i by $-y_i$, we conclude as well that

$$-\sum_{i=1}^{m} y_i f(t_i) \leqslant \rho_\infty(f, V)$$

whence

$$\left| \sum_{i=1}^{m} y_i f(t_i) \right| \leqslant \rho_\infty(f, V). \tag{6}$$

A direct proof of the last inequality can be given as follows: Let

$$v = \sum_{j=1}^{n} x_j v_j \in V$$

be chosen arbitrarily. Then from (5) we obtain

$$\pm \sum_{i=1}^{m} y_i f(t_i) = \pm \sum_{i=1}^{m} y_i [f(t_i) - v(t_i)] \leqslant \sum_{i=1}^{m} |y_i| \max |f(t_i) - v(t_i)| \leqslant \| v - f \|_\infty$$

and hence

$$\pm \sum_{i=1}^{m} y_i f(t_i) \leqslant \rho_\infty(f, V)$$

because v is arbitrary.

Problem I.3.2d If we choose $m = 4$, $t_1 = 0$, $t_2 = \frac{1}{4}$, $t_3 = \frac{3}{4}$, $t_4 = 1$ and $y_1 = \frac{1}{6}$, $y_2 = -\frac{1}{3}$, $y_3 = \frac{1}{3}$, $y_4 = -\frac{1}{6}$, then we obtain

$$\sum_{i=1}^{4} v_j(t_i) y_i = 0 \qquad \text{for } j = 1,2,3, \ \sum_{i=1}^{4} |y_i| = 1$$

where $v_1(t) = 1$, $v_2(t) = t$, $v_3(t) = t^2$. Therefore, by (6) with $f(t) = t^3$, one has

$$|0 \tfrac{1}{6} - \tfrac{1}{64} \tfrac{1}{3} - \tfrac{1}{6}| = \tfrac{1}{32} \leqslant \rho_\infty(f, V).$$

Problem I.3.3a If we put $f(t) = 0$ for all $t \in [0, 1]$ and $r = 1$, then we have

$$\int_0^1 t f(t)\, \mathrm{d}t + 2r = 2$$

which is the minimum value of the problem because

$$\int_0^1 t f(t)\, \mathrm{d}t \geqslant 0$$

for each $f \in L_2[0, 1]$ such that $f \geqslant 0$ a.e. and $r \geqslant 1$ which is a consequence of the first constraint of the problem.

Problem I.3.3b The adjoint mapping $A^* : F^* \to E^*$ is given by

$$A^*(y^*)(f, r) = y^*(A(f, r)) = \int_0^1 g(t) \left(\int_t^1 f(s)\, \mathrm{d}s + r \right) \mathrm{d}t$$

$$= \int_0^1 \int_0^t g(s)\, \mathrm{d}s\, [f(t) + g(t) r]\, \mathrm{d}t$$

where

$$y^*(y) = \int_0^1 g(t) y(t)\, \mathrm{d}t \qquad \text{for all } y \in F = L_2[0, 1].$$

Therefore $A^*(y^*) \leqslant c$ is equivalent to

$$\int_0^1 \left(t - \int_0^t g(s)\, \mathrm{d}s \right) f(t)\, \mathrm{d}t + \left(2 - \int_0^1 g(t)\, \mathrm{d}t \right) r \geqslant 0$$

for all $f \in L_2[0, 1]$ with $f \geqslant 0$ a.e. and all $r \geqslant 0$. This in turn is equivalent to

$$t - \int_0^t g(s)\, \mathrm{d}s \geqslant 0 \text{ for almost all } t \in [0, 1] \text{ and } 2 - \int_0^1 g(t)\, \mathrm{d}t \geqslant 0.$$

$y^* \geqslant \Theta_{F^*}$ is equivalent to $g \geqslant 0$ a.e. Finally, we have

$$y^*(b) = \int_0^1 g(t)\, \mathrm{d}t$$

which completes the proof.

Problem I.3.3c Apparently $\int_0^1 g(t)\, \mathrm{d}t \leqslant 2$ is a consequence of

$$\int_0^t g(s)\, \mathrm{d}s \leqslant t \text{ and } g(t) \geqslant 0 \qquad \text{for almost all } t \in [0, 1]. \tag{7}$$

Assume

$$\int_0^1 g(s)\, \mathrm{d}s > 1 \text{ for some } g \in L_2\,[0, 1] \text{ with } g \geqslant 0 \text{ a.e.}$$

Then, due to the continuity of the function $t \to \int_0^t g(s)\, \mathrm{d}s$, we would have

$$\int_0^t g(s)\, \mathrm{d}s > 1 \geqslant t \qquad \text{for all } t \in [1-\epsilon, 1] \text{ and some } \epsilon > 0.$$

Therefore (7) implies that $\int_0^1 g(s)\, \mathrm{d}s \leqslant 1$. If we put $g(t) = 1$ for all $t \in [0, 1]$, then (7) is satisfied and we obtain $\int_0^1 g(s)\, \mathrm{d}s = 1$. This implies that $g \equiv 1$ is optimal and $\beta = 1$.

Problem I.4.1 We have to show that $\overline{L_b \cap K(A, c)} \supseteq L_b \cap \overline{K(A, c)}$ which is an immediate consequence of $\overline{K(A, c)} = K(A, c)$.

Problem I.4.2

(1) Let the problem (P) be normal. If $v^*(A^*, b, c) < +\infty$, then by Theorem I.4.7*a* and Lemma I.4.3 it follows that $M \neq \phi$ and $v(A, c, b) = v^*(A^*, b, c)$.

(2) Assume the implication

$$v^*(A^*, b, c) < +\infty \Rightarrow (M \neq \phi \text{ and } v(A, c, b) = v^*(A^*, b, c)) \tag{8}$$

to be true. Consider $(\alpha, z) \in L_b \cap \overline{K(A, c)}$. Then $z = b$ and $(\alpha, b) \in \overline{K(A, c)}$. By Theorem I.4.6 it follows that $v^*(A^*, b, c) \leqslant \alpha < +\infty$. Because of (8) there exists a sequence (x_k) in M such that

$$v^*(A^*, b, c) = v(A, c, b) = \lim_{k \to \infty} c(x_k).$$

Put $y_k = A(x_k) - b$. Then $y_k \geqslant \Theta_F$ for all k. If $\alpha = v^*(A^*, b, c)$, then

$$(\alpha, z) = \lim_{k \to \infty} (c(x_k), A(x_k) - y_k) = \lim_{k \to \infty} (c(x_k), b)$$

which implies $(\alpha, z) \in \overline{L_b \cap K(A, c)}$, hence $L_b \cap \overline{K(A, c)} \subseteq \overline{L_b \cap K(A, c)}$ where in fact equality holds. If $\alpha > v^*(A^*, b, c)$, then we put $r_k = \alpha - c(x_k)$ such that $r_k \geqslant 0$ for k sufficiently large. Further,

$$(\alpha, z) = \lim_{k \to \infty} (c(x_k) + r_k, A(x_k) - y_k) = \lim_{k \to \infty} (c(x_k) + r_k, b)$$

which again implies $(\alpha, z) \in \overline{L_b \cap K(A, c)}$. This completes the proof.

Problem I.5.1a A pair $(x, \gamma) \in \mathbb{R}^n \times \mathbb{R}$ is a feasible solution of problem (P) if and only if

$$\left\| \sum_{j=1}^{n} v_j x_j - f \right\|_\infty \leqslant \gamma \tag{9}$$

which implies

$$\rho_\infty(f, V) \leqslant \inf \{\gamma : (x, \gamma) \in \mathbb{R}^{n+1} \text{ feasible for some } x \in \mathbb{R}^n\}. \tag{10}$$

If $(\hat{x}, \hat{\gamma}) \in \mathbb{R}^n \times \mathbb{R}$ is an optimal solution of problem (P), then by (9) we have

$$\left\| \sum_{j=1}^{n} v_j \hat{x}_j - f \right\|_\infty = \hat{\gamma}$$

and (10) implies $\hat{\gamma} = \rho_\infty(f, V)$.

Problem I.5.1b If $\sum_{j=1}^n v_j \hat{x}_j$ solves the approximation problem, then by (9) and (10) it follows that $(\hat{x}, \rho_\infty(f, V))$ solves problem (P).

Problem I.5.2 The proof of (*a*) is straightforward.

In (*b*) the implication (I.5.11), (I.5.12) $\Rightarrow$ (I.5.10) is trivial. In order to prove the converse implication we assume $y_1(v_{j_0}) - y_2(v_{j_0}) \neq 0$ for some $j_0 \in \{1, \ldots, n\}$. If we choose $x_{j_0} = \text{sgn}(y_1(v_{j_0}) - y_2(v_{j_0}))$, $x_j = 0$ for all $j \neq j_0$, and $\gamma = 0$, then (I.5.10) is violated. If $y_1^*(e) + y_2^*(e) > 1$, then we choose $x_j = 0$ for all j and $\gamma = 1$ and again (I.5.10) is violated. Hence (I.5.10) $\Rightarrow$ (I.5.11), (I.5.12) follows by contraposition.

Problem I.5.3a Let $\hat{v} \in V$ be a best approximation of f. Then by Theorem I.5.6 there exist $s \leqslant (n + 1)$ distinct points $\hat{t}_1, \ldots, \hat{t}_s \in E_{\hat{v}}$ (I.5.20) and numbers $\hat{y}_1, \ldots, \hat{y}_s \in \mathbb{R}$ such that (I.5.21), (I.5.22), and (I.5.23) hold. This implies

$$\rho_\infty(f, V) = \| f - \hat{v} \|_\infty = \| f - \hat{v} \|_\infty \sum_{i=1}^{s} |\hat{y}_i| = \sum_{i=1}^{s} [f(\hat{t}_i) - \hat{v}(\hat{t}_i)] = \sum_{i=1}^{s} \hat{y}_i f(\hat{t}_i).$$

Problem I.5.3b Under the assumption (I.5.22) the statement (I.5.23) implies

$$\sum_{i=1}^{s} \hat{y}_i f(\hat{t}_i) = \sum_{i=1}^{s} \hat{y}_i [f(\hat{t}_i) - \hat{v}(\hat{t}_i)] = \| f - \hat{v} \|_\infty \sum_{i=1}^{s} |\hat{y}_i|.$$

Conversely, if (I.5.22) and

$$\sum_{i=1}^{s} \hat{y}_i f(\hat{t}_i) = \| f - \hat{v} \| \sum_{i=1}^{s} |\hat{y}_i|$$

hold, then

$$\sum_{\hat{y}_i \neq 0} |\hat{y}_i| \{\| f - \hat{v} \|_\infty - \text{sgn}(\hat{y}_i) [f(\hat{t}_i) - \hat{v}(\hat{t}_i)]\} = 0$$

which implies (I.5.23).

Problem I.5.3c Put $\varphi(t) = f(t) - \hat{v}(t)$. Then φ is convex, $\varphi(0) = 0$, $\varphi(1) = 3 - 2\sqrt{2}$ and $\varphi'(\sqrt{2} - 1) = 0$, hence

$$\min_{t \in [0,1]} \varphi(t) = (\sqrt{2} - 1) = -(3 - 2\sqrt{2})$$

and

$$\max_{t \in [0,1]} \varphi(t) = \varphi(1) = 3 - 2\sqrt{2}$$

which implies $\| f - \hat{v} \|_\infty = 3 - 2\sqrt{2}$. Choose $t_1 = 1$ and $t_2 = \sqrt{2} - 1$ then (I.5.21) and (I.5.22) read

$$|\hat{y}_1| + |\hat{y}_2| = 1$$

and

$$\hat{y}_1 + \hat{y}_2(\sqrt{2} - 1) = 0.$$

A solution is given by $\hat{y}_1 = 1 - 1/\sqrt{2}$ and $\hat{y}_2 = -1/\sqrt{2}$. Hence $\hat{y}_1 > 0$ and $\hat{y}_2 < 0$ and

$$f(\hat{t}_i) - \hat{v}(\hat{t}_i) = \| f - \hat{v} \|_\infty \operatorname{sgn} \hat{y}_i \qquad \text{for } i = 1, 2.$$

By Theorem I.5.6 $\hat{v}$ is a best approximation of f in V.

Problem I.5.4 It suffices to give the proof for V^+. The closedness of V^+ in V follows from the closedness of V in $C(M)$ (as a finite dimensional subspace) and the implication

$$\left.\begin{array}{r} v_k \to v \text{ (uniformly)} \\ v_k(t) \geqslant f(t) \text{ for all } t \in M \end{array}\right\} \Rightarrow v(t) \geqslant f(t) \qquad \text{for all } t \in M.$$

The convexity of V^+ follows from the convexity of V and the implication

$$\left.\begin{array}{r} v_1(t) \geqslant f(t) \\ v_2(t) \geqslant f(t) \end{array}\right\} \text{for all } t \in M, \lambda \in [0, 1] \Rightarrow \lambda v_1(t) + (1 - \lambda) v_2(t) \geqslant f(t) \quad \text{for all } t \in M.$$

Problem I.5.5 The proof is completely analogous to the solution of Problem I.5.1.

Problem I.5.6 Let

$$\max_{t \in [0,1]} |f(t) - \varphi(t, 1)| \leqslant \gamma. \tag{11}$$

Then

$$\varphi(t, 1) - \gamma \leqslant f(t) \leqslant \varphi(t, 1) + \gamma \text{ for all } t \in [0, 1] \Leftrightarrow \varphi - \gamma \leqslant f \leqslant \varphi + \gamma \text{ on } \Gamma.$$

Further

$$\Delta(\varphi(t, s) - \gamma) = \Delta(\varphi(t, s) + \gamma) = 0.$$

Hence the maximum principle (I.2.20) implies

$$\varphi(t,s)-\gamma \leqslant \hat{u}(t,s) \leqslant \varphi(t,s)+\gamma \qquad \text{for all } (t,s)\in \bar{B}$$

which is equivalent to

$$\max_{(t,s)\in\bar{B}} |\hat{u}(t,s)-\varphi(t,s)| \leqslant \gamma.$$

According to the arguments preceding Problem I.5.6 the statement (11) is true for $\gamma = 3.12 \times 10^{-3}$ which completes the proof.

Problem I.5.7 By the definition of B^* we have

$$\| B^* y^* \| = \sup_{\substack{u\in C[0,T]\\ \|u\|_\infty \leqslant 1}} B^*(y^*)(u) = \sup_{\substack{u\in C[0,T]\\ \|u\|_\infty \leqslant 1}} y^*(B(u)).$$

Therefore (I.5.52) ⇒ (I.5.55) since the conditions $y^* \in C[-1,+1]^*$, $\| y^* \| \leqslant 1$ are satisfied for $-y^*$ as well.

Problem I.5.8a By Section I.1.2 the solution of (I.1.2), (I.1.3), and (I.1.5) for $u = 0$ is given by

$$y(t,x,0) = y_0 \sum_{k=1}^{\infty} B_k \cos \mu_k x \exp(-\mu_k^2 t) \qquad \text{for each } y_0 \in \mathbb{R}.$$

Assume $y_0 \neq 0$ and put $t = 0$. Then $y(0,x,0) = y_0$ and

$$y(0,x,0) = y_0 \sum_{k=1}^{\infty} B_k \cos \mu_k x$$

which implies

$$\sum_{k=1}^{\infty} B_k \cos \mu_k x = 1 \qquad \text{for all } x \in [0,1].$$

Problem I.5.8b By Problem I.1.1 the series

$$\sum_{k=1}^{\infty} B_k \cos \mu_k x$$

converges uniformly on $[0,1]$. Hence

$$1 = \int_0^1 \sum_{k=1}^{\infty} B_k \cos \mu_k x = \sum_{k=1}^{\infty} B_k \int_0^1 \cos \mu_k x \mathrm{d}x = \sum_{k=1}^{\infty} B_k \, [(\sin \mu_k)/\mu_k]$$

$$= \sum_{k=1}^{\infty} B_k \, [b(\cos \mu_k)/\mu_k^2]$$

which implies the assertion.

CHAPTER II

Convex Problems

II.1 EXAMPLES OF CONVEX APPROXIMATION AND OPTIMIZATION PROBLEMS

II.1.1 The general linear approximation problem

Let Z be a normed vector space over $\mathbb{R}$, V be a linear subspace of Z and z be an arbitrary element of Z. We seek a $\hat{v} \in V$ satisfying

$$\| \hat{v} - z \| \leqslant \| v - z \| \qquad \text{for all } v \in V \tag{II.1.1}$$

where $\| \ \|$ denotes the norm in Z.

Every $\hat{v} \in V$ satisfying (II.1.1) is called a best approximation of z in V and the quantity

$$\rho(z, V) = \inf_{v \in V} \| v - z \| \tag{II.1.2}$$

is the minimal deviation (or the distance) of the element z from the set V.

There are numerous examples of such approximation problems. In Sections I.2.1 and I.5.2, we treated the case of the uniform approximation of functions, and in that case Z was the vector space $C(M)$ of continuous real-valued functions equipped with the maximum norm (I.2.1) defined on a compact subset M of a normed space, and V was a finite dimensional linear subspace of Z. A further example occurs when Z is the vector space $C[a, b]$ $(a < b)$ of continuous real-valued functions defined on a compact interval $[a, b]$ and equipped with the Euclidean norm

$$\| x \| = \left(\int_a^b | x(t) |^2 \mathrm{d}t \right)^{1/2} \qquad \text{for } x \in C[a, b]. \tag{II.1.3}$$

If V is a finite dimensional subspace of Z, then the above approximation problem is the well known least squares problem. In fact $Z = C[a, b]$ with the Euclidean norm (II.1.3) is a unitary space with the usual scalar product. The approximation problem (II.1.1) can now be re-written in various ways as a problem of convex optimization. For example, if one defines the real-valued functional f on Z by

$$f(y) = \| y - z \| \qquad \text{for } y \in Z \tag{II.1.4}$$

then f is convex (see Section II.2.1).

The approximation problem (II.1.1) is equivalent to the optimization problem of minimizing f. That is a convex optimization problem without explicit side conditions. A problem with side conditions is obtained by introducing the topological dual space Z^* of Z (see Section IV.1.2). We norm Z^* by

$$\| L \| = \sup_{\| y \| \leqslant 1} | L(y) | \qquad \text{for } L \in Z^* \tag{II.1.5}$$

and define the unit sphere in Z^* by

$$B^* = \{L \in Z^* : \| L \| \leqslant 1\}. \tag{II.1.6}$$

Then B^* is a convex subset (see Section IV.3) of Z^* (Proof = Exercise). Of basic importance is the following lemma.

Lemma II.1.1 *For every $y \in Z$, one has*

$$\| y \| = \max_{L \in B^*} L(y). \tag{II.1.7}$$

Proof For $y = \Theta_Z$ the assertion is trivial. Suppose therefore that $y \neq \Theta_Z$ and $L \in B^*$ is arbitrary. The it follows that

$$L(y) \leqslant | L(y) | \leqslant \| L \| \, \| y \| \leqslant \| y \|.$$

By Theorem IV.1.4 there corresponds to y an $L_y \in Z^*$ with $\| L_y \| = 1$ and $L_y(y) = \| y \|$, whence

$$\| y \| = L_y(y) = \max_{L \in B^*} L(y).$$

In particular, by Lemma II.1.1,

$$\| v - z \| = \max_{L \in B^*} L(v - z). \tag{II.1.8}$$

Therefore the approximation problem (II.1.1) is equivalent to the problem of minimizing the functional $f(v, \gamma) = \gamma$ subject to the side conditions

$$(v, \gamma) \in V \times \mathbb{R}$$
$$\varphi_L(v, \gamma) = L(v) - \gamma - L(z) \leqslant 0 \qquad \text{for all } L \in B^* \tag{II.1.9}$$

(Proof = Exercise). Here $V \times \mathbb{R}$ is a linear subspace of $Z \times \mathbb{R}$, φ_L is an affine linear functional for each L (see Section II.2.1), and f is obviously a linear functional. In Section II.5.4 we shall show that this problem can be regarded as a linear optimization problem in the sense of Section I.3.

In special cases it is often possible to make the maximum statement (II.1.8) even for a subset of B^*. In the case of uniform approximation in Section I.2.1, the statement (II.1.8) obviously holds for all $L \in B^*$ which are so-called point functionals of the form

$$L(y) = y(t) \quad \text{or} \quad L(y) = -y(t) \qquad \text{for } t \in M$$

and the side conditions (II.1.9) go over into

$$\begin{aligned} &(v, \gamma) \in V \times \mathbb{R} \\ &v(t) - \gamma - z(t) \leqslant 0 \\ &-v(t) - \gamma + z(t) \leqslant 0 \end{aligned} \qquad \text{for all } t \in M \qquad \text{(II.1.10)}$$

(see (I.2.5)). In this case we also have a linear optimization problem.

II.1.2 Optimal error estimates for linear operator equations

II.1.2.1 Defect estimates

Suppose X is a complete normed vector space over $\mathbb{R}$ and A is an arbitrary linear mapping of X into itself (see Section IV.1.3). Suppose further that $y \in X$ is a fixed element. Then we consider the linear operator equation

$$x - A(x) = y. \qquad \text{(II.1.11)}$$

If one assumes that

$$\| A \| = \sup_{\|x\| \leqslant 1} \| A(x) \| \leqslant \alpha < 1 \qquad \text{(II.1.12)}$$

then, as is well known, the equation (II.1.11) has for each $y \in X$ precisely one solution $x_y \in X$, which can be computed iteratively by means of the method of successive approximations (see for example Kantorowitsch and Akilov [64]).

For this one chooses as a starting element $x_0 = y$ and determines accordingly a sequence $\{x_k\}$ of approximations $x_k \in X$ for x_y by the scheme

$$x_{k+1} = A(x_k) + y \qquad \text{for } k = 0, 1, 2, \ldots.$$

The sequence $\{x_k\}$ defined in this way converges to x_y. However, the numerical execution of this procedure is usually very tedious. Therefore one often tries to select from some given class V of approximations for x_y an element $v \in V$ such that the deviation $\| x_y - v \|$ is as small as possible.

As a rule V is a finite dimensional linear subspace of X. If any $v \in V$ is given, then the substitution of v into the left-hand side of equation (II.1.11) yields

$$w = v - A(v) \qquad \text{(II.1.13)}$$

and we obtain the so-called defect

$$y - w = (x_y - v) - A(x_y - v)$$

where we again denote by x_y the unique solution of (II.1.11) (under the assumption (II.1.12)).

Intuitively one will now choose $v \in V$ so that

$$\| y - w \| = \| y - [v - A(v)] \|$$

is as small as possible. If one sets

$$W = \{w = v - A(v) : v \in V\} \tag{II.1.14}$$

then W as well as V is a finite dimensional linear subspace of X and we have the approximation problem of determining $\hat{w} = \hat{v} - A(\hat{v})$ with $\hat{v} \in V$ such that

$$\| \hat{w} - y \| \leqslant \| w - y \| \qquad \text{for all } w \in W. \tag{II.1.15}$$

This procedure is justified by the following theorem.

Theorem II.1.2 *Under the assumption (II.1.12) the error estimate*

$$\frac{\| y - w \|}{1 + \alpha} \leqslant \| x_y - v \| \leqslant \frac{\| y - w \|}{1 - \alpha} \tag{II.1.16}$$

holds for every $v \in V$ with w defined by (II.1.13), so that by minimizing $\| y - w \|$ (i.e. by solving (II.1.15)) the error $\| x_y - v \|$ is in general also made small.

Proof First

$$\begin{aligned} \| y - w \| &= \| x_y - v - A(x_y - v) \| \geqslant \| x_y - v \| - \| A(x_y - v) \| \\ &\geqslant \| x_y - v \| - \| A \| \, \| x_y - v \| \geqslant \| x_y - v \| \{1 - \alpha\} \end{aligned}$$

from which the right-hand inequality of (II.1.16) follows. Furthermore

$$\begin{aligned} \| y - w \| &= \| x_y - v - A(x_y - v) \| \leqslant \| x_y - v \| + \| A(x_y - v) \| \\ &\leqslant \| x_y - v \| + \| A \| \, \| x_y - v \| \leqslant \| x_y - v \| \{1 + \alpha\} \end{aligned}$$

which yields the left-hand inequality of (II.1.16).

As an example we consider a Fredholm integral equation of the second kind

$$x(t) - \int_a^b K(t, s) x(s) \, ds = y(t) \qquad \text{for } t \in [a, b]. \tag{II.1.11$'$}$$

Let X be the vector space $C[a, b]$ of continuous real-valued functions on $[a, b]$ with $a < b$, equipped with the maximum norm

$$\| z \|_\infty = \max_{t \in [a, b]} | z(t) | \qquad \text{for } z \in C[a, b].$$

Let $y \in C[a, b]$ be given and suppose that the kernel K is a continuous real-valued function defined on $[a, b] \times [a, b]$. If one defines for every $z \in C[a, b]$

$$A(z)(t) = \int_a^b K(t, s) z(s) \, ds \tag{II.1.17}$$

then A is a linear mapping of $C[a, b]$ into itself.

Furthermore

$$\| A(z) \| \leqslant \max_{t \in [a, b]} \int_a^b | K(t, s) | \, ds \, \| z \|_\infty$$

holds for all $z \in C[a, b]$. By Theorem IV.1.6, A is therefore continuous and

$$\| A \| \leqslant \alpha = \max_{t \in [a,b]} \int_a^b | K(t, s) | \, ds.$$

In fact $\| A \| = \alpha$ holds (see Ljusternik and Sobolew [68]). With A defined by (II.1.17) the integral equation (II.1.11′) goes over into the general linear operator equation (II.1.11). On the basis of the above remarks its unique solution is guaranteed, if the kernel K in (II.1.11′) satisfies

$$\alpha = \max_{t \in [a,b]} \int_a^b | K(t, s) | \, ds < 1. \tag{II.1.12′}$$

Minimizing the defect leads in this case to the uniform linear approximation problem of seeking a $\hat{w} \in W$ (according to (II.1.14)) satisfying

$$\| \hat{w} - y \|_\infty \leqslant \| w - y \|_\infty \qquad \text{for all } w \in W. \tag{II.1.15′}$$

Subject to the assumption (II.1.12′), we obtain for every v of the linear subspace V of $C[a, b]$ the error estimate

$$\frac{\| y - [v - A(v)] \|_\infty}{1 + \alpha} \leqslant \| x_y - v \|_\infty \leqslant \frac{\| y - [v - A(v)] \|_\infty}{1 - \alpha} \tag{II.1.16′}$$

whereby x_y is the unique solution of (II.1.11′). By (II.1.16′) in turn the minimization of the norm $\| y - [v - A(v)] \|_\infty$ of the defect $y - [v - A(v)]$ is justified.

The right-hand inequality of (II.1.16′) has already been derived in Section I.3.3.2 for the linear boundary value problem

$$L[x](s) = -x''(s) + (1 + s^2)x(s) = s^2 \qquad \text{for } s \in [-1, +1], x(-1) = x(+1) = 0. \tag{II.1.18}$$

If $G = G(t, s)$ is the Green's function for the boundary value problem

$$-x''(s) = r(s) \qquad \text{for } s \in [-1, +1], x(-1) = x(+1) = 0$$

where $r \in C[-1, +1]$, then by Section I.2.4.1 problem (II.1.18) is equivalent to the linear integral equation

$$x(t) + \int_{-1}^{+1} K(t, s)x(s) \, ds = y(t) \qquad \text{for } t \in [-1, +1] \tag{II.1.11″}$$

with

$$K(t, s) = G(t, s)(1 + s^2) \qquad y(t) = \int_{-1}^{+1} G(t, s)s^2 \, ds.$$

In this case $G(t, s)$ is given by

$$G(t, s) = \begin{cases} \frac{1}{2}(s + 1)(1 - t) & \text{for } -1 \leqslant s \leqslant t \leqslant +1 \\ \frac{1}{2}(t + 1)(1 - s) & \text{for } -1 \leqslant t \leqslant s \leqslant +1. \end{cases}$$

Furthermore we have

$$\alpha = \max_{t\in[-1,+1]} \int_{-1}^{+1} |K(t,s)|\,ds = \max_{t\in[-1,+1]} \int_{-1}^{+1} G(t,s)(1+s^2)\,ds = \tfrac{7}{12}.$$

Let $v \in C[-1,+1]$ be any approximate solution of (II.1.11″),

$$z(t) = v(t) + \int_{-1}^{+1} K(t,s)v(s)\,ds$$

and let x_y be the solution of (II.1.11″) which is equivalent to the boundary value problem (II.1.18). Then (II.1.16′) yields the following error estimate:

$$\tfrac{12}{19}\,\|y - z\|_\infty \leqslant \|x_y - v\|_\infty \leqslant \tfrac{12}{5}\,\|y - z\|_\infty. \tag{II.1.19}$$

If one chooses in particular, as in Section I.3.3.2,

$$v(s) = -\tfrac{1}{34}(1-s^2) + \tfrac{3}{34}(1-s^4) \quad (\Rightarrow v(-1) = v(+1) = 0)$$

and sets

$$r(s) = -v''(s) + (1+s^2)v(s) \tag{II.1.20}$$

then it follows that

$$\max_{s\in[-1,+1]} |r(s) - s^2| = \tfrac{200}{4131} \approx 0.0485.$$

Furthermore it follows from (II.1.20) that

$$z(t) = v(t) + \int_{-1}^{+1} G(t,s)(1+s^2)v(s)\,ds = \int_{-1}^{+1} G(t,s)r(s)\,ds$$

and consequently

$$\|y - z\|_\infty = \max_{t\in[-1,+1]} \left| \int_{-1}^{+1} G(t,s)[s^2 - r(s)]\,ds \right|$$

$$\leqslant \max_{t\in[-1,+1]} \int_{-1}^{+1} G(t,s)\,ds \max_{s\in[-1,+1]} |r(s) - s^2| = \tfrac{1}{2}\,\tfrac{200}{4131}.$$

The right-hand inequality of (II.1.19) therefore yields the estimate

$$\|x_y - v\|_\infty \leqslant \tfrac{12}{5}\,\tfrac{1}{2}\,\tfrac{200}{4131} \approx 0.058$$

which we have also already derived in Section I.3.3.2.

II.1.2.2 Operator estimates

We think of the mapping A in the operator equation (II.1.11) replaced by a continuous linear mapping $\hat{A}$ of X into itself, such that

$$\|\hat{A}\| = \sup_{\|x\|\leqslant 1} \|\hat{A}(x)\| \leqslant \hat{\alpha} < 1. \tag{II.1.21}$$

We suppose that the equation

$$x - \hat{A}(x) = y \tag{II.1.22}$$

which is uniquely solvable under the assumption (II.1.21), can be solved explicitly.

Let $\hat{x}_y$ be the associated solution, which we now use as an approximate solution for the equation (II.1.11). Setting $\hat{x}_y$ into (II.1.11) yields

$$\hat{y} = \hat{x}_y - A(\hat{x}_y). \tag{II.1.23}$$

So we obtain the defect

$$y - \hat{y} = (x_y - \hat{x}_y) - A(x_y - \hat{x}_y) \tag{II.1.24}$$

where x_y again denotes the unique solution of (II.1.11) (under the assumption (II.1.12)). From the right-hand inequality in (II.1.16) we obtain the estimate

$$\| x_y - \hat{x}_y \| \leqslant \frac{\| y - \hat{y} \|}{1 - \alpha}. \tag{II.1.25}$$

From (II.1.23) and $y = \hat{x} - \hat{A}(\hat{x}_y)$ we get by subtraction

$$y - \hat{y} = A(\hat{x}_y) - \hat{A}(\hat{x}_y) = (A - \hat{A})(\hat{x}_y)$$

and from this

$$\| y - \hat{y} \| \leqslant \| A - \hat{A} \| \, \| \hat{x}_y \|. \tag{II.1.26}$$

Now

$$\begin{aligned} \| y \| &= \| \hat{x}_y - \hat{A}(\hat{x}_y) \| \geqslant \| \hat{x}_y \| - \| \hat{A}(\hat{x}_y) \| \\ &\geqslant \| \hat{x}_y \| - \| \hat{A} \| \, \| \hat{x}_y \| \geqslant \| \hat{x}_y \| (1 - \hat{\alpha}) \end{aligned} \tag{II.1.27}$$

with $\hat{\alpha}$ as in (II.1.21). Combining (II.1.25), (II.1.26), and (II.1.27) we get the estimate

$$\| x_y - \hat{x}_y \| \leqslant \frac{\| A - \hat{A} \| \, \| y \|}{(1 - \alpha)(1 - \hat{\alpha})}. \tag{II.1.28}$$

This suggests varying the operator $\hat{A}$ in a suitable class V of continuous linear mappings under the side conditions (II.1.21) in such a way that the deviation $\| A - \hat{A} \|$ is as small as possible. V should consist of such mappings which admit an explicit solution of the equation (II.1.22). We seek therefore a mapping $\hat{A}_0 \in V_{\hat{\alpha}}$, for $\hat{\alpha} \in (0, 1)$, with

$$\| A - \hat{A}_0 \| = \rho(A, V_{\hat{\alpha}}) = \inf_{A \in V_{\hat{\alpha}}} \| A - \hat{A} \| \tag{II.1.29}$$

whereby

$$V_{\hat{\alpha}} = \{ \hat{A} \in V : \| \hat{A} \| \leqslant \hat{\alpha} \}. \tag{II.1.30}$$

If one chooses in particular for V a finite dimensional linear subspace of the vector space $L(X, X)$ of all continuous linear mappings of X into itself with the norm (II.1.12), then $V_{\hat{\alpha}}$ is a non-empty, convex (see Section IV.3.1), closed subset of V which generates all of V and we are confronted with a convex approximation problem in the sense of Section II.5.1. By Theorem II.5.2 this is always solvable. We shall demonstrate this procedure again by the Fredholm integral equation (II.1.11′). We make the same requirements as in Section II.1.2.1 and assume in addition that the kernel $K = K(t, s)$ of the integral operator A defined by (II.1.17) is symmetric. To obtain approximation operators $\hat{A}$ of A, we choose an orthonormal system of function $\{\varphi_1, \ldots, \varphi_n\}$ in $X = C[a, b]$ (for example, a system of orthonormal eigenfunctions of A, if these are known) and define for every $\lambda \in \mathbb{R}^n$ a so-called degenerate kernel

$$K(\lambda, t, s) = \sum_{j=1}^{n} \lambda_j \varphi_j(t)\varphi_j(s) \qquad \text{for } a \leqslant t, s \leqslant b. \tag{II.1.31}$$

We replace the integral equation (II.1.11′) then by the approximation equation

$$x(t) - \int_a^b K(\lambda, t, s)x(s)\, ds = y(t). \tag{II.1.32}$$

If one defines for every λ the integral operator

$$\hat{A}(x) = A_\lambda(x) = \int_a^b K(\lambda, t, s)\, x(s)\, ds \qquad \text{for } x \in C[a, b] \tag{II.1.33}$$

then

$$\| \hat{A} \| = \| A_\lambda \| = \max_{t \in [a, b]} \int_a^b | K(\lambda, t, s) |\, ds$$

and the condition (II.1.21) goes over into

$$\max_{t \in [a, b]} \int_a^b |K(\lambda, t, s) |\, ds \leqslant \hat{\alpha} < 1. \tag{II.1.34}$$

Under this condition (II.1.32) is uniquely solvable and the solution $x_{\lambda, y} = x_{\lambda, y}(t)$ necessarily has the form

$$x_{\lambda, y}(t) = y(t) + \sum_{j=1}^{n} \rho_j(\lambda)\varphi_j(t) \tag{II.1.35}$$

with

$$\rho_j(\lambda) = \int_a^b \varphi_j(s) x_{\lambda, y}(s)\, ds\, \lambda_j. \tag{II.1.36}$$

Setting $x_{\lambda, y}$ into (II.1.32) yields

$$\sum_{j=1}^{n} \{\rho_j(\lambda) - \lambda_j(b_j(y) + \rho_j(\lambda))\}\, \varphi_j(t) = 0 \qquad \text{for all } t \in [a, b] \tag{II.1.37}$$

with

$$b_j(y) = \int_a^b \varphi_j(s) y(s)\, ds \qquad \text{for } j = 1, \ldots, n \tag{II.1.38}$$

(Proof = Exercise). Because of the linear independence of the functions $\varphi_1, \ldots, \varphi_n$ (which follows from the orthonormality-exercise) we conclude further that

$$(1 - \lambda_j)\rho_j(\lambda) = \lambda_j b_j(y) \qquad \text{for } j = 1, \ldots, n. \tag{II.1.39}$$

Under the assumption

$$\lambda_j \neq 1 \qquad \text{for } j = 1, \ldots, n \tag{II.1.40}$$

the solution $x_{\lambda,y}$ ($\lambda = (\lambda_1, \ldots, \lambda_n)$) of (II.1.32) necessarily has the form (II.1.35) with

$$\rho_j(\lambda) = \frac{\lambda_j b_j(y)}{1 - \lambda_j} \qquad \text{for } j = 1, \ldots, n \tag{II.1.41}$$

Problem II.1.1a Show without using (II.1.34) that the converse holds, i.e. for every $\lambda \in \mathbb{R}^n$ with $\lambda_j \neq 1$, for all $j = 1, \ldots, n$, a solution $x_{\lambda,y}$ of (II.1.32) is given by (II.1.35) with $\rho_j(\lambda)$, $j = 1, \ldots, n$, defined according to (II.1.41).

Problem II.1.1b Making use of

$$\int_a^b K(\lambda, t, s)\varphi_j(s)\, ds = \lambda_j \varphi_j(t) \tag{II.1.42}$$

for all $t \in [a, b]$, $j = 1, \ldots, n$ and $\lambda \in \mathbb{R}^n$, prove that the condition (II.1.40) follows from (II.1.34).

The approximation problem (II.1.29), (II.1.30) in this case therefore consists of minimizing the deviation

$$\| A - A_\lambda \| = \max_{t \in [a,b]} \int_a^b | K(t, s) - K(\lambda, t, s) |\, ds \tag{II.1.43}$$

subject to the side conditions (II.1.34), i.e. subject to

$$\int_a^b | K(\lambda, t, s) |\, ds \leqslant \hat{\alpha}(<1) \qquad \text{for all } t \in [a, b] \tag{II.1.34$'$}$$

with $K(\lambda, t, s)$ defined by (II.1.31). Since $V = \{A_\lambda$ defined by (II.1.33) $: \lambda \in \mathbb{R}^n\}$ is an n-dimensional subspace of $L(X, X)$, $X = C[a, b]$, which is generated by the convex closed set $V_{\hat{\alpha}} = \{A_\lambda$ defined by (II.1.33) $: \lambda \in \mathbb{R}^n$ satisfies (II.1.34)$\}$, then this convex approximation problem is again solvable by Theorem II.5.2.

One can also regard it as an approximation problem in $C([a, b] \times [a, b])$, equipped with the mixed norm

$$\| g \| = \max_{t \in [a,b]} \int_a^b | g(t, s) |\, ds.$$

We shall go into such problems more deeply in Section II.5.3. If the deviation (II.1.43) for $\lambda = \hat{\lambda} \in \mathbb{R}^n$ satisfying (II.1.34) is minimal, then one obtains from (II.1.28) an optimal estimate of the form

$$\| x_y - x_{\hat{\lambda},y} \| \leqslant \frac{\| A - A_{\hat{\lambda}} \| \, \| y \|}{(1-\alpha)(1-\hat{\alpha})} \tag{II.1.44}$$

with α defined by (II.1.12′) for the deviation of the solution

$$x_{\hat{\lambda},y}(t) = y(t) + \sum_{j=1}^{n} \frac{\hat{\lambda}_j b_j(y)}{1-\hat{\lambda}_j} \varphi_j(t) \tag{II.1.45}$$

of (II.1.32) (for $\lambda = \hat{\lambda}$) from the desired solution x_y of the equation (II.1.11′). The estimate (II.1.44) can be found in a somewhat more general form also by Mikhlin and Smolitskiy [67]. A similar estimate was further set up under more general hypotheses by Kantorowitsch and Krylow [56], which, applied to the above situation, is in fact somewhat sharper than (II.1.44).

The estimate (II.1.16) was set up by Barrodale and Young [70] and applied to the approximate solution of linear operator equations with approximation and optimization methods.

II.1.3 A problem of optimal control

We consider a forced vibration process, which is described by a linear differential equation of second order

$$\ddot{x}(t) + a(t)\dot{x}(t) + b(t)x(t) = u(t) \qquad \text{for } t \in [0, T] \tag{II.1.46}$$

with the initial conditions

$$x(0) = x_0 \qquad \dot{x}(0) = \dot{x}_0. \tag{II.1.47}$$

The vibration described by $x = x(t)$ will be controlled by the function $u = u(t)$. Let $C[0, T]$ be the vector space of continuous real-valued functions on $[0, T]$. We assume that $a, b, u \in C[0, T]$. Then on the basis of the theory of ordinary linear differential equations (see, for example, Coddington and Levinson [55]) for each $u \in C[0, T]$ there is precisely one solution $x_u = x_u(t)$ of the initial value problem (II.1.46), (II.1.47). Furthermore x_u depends linearly on u, i.e. for every pair λ, $\mu \in \mathbb{R}$ and $u_1, u_2 \in C[0, T]$

$$x_{\lambda u_1 + \mu u_2} = \lambda x_{u_1} + \mu x_{u_2}.$$

It makes sense to suppose that the control functions which one admits (i.e. which are technically realizable), do not take on arbitrarily large values, i.e. that there is a non-negative function $c \in C[0, T]$ such that

$$|u(t)| \leqslant c(t) \qquad \text{for all } t \in [0, T].$$

Therefore let

$$\Omega = \{u \in C[0, T] : |u(t)| \leqslant c(t) \qquad \text{for all } t \in [0, T]\}. \tag{II.1.48}$$

The set Ω is convex (see Section IV.3; Proof = Exercise). Now we imagine $u \in \Omega$ chosen so that the associated trajectory $x_u = x_u(t)$ deviates as little as possible from a given trajectory $k = k(t)$ in the sense of the Euclidean norm in $C[0, T]$, i.e. that

$$\| x_u - k \|_2 = \left(\int_0^T | x_u(t) - k(t) |^2 \, dt \right)^{1/2}$$

is minimized.

We are faced therefore with the problem of approximating a function $k \in C[0, T]$ as closely as possible by means of functions x_u from the (likewise convex) subset $\{x_u : u \in \Omega$ of $C[0, T]\}$ in the sense of the Euclidean norm of $C[0, T]$. This is a convex approximation problem (see Sections II.5.1 and II.5.4). The functions x_u are given here only implicitly, i.e. as solutions of the initial value problem (II.1.46), (II.1.47). Thus also the (convex) functional $\varphi(u) = \| x_u - k \|_2$, which is to be minimized on Ω, is given only implicitly. However, if one knows a fundamental solution system for the associated homogeneous differential equation, then every x_u can be given explicitly.

II.2 CONVEX FUNCTIONS

II.2.1 Convex functionals

Let E be a linear vector space and X a non-empty convex subset of E (see Section IV.3.1).

Definition A functional $f: X \to \mathbb{R}$ is said to be convex if

$$f(\lambda x + (1 - \lambda)y) \leqslant \lambda f(x) + (1 - \lambda) f(y) \tag{II.2.1}$$

for all $\lambda \in [0, 1]$ and $x, y \in X$.

Example If $X = E$, then $f(x) = \| x \|, x \in E$ is a convex functional.

We shall encounter further examples.

A functional $f: X \to \mathbb{R}$ is said to be concave, if $-f$ is convex. It is affine linear if it is both concave and convex. That is precisely the case when the equality sign holds in (II.2.1).

Moreover, we have the lemma now given.

Lemma II.2.1 *If X is a linear manifold in E (and hence convex, see Section IV.3.1), then a functional $f: X \to \mathbb{R}$ is affine linear, if and only if for all $x, y \in X$ and $\lambda \in \mathbb{R}$ the following equality holds*

$$f(\lambda x + (1 - \lambda)y) = \lambda f(x) + (1 - \lambda) f(y). \tag{II.2.2}$$

Proof If (II.2.2) holds for all $x, y \in X$ and $\lambda \in \mathbb{R}$, then f is certainly affine linear on X. If the reverse is the case, then (II.2.2) holds for all $x, y \in X$ and $\lambda \in [0, 1]$.

If now, say, $\lambda < 0$, then we set $z = \lambda x + (1 - \lambda)y$. Then it follows that

$$y = \frac{1}{1-\lambda} z + \frac{(-\lambda)}{1-\lambda} x \quad \text{with} \quad \frac{1}{1-\lambda} \in (0,1) \quad \text{and} \quad \frac{(-\lambda)}{1-\lambda} = 1 - \frac{1}{1-\lambda}.$$

Thus

$$f(y) = \frac{1}{1-\lambda} f(z) + \frac{(-\lambda)}{1-\lambda} f(x) \Rightarrow f(z) = \lambda f(x) + (1-\lambda) f(y).$$

Analogously, one comes to the same conclusion for $\lambda > 1$.

If X is a linear subspace of E, $c : X \to \mathbb{R}$ is a linear form and $\beta \in \mathbb{R}$ is a fixed number, then $f(x) = c(x) + \beta$ is an affine linear functional on X. If, on the other hand, $f : X \to \mathbb{R}$ is affine linear and X is a linear subspace of E, then there is a linear form $c : X \to \mathbb{R}$ and a $\beta \in \mathbb{R}$ with $f(x) = c(x) + \beta$ for all $x \in X$ (Proof = Exercise).

A convex functional is not necessarily continuous. However, we have the next theorem.

Theorem II.2.2 *If E is a finite dimensional normed vector space and X is a convex subset with non-empty interior $\mathring{X}$, then every convex functional $f : X \to \mathbb{R}$ is continuous on $\mathring{X}$.*

We shall not give the proof here, but for it refer the reader to Collatz and Wetterling [71].

Lemma II.2.3 *If $f : X \to \mathbb{R}$ is a convex functional on the convex subset X of a linear vector space, then for every $\alpha \in \mathbb{R}$, the set*

$$X_\alpha = \{x \in X : f(x) \leqslant \alpha\}$$

is convex.

Proof Let $x, y \in X_\alpha$ and $\lambda \in [0, 1]$; then

$$z = \lambda x + (1-\lambda)y \in X \qquad \text{and} \qquad f(z) \leqslant \lambda f(x) + (1-\lambda) f(y) \leqslant \alpha$$

and hence $z \in X_\alpha$.

The converse of Lemma II.2.3 is in general not true and gives rise to the definition of the concept of quasi-convexity (see Mangasarian [69]).

Definition A functional $f : X \to \mathbb{R}$ defined on a non-empty subset X of a normed vector space is said to be weakly semi-continuous from below on X, if

$$x_k \rightharpoonup x, x_k, x \in X \Rightarrow \liminf_{k \to \infty} f(x_k) \geqslant f(x). \tag{II.2.3}$$

(Here $x_k \rightharpoonup x$ signifies the weak convergence of Section IV.3.3.)

The significance of the weak semi-continuity from below is clear from the following generalization of a well known theorem of Weierstrass.

Theorem II.2.4 *Let X be a non-empty weakly sequentially compact subset of a normed vector space (see Section IV.3.3) and $f : X \to \mathbb{R}$ a weakly semi-continuous* *functional from below. Then there is an* $\hat{x} \in X$ *with*

$$f(\hat{x}) = \inf_{x \in X} f(x). \tag{II.2.4}$$

Proof Let $\{x_k\}$ be a minimizing sequence in X, i.e. a sequence with

$$\lim_{k \to \infty} f(x_k) = \inf_{x \in X} f(x).$$

Since X is weakly sequentially compact, there is a subsequence $\{x_{k_i}\}$ and an $\hat{x} \in X$ with $x_{k_i} \rightharpoonup \hat{x}$. From (II.2.3) there consequently follows

$$f(\hat{x}) \leqslant \liminf_{i \to \infty} f(x_{k_i}) = \inf_{x \in X} f(x)$$

which yields (II.2.4).

Problem II.2.1a Suppose E is a finite dimensional normed vector space and X a non-empty subset of E. Show that each continuous functional $f: X \to \mathbb{R}$ is also weakly semi-continuous from below.

Problem II.2.1b Making use of Theorem II.2.4, show further that f assumes its infimum on X if X is compact.

This statement is also true if E is of infinite dimension, as one proves directly (Proof = Exercise).

Lemma II.2.5 *Suppose X is a non-empty closed convex subset of a normed vector space. Then every continuous convex functional f on X is weakly semi-continuous from below.*

Proof Since f is continuous and X is closed, the set

$$X_\alpha = \{x \in X : f(x) \leqslant \alpha\}$$

is closed for each $\alpha \in \mathbb{R}$. Since f and X are convex, X_α is also convex for each $\alpha \in \mathbb{R}$ by Lemma II.2.3. Consequently by Theorem IV.3.7, X_α is weakly sequentially closed.

Suppose for the sake of contradiction that f were not weakly semi-continuous from below. Then there would be a sequence $\{x_k\}$ in X with $x_k \rightharpoonup x \in X$ and $\liminf_{k \to \infty} f(x_k) < f(x)$. If one chooses $\alpha \in \mathbb{R}$ with $\liminf_{k \to \infty} f(x_k) < \alpha < f(x)$, then there is a subsequence $\{x_{k_i}\}$ with $x_{k_i} \in X_\alpha$ for all i and $x_{k_i} \rightharpoonup x$. Since X_α is weakly closed, it follows that $x \in X_\alpha$ which implies $f(x) \leqslant \alpha$, contradicting the choice of α.

Summing up, we obtain from Theorem IV.3.10, Theorem II.2.4 and Lemma II.2.5 the following theorem.

Existence theorem *Suppose X is a non-empty, closed, bounded and convex subset of a reflexive Banach space and* $f: X \to \mathbb{R}$ *is a continuous convex functional. Then f assumes its infimum on X.*

Proof By Theorem IV.3.10, X is weakly sequentially compact and, by Lemma

II.2.5, f is weakly semi-continuous from below on X, so that the assertion follows from Theorem II.2.4.

Now let X be a non-empty subset of a normed vector space E and $f : X \to \mathbb{R}$ a functional. Every $\hat{x} \in X$ satisfying $f(\hat{x}) \leqslant f(x)$ for all $x \in X$ is referred to as a minimum of f in X.

Theorem II.2.6 *The set of minima of a convex functional defined on a convex set X is convex.*

Proof Let $\hat{x}, \hat{x}^* \in X$ be minima of f on X. Then $\lambda\hat{x} + (1 - \lambda)\hat{x}^* \in X$ for every $\lambda \in [0, 1]$ by the convexity of X and, consequently,

$$f(\lambda\hat{x} + (1 - \lambda)\hat{x}^*) \leqslant \lambda f(\hat{x}) + (1 - \lambda)f(\hat{x}^*) = f(\hat{x}) = f(\hat{x}^*)$$

by the convexity of f, which implies

$$f(\lambda\hat{x} + (1 - \lambda)\hat{x}^*) = f(\hat{x}) = f(\hat{x}^*)$$

whence the assertion follows.

An $\hat{x} \in X$ is called a local minimum of f in X if there is a ball $B(\hat{x}, \rho)$ with radius $\rho > 0$ centred at $\hat{x}$ such that

$$f(\hat{x}) \leqslant f(x) \qquad \text{for all } x \in B(\hat{x}, \rho) \cap X.$$

Theorem II.2.7 *Every local minimum of a convex functional f on a convex set X is also a minimum (the converse is trivial).*

Proof Let $\hat{x} \in X$ be a local minimum of f in X and let $x \in X$ be arbitrary. If $x \in B(\hat{x}, \rho)$, then by assumption $f(\hat{x}) \leqslant f(x)$. Suppose, therefore, that $x \notin B(\hat{x}, \rho)$, i.e. $\| x - \hat{x} \| > \rho$. Choose λ so that

$$0 < \lambda \leqslant \frac{\rho}{\| x - \hat{x} \|} (<1).$$

Then $x_\lambda = \lambda x + (1 - \lambda)\hat{x} \in X$ and $\| x_\lambda - \hat{x} \| = \lambda \| x - \hat{x} \| \leqslant \rho$, i.e. from $x_\lambda \in X \cap B(\hat{x}, \rho)$ follows

$$f(\hat{x}) \leqslant f(x_\lambda) \leqslant \lambda f(x) + (1 - \lambda)f(\hat{x})$$

which implies $f(x) - f(\hat{x}) \geqslant 0$.

II.2.2 Convex mappings

Let E be a linear vector space and X a non-empty convex subset of E. Further, let F be a partially ordered linear vector space with Y as the positive cone (see Section IV.1.1).

Definition A mapping $g : X \to F$ is called convex if for all $x, y \in X$ and $\lambda \in [0, 1]$

$$g(\lambda x + (1 - \lambda)y) \leqslant \lambda g(x) + (1 - \lambda)g(y)$$

which, by definition (IV.1.2), is equivalent to

$$\lambda g(x) + (1 - \lambda)g(y) - g(\lambda x + (1 - \lambda)y) \in Y.$$

If, in particular, $F = \mathbb{R}$, then we always think of F in the sequel as being partially ordered by the natural cone $Y = \{y \in \mathbb{R} : y \geqslant 0\}$ or the trivial cone $Y = \{0\}$, so that a convex mapping $g : X \to \mathbb{R}$ is a convex or an affine linear function in the sense of Section II.2.1.

A mapping $g : X \to F$ is said to be concave if the mapping $(-g)$ is convex. It is affine linear if g is both convex and concave. In the latter case we have

$$g(\lambda x + (1 - \lambda)y) = \lambda g(x) + (1 - \lambda)g(y) \tag{II.2.5}$$

for all $x, y \in X$ and $\lambda \in [0, 1]$.

If, in particular, X is a linear manifold and g is affine linear on X, then (II.2.5) in fact holds for all $x, y \in X$ and $\lambda \in \mathbb{R}$ which one proves in a manner analogous to the proof of Lemma II.2.1.

If F is partially ordered by the trivial cone $Y = \{\Theta_F\}$, then every convex (concave) mapping is affine linear. Let $F = \mathbb{R}^m$ be equipped with the positive cone

$$Y = \{y \in \mathbb{R}^m : y_1 \geqslant 0, \ldots, y_r \geqslant 0, y_{r+1} = \ldots = y_m = 0\}$$

for some $r \in \{0, \ldots, m\}$. A mapping $g : X \to F, g(x) = (g_1(x), \ldots, g_m(x))^T$, is convex if and only if the functionals $g_j : X \to \mathbb{R}$ are convex and for $j = 1, \ldots, r$ and affine linear for $j = (r + 1), \ldots, m$.

Lemma II.2.8 *Let E, F and G be linear vector spaces. Further, let F and G be partially ordered and X be a non-empty convex subset of E.*
(a) If $g : X \to F$ is affine linear and $h : F \to G$ is convex, then $h \circ g : X \to G$ is convex.
(b) If $F = G = \mathbb{R}$, $g : X \to F$ is convex, and $h : F \to G$ is monotonically non-decreasing and convex, then $h \circ g : X \to G$ is convex.

Proof Let $x, y \in X$ and $\lambda \in [0, 1]$ be given. Then, in case (*a*),

$$\begin{aligned}(h \circ g)(\lambda x + (1 - \lambda)y) &= h(\lambda g(x) + (1 - \lambda)g(y)) \\ &\leqslant \lambda(h \circ g)(x) + (1 - \lambda)(h \circ g)(y)\end{aligned}$$

and, in case (*b*),

$$g(\lambda x + (1 - \lambda)y \leqslant \lambda g(x) + (1 - \lambda)g(y)$$

which implies

$$\begin{aligned}(h \circ g)(\lambda x + (1 - \lambda)y) &\leqslant h(\lambda g(x) + (1 - \lambda)g(y)) \\ &\leqslant \lambda(h \circ g)(x) + (1 - \lambda)(h \circ g)(y).\end{aligned}$$

Problem II.2.2 Show by giving a counter-example that in the case (*b*) the monotonicity cannot be dispensed with.

Lemma II.2.9 *Suppose E is a linear vector space and X is a non-empty convex subset in E. Suppose, further, that F is a partially ordered normed vector space with*

Y as a positive cone and $g : X \to F$ is a convex mapping. Then for every continuous linear form $y^* \geqslant \Theta_{F^*}$ (in the sense of the ordering induced by (IV.1.14) in the topological dual space F^* of F), the functional $f : X \to \mathbb{R}$ defined by

$$f(x) = y^*(g(x)) \qquad \text{for } x \in X$$

is convex.

Proof Let $x, y \in X$ and $\lambda \in [0, 1]$ be given. Then, because of the convexity of g, one has

$$\lambda g(x) + (1 - \lambda)g(y) - g(\lambda x + (1 - \lambda)y) \in Y$$

and since $y^* \geqslant \Theta_{F^*}$ (according to (IV.1.14))

$$\begin{aligned} &\lambda f(x) + (1 - \lambda)f(y) - f(\lambda x + (1 - \lambda)y) = \lambda y^*(g(x)) + (1 - \lambda)y^*(g(y)) \\ &\quad -y^*[g(\lambda x + (1 - \lambda)y)] = y^*[\lambda g(x) + (1 - \lambda)g(y) - g(\lambda x + (1 - \lambda)y)] \\ &\quad \geqslant 0. \end{aligned}$$

Problem II.2.3 How does Lemma II.2.9 read for $F = \mathbb{R}^m$ with $Y = K_m^m$ (see (IV.1.7')) as the positive cone?

Lemma II.2.10 *Let E be a linear vector space and X a non-empty convex subset of E. Further let F and G be two partially ordered linear vector spaces and $f : X \to F$, $g : X \to G$ be two given mappings. Then we have the assertion that if f is convex and g concave, the set*

$$K = \bigcup_{x \in X} \{(y, z) \in F \times G : y \geqslant f(x), z \leqslant g(x)\}$$

is convex.

Proof Let $(y, z), (\hat{y}, \hat{z}) \in K$ and $\lambda \in [0, 1]$ be given. Then there are points $x, \hat{x} \in X$ with

$$y \geqslant f(x), \hat{y} \geqslant f(\hat{x}) \Rightarrow \lambda y + (1 - \lambda)\hat{y} \geqslant \lambda f(x) + (1 - \lambda)f(\hat{x}) \geqslant f(\lambda x + (1 - \lambda)\hat{x})$$

since $\lambda x + (1 - \lambda)\hat{x} \in X$ and

$$z \leqslant g(x), \hat{z} \leqslant g(\hat{x}) \Rightarrow \lambda z + (1 - \lambda)\hat{z} \leqslant \lambda g(x) + (1 - \lambda)g(\hat{x}) \leqslant g(\lambda x + (1 - \lambda)\hat{x})$$

since $\lambda x + (1 - \lambda)\hat{x} \in X$, which implies $\lambda(y, z) + (1 - \lambda)(\hat{y}, \hat{z}) = (\lambda y + (1 - \lambda)\hat{y}, \lambda z + (1 - \lambda)\hat{z}) \in K$.

Problem II.2.4a Let E be a linear vector space, X a non-empty convex subset of E and F a partially ordered linear vector space. Show that a mapping $g : X \to F$ is convex, concave, respectively, if and only if its supergraph

$$E_g = \{(x, y) \in X \times F : g(x) \leqslant y\}$$

or its subgraph

$$H_g = \{(x, y) \in X \times F : g(x) \geqslant y\}$$

respectively, is convex (see Figure II.2.1).

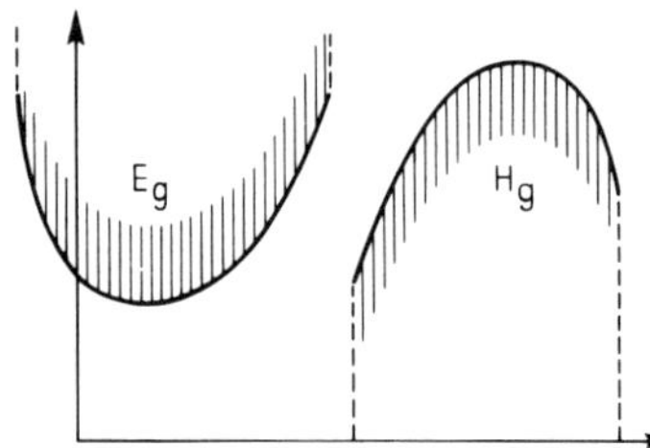

Figure II.2.1 The supergraph E_g and the subgraph H_g of a convex and concave function, respectively

Problem II.2.4b Let $\{f_i\}_{i\in I}$ be a family of convex, or concave, functionals, respectively on a non-empty convex subset X of a linear vector space such that for every $x \in X$ the family $\{f_i(x)\}_{i\in I}$ is bounded above, or below, respectively. Then the functional

$$f(x) = \sup_{i\in I} f_i(x) \qquad \text{or} \qquad f(x) = \inf_{i\in I} f_i(x)$$

is also convex, or concave, respectively (see Figure II.2.2).

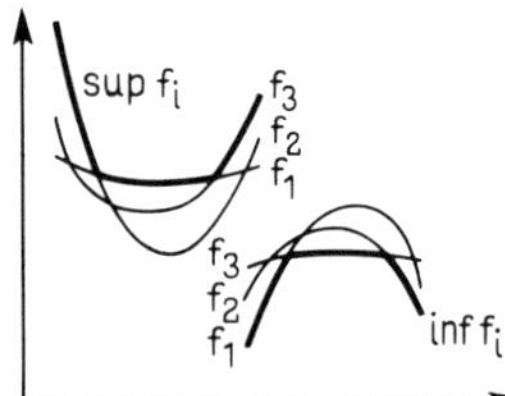

Figure II.2.2 The convexity and concavity of the supremum and the infimum of convex and concave functions respectively

II.2.3 Existence statements for linear control problems

We take up the problem of Section I.2.5 again in a somewhat more general form. Let $C[0, 1]^n$ be the vector space of the vector-valued functions $x = x(t)$ with n components which are defined and continuous on $[0, 1]$ and let $L_2[0, 1]^r$ be the vector space of vector-valued functions $u = u(t) = (u_1(t), \ldots, u_r(t))$ which are defined on $[0, 1]$ and whose component functions $u_k = u_k(t)$ for $k = 1, \ldots, r$ are measurable and quadratically integrable on $[0, 1]$ (more precisely, one should consider classes of such functions, which differ at most on a set of measure zero). With the scalar product

$$\langle u, v \rangle = \sum_{k=1}^{r} \int_0^1 u_k(t)v_k(t)\mathrm{d}t \qquad \text{for } u, v \in L_2[0, 1]^r$$

$L_2[0, 1]^r$ becomes a Hilbert space, which contains the vector space $C[0, 1]^r$ as a subspace.

Now let $A = A(t), B = B(t)$, be continuous $n \times n, n \times r$ matrix functions, respectively, on $[0, 1]$, $x_0 \in \mathbb{R}^n$ a given vector, Ω a closed, bounded and convex subset of $L_2[0, 1]^r$ and $\varphi : C[0, 1]^n \times L_2[0, 1]^r \to \mathbb{R}$ a continuous convex functional. Then we consider the following initial value problem. For every

$u \in L_2[0, 1]^r$ an $x \in C[0, 1]^n$ is sought with

$$\dot{x}(t) = \frac{dx}{dt}(t) = A(t)x(t) + B(t)u(t) \tag{II.2.6}$$

for almost all $t \in [0, 1]$ and

$$x(0) = x_0. \tag{II.2.7}$$

Physically, a motion $x = x(t)$ with x_0 as a starting point is described by (II.2.6), (II.2.7), which is controlled by $u = u(t)$. From the theory of differential equations (see for example Coddington and Levinson [55]), it is known that for every $u \in L_2[0, 1]^r$ precisely one absolutely continuous solution $x_u = x_u(t)$ of the initial value problem (II.2.6), (II.2.7) exists which is given by

$$x_u(t) = Y(t)\left(x_0 + \int_0^t Y(s)^{-1}B(s)u(s)ds\right). \tag{II.2.8}$$

Here $Y = Y(t)$ is the differentiable $n \times n$ matrix function, which for every $t \in [0, 1]$ is non-singular, with

$$\dot{Y}(t) = A(t)Y(t) \qquad \text{for all } t \in [0, 1]$$

and $Y(0)$ is the $n \times n$ unit matrix.

The control problem consists now in selecting a $u \in \Omega$ so that with x_u defined by (II.2.8) the functional value $\varphi(x_u, u)$ is minimized.

In Section I.2.5, Ω is given by

$$\Omega = \{u \in U : \| u(t) \|_\infty \leqslant \gamma \text{ for almost all } t \in [0, 1]\}. \tag{II.2.9}$$

Here $\gamma > 0$ is a given constant, U is a linear subspace of $C[0, 1]^r$, and $\| \ \|_\infty$ denotes the maximum norm in $\mathbb{R}^r$.

The set Ω defined by (II.2.9) is convex and bounded in $L_2[0, 1]^r$. In general, however, it is not closed in the sense of the L_2 norm, even if U is a linear subspace of $C[0, 1]^r$ which is closed in the sense of the maximum norm.

Problem II.2.5 Show that Ω is convex and closed in $L_2[0, 1]^r$ and closed if U is a finite dimensional subspace of $C[0, 1]^r$.

The functional $\varphi : C[0, 1]^n \times L_2[0, 1]^r \to \mathbb{R}$ was given in Section I.2.5 by

$$\varphi(x, u) = \max_{t \in [0, 1]} \| x(t) - \hat{x}(t) \|_\infty$$

where $\hat{x} \in C[0, 1]^n$ is given and fixed and $\| \ \|_\infty$ denotes the maximum norm in $\mathbb{R}^n$. φ is convex and continuous (Proof = Exercise).

We again consider the general problem and define a functional $f : L_2[0, 1]^r \to \mathbb{R}$ by

$$f(u) = \varphi(x_u, u) \qquad \text{for } u \in L_2[0, 1]^r. \tag{II.2.10}$$

Since the mapping $u \to (x_u, u)$ is affine linear and continuous (Proof = Exercise) and φ is convex and continuous, f (by Lemma II.2.8*a*) is convex and continuous.

The above general control problem now consists in minimizing the convex continuous functional f defined by (II.2.10) with x_u defined according to (II.2.8) on a closed, bounded and convex subset Ω of the Hilbert space $L_2[0, 1]^r$. The existence theorem in Section II.2.1 then yields the following theorem.

Theorem II.2.11 *If the set Ω of control functions is not empty, then the control problem is always solvable.*

In a similar way, the problem of optimal control in Section II.1.3 can be treated.

II.3. THE GENERAL CONVEX OPTIMIZATION PROBLEM: EXISTENCE AND DUALITY THEOREMS

II.3.1 Posing the problem, existence theorems, subconsistency

Suppose we are given a linear vector space E, a non-empty convex subset X of E, a partially ordered normed vector space F with Y as the positive cone (see Sections IV.1.1 and IV.3.1), a convex functional $f : X \to \mathbb{R}$, a concave mapping $g : X \to F$ (see Sections II.2.1 and II.2.2) and a fixed element $b \in F$. Suppose the set

$$S(X, g, b) = \{x \in X : g(x) \geqslant b\} \tag{II.3.1}$$

is not empty. Here $g(x) \geqslant b$ is equivalent to $g(x) - b \in Y$ or $g(x) \in (Y + b)$. The concavity of g refers to the partial ordering in F induced by Y. The set $S(X, g, b)$ in (II.3.1) is convex (Proof = Exercise), and the general convex optimization problem is given as problem (P): we seek an $\hat{x} \in S(X, g, b)$ such that

$$f(\hat{x}) \leqslant f(x) \qquad \text{for all } x \in S(X, g, b). \tag{II.3.2}$$

Every $x \in S(X, g, b)$ is called consistent and every $\hat{x} \in S(X, g, b)$ satisfying (II.3.2) is called optimal.

We shall state the problem somewhat more geometrically. By analogy to (I.4.4) we define the set

$$K(g, f) = \{(f(x) + r, g(x) - y) : x \in X, r \geqslant 0, r \in Y\} \tag{II.3.3a}$$

which one can also represent as follows:

$$K(g, f) = \bigcup_{x \in X} \{(\alpha, z) \in \mathbb{R} \times F : \alpha \geqslant f(x), z \leqslant g(x)\}. \tag{II.3.3b}$$

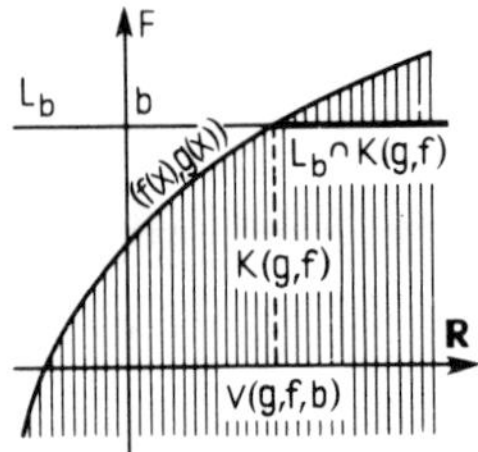

Figure II.3.1 The set $L_b \cap K(g, f)$

By Lemma II.2.10 the set $K(g,f)$ in $\mathbb{R} \times F$ is convex. If one defines further as in (I.4.5)

$$L_b = R \times \{b\} \tag{II.3.4}$$

then the set $S(X, g, b)$ is non-empty if and only if the intersection $L_b \cap K(g,f)$ is non-empty. Furthermore, for every $(\alpha, b) \in K(g,f)$ there is an $x \in X$ with $b \leqslant g(x)$ and $\alpha \geqslant f(x)$. The extremal value $v(g, f, b)$ of problem (P) is given by

$$v(g,f,b) = \begin{cases} \inf\limits_{X \in S(X,g,b)} f(x) & \text{if } S(X,g,b) \neq \phi \\ +\infty & \text{otherwise} \end{cases} \tag{II.3.5}$$

$$= \begin{cases} \inf\limits_{(\alpha,z) \in L_b \cap K(g,f)} \alpha & \text{if } L_b \cap K(g,f) \neq \phi \\ +\infty & \text{otherwise} \end{cases}$$

(Proof = Exercise), and problem (P) is equivalent to seeking an $(\hat{\alpha}, \hat{z}) \in L_b \cap K(g,f)$ such that

$$\hat{\alpha} \leqslant \alpha \qquad \text{for all } (\alpha, z) \in L_b \cap K(g,f). \tag{II.3.6}$$

Stated geometrically, we seek the point of the set $L_b \cap K(g,f)$ lying farthest to the left (see Figure II.3.1). If $\hat{x} \in S(X, g, b)$ is a solution of problem (P), then $(f(\hat{x}), b) \in L_b \cap K(g,f)$ is the farthest left point of $L_b \cap K(g,f)$. Conversely if $(\hat{\alpha}, \hat{z}) \in L_b \cap K(g,f)$ is the farthest left point of $L_b \cap K(g,f)$, then necessarily $\hat{z} = b$, $\hat{\alpha} = v(g,f,b)$ and there is an $\hat{x} \in X$ with $g(\hat{x}) \geqslant b$ and $f(\hat{x}) = \hat{\alpha}$, i.e. $\hat{x}$ is optimal. Certainly there is a farthest left point of the set $L_b \cap K(g,f)$ if it is closed and not empty, which is obviously the case if it is not empty and the set $K(g,f)$ is closed. More generally, the following lemma concerns this.

Lemma II.3.1 *Suppose K is a non-empty closed subset of* $\mathbb{R} \times F$ *and* $b \in F$ *is an element such that the intersection* $L_b \cap K$ *is not empty and bounded on the left (with* L_b *given by (II.3.4)). Then the set* $L_b \cap K$ *has a point lying farthest to the left which is given by* $(\hat{\alpha}, b)$ *with* $\hat{\alpha} = \min\limits_{(\alpha,b) \in K} \alpha$ *(Proof = Exercise).*

From this lemma follows immediately the next lemma.

Lemma II.3.2 *If the set* $K(g,f)$ *given by (II.3.3) is closed, then for every* $b \in F$ *with*

$$L_b \cap K(g,f) \neq \phi \qquad \text{that is} \qquad S(X,g,b) \neq \phi \tag{II.3.7}$$

the problem (P) is solvable if f is bounded from below (even without the convexity of X and f and without the concavity of g).

Sufficient conditions for Lemma II.3.2 to be satisfied are contained in the following.

Lemma II.3.3 *If E is a normed vector space, X a non-empty compact subset of E, Y a non-empty closed subset of F,* $g: X \to F$ *a continuous mapping (see Section IV.1.3) and* $f: X \to \mathbb{R}$ *a continuous functional (see Section IV.1.2), then the set* $K(g,f)$ *defined by (II.3.3) is closed in* $\mathbb{R} \times F$ *if one norms* $\mathbb{R} \times F$*, say, by* $\|(\alpha, z)\| =$

$|\alpha| + \|z\|$. *Thus, problem (P) is solvable for all $b \in F$ satisfying (II.3.7) (even without convexity and concavity requirements).*

Proof Let $\{(\alpha_k, z_k)\}$ be a sequence in $K(g,f)$ with $\lim_{k\to\infty} (\alpha_k, z_k) = (\alpha, z)$. By assumption, we have for all k

$$\alpha_k = f(x_k) + r_k \qquad z_k = g(x_k) - y_k$$

with $x_k \in X, r_k \geqslant 0, y_k \in Y$. Since X is compact, there is a subsequence $\{x_{k_i}\}$ and an $x \in X$ with $x = \lim_{i\to\infty} x_{k_i}$, which implies $\lim_{i\to\infty} f(x_{k_i}) = f(x)$ and since $\alpha = \lim_{i\to\infty} \alpha_{k_i}$ exists, also

$$r = \lim_{i\to\infty} r_{k_i} = \lim_{i\to\infty} \alpha_{k_i} - \lim_{i\to\infty} f(x_{k_i}) = \alpha - f(x)$$

and $r \geqslant 0$ since all $r_{k_i} \geqslant 0$.

Consequently $\alpha = f(x) + r$ with $x \in X$ and $r \geqslant 0$. Furthermore, it follows that $\lim_{i\to\infty} g(x_{k_i}) = g(x)$ exists and since $z = \lim_{i\to\infty} z_{k_i}$ exists, also

$$y = \lim_{i\to\infty} y_{k_i} = \lim_{i\to\infty} g(x_{k_i}) - \lim_{i\to\infty} z_{k_i} = g(x) - z.$$

Since $y_{k_i} \in Y$ for all i and because Y is closed, it follows also that $y \in Y$, which implies $z = g(x) - y$ with $x \in X$ and $y \in Y$.

As in Section I.4.2, we introduce again the concept of subconsistency of a problem.

Definition The problem (P) is called consistent, or subconsistent, respectively, if the intersection $L_b \cap K(g,f)$, or $L_b \cap \overline{K(g,f)}$, is not empty. (Here L_b is given by (II.3.4) and $K(g,f)$ by (II.3.3) and $\overline{K(g,f)}$ is the closure of $K(g,f)$).

We define the subvalue of the problem by

$$v_s(g,f,b) = \begin{cases} \inf\limits_{(\alpha,\, b)\in \overline{K(g,f)}} \alpha & \text{if } L_b \cap \overline{K(g,f)} \neq \phi \\ +\infty & \text{otherwise.} \end{cases} \tag{II.3.8}$$

As an extension of problem (P), we have problem (P_s).

Problem P_s We seek a point $(\hat{\alpha}, b) \in \overline{K(g,f)}$ with $\hat{\alpha} \leqslant \alpha$ for all points $(\alpha, b) \in \overline{K(g,f)}$.

Since $\overline{K(g,f)}$ is closed, problem (P_s) is solvable for every $b \in F$ with $L_b \cap \overline{K(g,f)}$ non-empty, bounded from the left, by Lemma II.3.1. Thus, we have

$$v_s(g,f,b) = \min_{(\alpha,z)\in L_b \cap \overline{K(g,f)}} \alpha. \tag{II.3.9}$$

A direct consequence of the above definition is the next lemma.

Lemma II.3.4 *(See Lemma I.4.2.) If problem (P) is consistent, then it is also subconsistent and*

$$v_s(g,f,b) \leqslant v(g,f,b) < +\infty$$

holds with v_s defined according to (II.3.8) and v according to (II.3.5).

Definition The problem (P) is called normal if

$$\overline{L_b \cap K(g,f)} = L_b \cap \overline{K(g,f)}. \tag{II.3.10}$$

Lemma II.3.5 (See Lemma I.4.4). Let problem (P) be normal. Then the following two statements hold:
(a) The problem (P) is consistent if and only if it is subconsistent.
(b) $v(g,f,b) = v_s(g,f,b)$.

The proof is the same as that for Lemma I.4.3.

If the problem (P) is not normal, then in general one has

$$v_s(g,f,b) < v(g,f,b)$$

which was shown in Section I.4.2 by means of an example from Section I.3.4.2.

II.3.2 The dual problem

In order to set up a problem dual to our problem, we define in $\mathbb{R} \times F^*$, where F^* is the topological dual space of F (see Section IV.1.2), the set

$$S^* = \{(\beta, y^*) \in \mathbb{R} \times F^* : f(x) - y^*(g(x)) \geqslant \beta - y^*(y) \text{ for all } x \in X, y \in Y\} \tag{II.3.11}$$

and consider the dual problem to be problem (D): we seek a pair $(\hat{\beta}, \hat{y}^*) \in S^*$ such that

$$\hat{\beta} + \hat{y}^*(b) \geqslant \beta + y^*(b) \qquad \text{for all } (\beta, y^*) \in S^*. \tag{II.3.12}$$

Each pair $(\beta, y^*) \in S^*$ is said to be dually consistent and each pair $(\hat{\beta}, \hat{y}^*) \in S^*$ satisfying (II.3.12) dually optimal.

The problem (D) is called consistent if S^* is not empty. Finally, we define the extremal value of the problem (D)

$$v^*(D) = \begin{cases} \sup\limits_{(\beta, y^*) \in S^*} \beta + y^*(b) & \text{if } S^* \neq \phi \\ -\infty & \text{otherwise} \end{cases} \tag{II.3.13}$$

The dual problem also has a geometric meaning which we shall now go into. By (IV.1.18) every continuous linear form z^* on $\mathbb{R} \times F$ is given by

$$z^*(\rho, y) = \lambda\rho + y^*(y) \qquad \text{for all } (\rho, y) \in \mathbb{R} \times F.$$

Here $\lambda \in \mathbb{R}$ and $y^* \in F^*$ are determined uniquely by z^*. A (closed) hyperplane in $\mathbb{R} \times F$ is therefore given in the form

$$H((\lambda, y^*), \alpha) = \{(\rho, y) \in \mathbb{R} \times F : \lambda\rho + y^*(y) = \alpha\}.$$

We call a hyperplane $H((\lambda, y^*), \alpha)$ in $\mathbb{R} \times F$ non-horizontal if $\lambda \neq 0$. Since

$$H((\lambda, y^*), \alpha) = H((1, y^*/\lambda), \alpha/\lambda)$$

we can assume without loss of generality that $\lambda = 1$. If one associates with each non-horizontal hyperplane $H((1, y^*), \alpha)$ in $\mathbb{R} \times F$ the half space

$$R((1, y^*), \alpha) = \{(\rho, y) \in \mathbb{R} \times F : \rho + y^*(y) \geqslant \alpha\}$$

then we have the following lemma.

Lemma II.3.6 *We have* $(\beta, y^*) \in S^*$, *with* S^* *given by (II.3.11) if and only if*

$$K(g, f) \subseteq R((1, -y^*), \beta). \tag{II.3.14}$$

Proof

1. Let $(\beta, y^*) \in S^*$. If $(\alpha, z) \in K(g, f)$ is given, then $\alpha = f(x) + r$, $z = g(x) - y$ for some $x \in X, r \geqslant 0$ and $y \in Y$. From (II.3.11) it follows that

$$\alpha - y^*(z) \geqslant f(x) - y^*(g(x)) + y^*(y) \geqslant \beta$$

which proves that $(\alpha, z) \in R((1, -y^*), \beta)$.

2. Let (II.3.14) be satisfied. Then it follows that

$$f(x) - y^*(g(x)) + y^*(y) \geqslant \beta \qquad \text{for all } x \in X, y \in Y$$

which implies $(\beta, y^*) \in S^*$. Furthermore, we obviously have

$$H((1, -y^*), \beta) \cap L_b = \{(y^*(b) + \beta, b)\}.$$

The dual problem (D) consists, therefore, in finding a non-horizontal hyperplane $H((1, -y^*), \beta)$ such that the set $K(g, f)$ given by (II.3.3) is entirely contained in the half space $R((1, -y^*), \beta)$ and the point of intersection of $H((1, -y^*), \beta)$ with the (horizontal) straight line L_b in $\mathbb{R} \times F$ given by (II.3.4) lies as far as possible to the right (see Figure II.3.2). Since the half space $R((1, -y^*), \beta)$ is closed, it follows from (II.3.14) also that

$$\overline{K(g, f)} \subseteq R((1, -y^*), \beta) \tag{II.3.14$'$}$$

An initial connection between the dual problem (D) and the problem (P_s) is given by the lemma now stated.

Lemma II.3.7 *(See Lemma I.4.4.) If the problem (P) is subconsistent (i.e.* $L_b \cap \overline{K(g, f)} \neq \phi$*) and the problem (D) is consistent (i.e.* $S^* \neq \phi$*), then it follows that*

$$-\infty < v^*(D) \leqslant v_s(g, f, b) < +\infty. \tag{II.3.15}$$

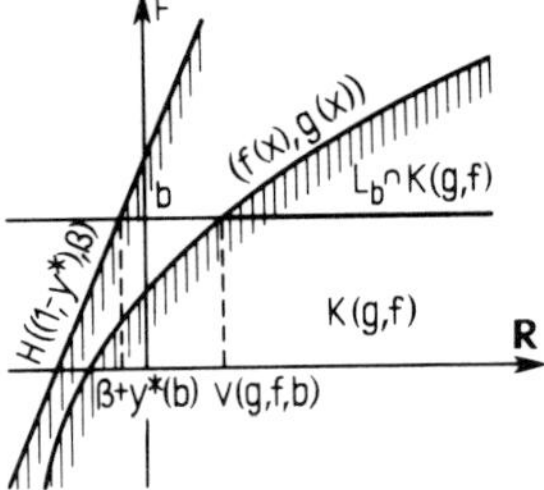

Figure II.3.2 The dual problem

Proof Let $(\alpha, b) \in \overline{K(g,f)}$ and $(\beta, y^*) \in S^*$ be given. From Lemma II.3.6 follows (II.3.14) and hence (II.3.14′) which in turn implies $\alpha - y^*(b) \geqslant \beta$ and thus $\alpha \geqslant \beta + y^*(b)$. Since $(\alpha, b) \in \overline{K(g,f)}$ and $(\beta, y^*) \in S^*$ are arbitrary, (II.3.15) follows.

By combining Lemmas II.3.4 and II.3.7 we obtain Lemma II.3.8.

Lemma II.3.8 *If the problem (P) is consistent and the problem (D) is consistent, it follows that*

$$-\infty < v^*(D) \leqslant v_S(g,f,b) \leqslant v(g,f,b) < +\infty.$$

II.3.3 General existence and duality theorems

As in Section I.4.3 we shall also show here that Lemma II.3.7 can be significantly sharpened. First we prove the next theorem.

Theorem II.3.9 *Let S^* (see (I.3.11)) be non-empty, i.e. the problem (D) is consistent. For every pair $(\alpha, b) \in \mathbb{R} \times F$, we then have*

$$(\alpha, b) \in \overline{K(g,f)} \Leftrightarrow \alpha \geqslant v^*(D) \tag{II.3.16}$$

with $K(g,f)$ defined by (II.3.3) and $v^*(D)$ by (II.3.13).

This theorem corresponds to Theorem I.4.6 and will also be proven completely analogously.

Proof

1. Let $(\alpha, b) \in \overline{K(g,f)}$. For every pair $(\beta, y^*) \in S^*$ it follows then as in the proof of Lemma II.3.7 that $\alpha \geqslant \beta + y^*(b)$ which implies $\alpha \geqslant v^*(D)$.

Remark The implication '$\Rightarrow$' is, by the way, trivially true if S^* is empty, since then by definition $v^*(D) = -\infty$.

2. Let $(\alpha, b) \notin \overline{K(g,f)}$. Since $\overline{K(g,f)}$ is convex by Lemma II.2.10 and Theorem IV.3.1 and is also closed, the separation theorem 2 in Section IV.3.2 and the representation formula (IV.1.18) (for $E = \mathbb{R}$) imply the existence of $(\lambda, y^*) \in \mathbb{R} \times F^*$ and $\rho \in \mathbb{R}$ with

$$\lambda\alpha + y^*(b) < \rho \leqslant \lambda\beta + y^*(z) \qquad \text{for all } (\beta, z) \in \overline{K(g,f)}.$$

Hence, in particular, we find

$$\lambda\alpha + y^*(b) < \rho \leqslant \lambda\{f(x) + r\} + y^*(g(x) - y) \tag{II.3.17}$$

for all $x \in X$, $r \geqslant 0$ and $y \in Y$.

From this follows first that $\lambda \geqslant 0$. Otherwise with any fixed choice of $x \in X$ and $y \in Y$ one could make the expression $\lambda\{f(x) + r\} + y^*(g(x) - y)$ arbitrarily small for sufficiently large $r > 0$, a contradiction to the right-hand inequality (II.3.17).

We now distinguish two cases.

(α) $\lambda > 0$. We can assume without loss of generality that $\lambda = 1$ and obtain from (II.3.17) with $\hat{y}^* = -y^*$ the conclusions

$$\alpha < \rho + \hat{y}^*(b) \qquad \text{and} \qquad -\hat{y}^*(y) \leqslant f(x) - \hat{y}^*(g(x))$$

for all $x \in X$ and $y \in Y$ which implies $(\rho, \hat{y}^*) \in S^*$ and $\alpha < v^*(D)$.
(β) $\lambda = 0$. From (II.3.17) follows

$$y^*(b) < \rho \leqslant y^*(g(x) - y) \qquad \text{for all } x \in X \text{ and } y \in Y. \tag{II.3.18}$$

Since S^* is not empty, there is a pair $(\beta, \hat{y}^*) \in \mathbb{R} \times F^*$ with

$$\beta - \hat{y}^*(y) \leqslant f(x) - \hat{y}^*(g(x)) \qquad \text{for all } x \in X \text{ and } y \in Y \tag{II.3.19}$$

From (II.3.18) and (II.3.19) it follows that

$$\beta + \sigma\rho - (\hat{y}^* - \sigma y^*)(y) \leqslant f(x) - (\hat{y}^* - \sigma y^*)(g(x))$$

for all $x \in X, y \in Y$, and all $\sigma > 0$. Hence

$$(\beta + \sigma\rho, \hat{y}^* - \sigma y^*) \in S^* \qquad \text{for all } \sigma > 0.$$

Since $y^*(b) < \rho$, we also have

$$\lim_{\sigma \to \infty} \{\beta + \sigma\rho + (\hat{y}^* - \sigma y^*)(b)\} = \lim_{\sigma \to \infty} \{\beta + \hat{y}^*(b) + \sigma(\rho - y^*(b))\} = +\infty$$

which implies $v^*(D) = +\infty$. Thus $\alpha \geqslant v^*(D)$ is impossible, which proves the implication in the direction '$\Leftarrow$' in (II.3.16).

Here also the Theorem I.4.7 is decisive for the theory. It will be reformulated as follows.

Theorem II.3.10(a) *The problem (P) is subconsistent and its subvalue $v_s(g, f, b)$ (see (II.3.8)) is finite if and only if the problem (D) is consistent and the dual extreme value $v^*(D)$ (see (II.3.13)) is finite. In both cases*

$$-\infty < v_s(g, f, b) = v^*(D) < +\infty.$$

(*b*) *If the problem (P) is not subconsistent and the dual problem (D) is consistent, then*

$$v_s(g, f, b) = v^*(D) = +\infty.$$

(*c*) *If the problem (P) is subconsistent and the dual problem (D) is not consistent, there follows*

$$v_s(g, f, b) = v^*(D) = -\infty.$$

Proof The statements (*b*) and (*c*) are a direct consequence of (*a*). For the proof of (*a*) we assume the following.

(α) Suppose that problem (D) is consistent and $v^*(D) < +\infty$. Then, by Theorem II.3.9, one has $(v^*(D), b) \in \overline{K(g, f)}$, i.e. the problem (P) is subconsistent and thus $v^*(D) \geqslant v_s(g, f, b)$. The equality follows from Lemma II.3.7.

(β) Suppose that problem (P) is subconsistent and $v_s(g, f, b) > -\infty$. If one chooses $\beta < v_s(g, f, b)$ arbitrarily, then $(\beta, b) \notin \overline{K(g, f)}$. By the separation theorem 2 in Section IV.3.2 and from the representation formula (IV.1.18) therefore follows the existence of $(\lambda, y^*) \in \mathbb{R} \times F^*$ and $\rho \in \mathbb{R}$ with

$$\lambda\beta + y^*(b) < \rho \leqslant \lambda\alpha + y^*(z) \qquad \text{for all } (\alpha, z) \in \overline{K(g, f)}. \tag{II.3.17'}$$

Since $(v_s(g, f, b), b) \in \overline{K(g,f)}$ this implies

$$\lambda\{v_s(g, f, b) - \beta\} > 0$$

and hence $\lambda > 0$. Furthermore, it follows from (II.3.17′) that

$$\lambda\{f(x) - r\} + y^*(g(x) - y) \geqslant \rho \qquad \text{for all } x \in X, r \geqslant 0, y \in Y$$

hence,

$$f(x) - \hat{y}^*(g(x)) \geqslant \rho/\lambda - \hat{y}^*(y) \qquad \text{for all } x \in X, y \in Y$$

if $\hat{y}^* = (1/\lambda)y^*$. Hence $(\rho/\lambda, \hat{y}^*) \in S^*$ (see (II.3.11)), i.e. this implies the consistency of problem (D). It follows further that

$$\rho/\lambda + \hat{y}^*(b) \leqslant v^*(D)$$

and hence

$$\rho \leqslant \lambda v^*(D) + y^*(b).$$

By using the left-hand inequality of (II.3.17′) we conclude

$$\lambda\{v^*(D) - \beta\} > 0$$

which implies $\beta < v^*(D)$. Since $\beta < v_s(g, f, b)$ was chosen arbitrarily, we necessarily have $v_s(g, f, b) \leqslant v^*(D)$ and the equality is obtained from Lemma II.3.7.

Corollary *If the problem (P) is normal, then, by Lemma II.3.5, in all the statements in Theorem II.3.10 the word subconsistent can be replaced by the word consistent, and the subvalue $v_s(g, f, b)$ by the value $v(g, f, b)$ of problem (P).*

By problem I.4.1, one has normality if the convex set $K(g, f)$ defined by (II.3.3) is closed. There is, however, still another important sufficient condition for normality, the so-called (generalized) Slater condition: let the positive cone Y of F have a non-empty interior $\mathring{Y}$ and let there be an $x_0 \in X$ with $g(x_0) \in \mathring{Y} + b$.

Lemma II.3.11 *If problem (P) satisfies the Slater condition, then it is normal.*

Proof From the Slater condition it follows first that the set $K(g, f)$ defined by (II.3.3) also has a non-empty interior $\mathring{K}(g, f)$ and that

$$(f(x_0) + r, b) \in \mathring{K}(g, f) \qquad \text{for all } r > 0$$

(Proof = Exercise). Since $\overline{L_b \cap K(g, f)} \subseteq L_b \cap \overline{K(g, f)}$ we must show that $L_b \cap \overline{K(g, f)} \subseteq \overline{L_b \cap K(g, f)}$. Let therefore $(\alpha, b) \in \overline{K(g, f)}$ be given. Fix $r > 0$ and let $\alpha_0 = f(x_0) + r$. Then $(\alpha_0, b) \in \mathring{K}(g, f)$ and Theorem IV.3.3 yields

$$B = \{\lambda(\alpha, b) + (1 - \lambda)(\alpha_0, b) : 0 \leqslant \lambda < 1\} \subseteq K(g, f)$$

which implies $L_b \cap B \subseteq L_b \cap K(g, f)$, hence $\overline{L_b \cap B} \subseteq \overline{L_b \cap K(g, f)}$. Since $B \subseteq L_b$ and $(\alpha, b) \in \overline{B}$, it follows that $(\alpha, b) \in \overline{L_b \cap B} \subseteq \overline{L_b \cap K(g, f)}$.

By the corollary to Theorem II.3.10 the normality assures the equality of the extremal values of the problems (P) and (D), i.e. the non-occurrence of a duality gap, if the extremal values are finite. Conversely, the normality is also necessary for the non-occurrence of a duality gap. By analogy to Theorem I.4.9 the following theorem also holds here.

Theorem II.3.12 *Suppose that the dual problem (D) is consistent, i.e. the set S^* defined by (II.3.11) is not empty. If the implication*

$$v^*(D) < +\infty \Rightarrow L_b \cap K(g,f) \neq \phi \qquad \text{and} \qquad v^*(D) = v(g,f,b) \tag{II.3.20}$$

is valid, then the problem (P) is normal.

Proof We have to show that $L_b \cap \overline{K(g,f)} \subseteq \overline{L_b \cap K(g,f)}$. Therefore let $(\hat{\alpha}, b) \in L_b \cap \overline{K(g,f)}$ be given. Then the problem (P) is subconsistent and by Lemma II.3.7 one has $v^*(D) \leqslant v_s(g,f,b) \leqslant \hat{\alpha} < +\infty$. On the basis of the implication (II.3.20), one also has

$$v^*(D) = v(g,f,b) = \min_{(\alpha,z)\in \overline{L_b \cap K(g,f)}} \alpha.$$

Hence $(v^*(D), b \in \overline{L_b \cap K(g,f)}$ which, because of $v^*(D) \leqslant \hat{\alpha}$ and the definition of $K(g,f)$, implies $(\hat{\alpha}, b) \in \overline{L_b \cap K(g,f)}$.

From the corollary to Theorem II.3.10 and Lemma II.3.2 we also obtain our next theorem.

Theorem II.3.13 *(General existence and duality theorem.) Let the convex set $K(g,f)$ defined by (II.3.3) be closed. Then the following assertions hold:*
(a) Problem (P) is consistent and its value $v(g,f,b)$ is finite if and only if the dual problem (D) is consistent and its value $v^*(D)$ is finite. In both cases, problem (P) is solvable and

$$-\infty < v^*(D) = v(g,f,b) < +\infty.$$

(b) If problem (P) is not consistent and problem (D) is consistent, then

$$v(g,f,b) = v^*(D) = +\infty.$$

(c) If problem (P) is consistent and problem (D) is not consistent, then

$$v(g,f,b) = y^*(D) = -\infty.$$

The analogue of Theorem I.4.8 also holds, which says that the closedness of $K(g,f)$ is in a certain sense characteristic of the solvability of problem (P). We shall not go into this any further.

II.3.4 The linear case

We specialize the problem posed in Section II.3.1 and assume that E (as well as F) is a partially ordered normed vector space with X as the positive cone, $f: E \to \mathbb{R}$ is a continuous linear form (see Section IV.1.2) and $g: E \to F$ is a continuous linear mapping (see Section IV.1.3). Then f, and g are defined on X and are convex, and concave, respectively, there. The problem (P) in Section II.3.1 now consists of minimizing the functional $f = f(x)$ subject to the side conditions

$$g(x) \geqslant b \qquad \text{for } x \in X (\Leftrightarrow x \geqslant \Theta_E).$$

This is precisely the linear optimization problem treated in Sections I.3 and I.4.

We now want to show that the dual problem introduced in Section II.3.2 is equivalent to that defined in Section I.3.1. For this purpose we define (by analogy to the set N of (I.3.4)) the set

$$S(Y^*, g^*, f) = \{y^* \in Y^* : g^*(y^*) \leqslant f\}. \tag{II.3.21}$$

Here Y^* is the positive cone in F^* adjoint to Y (see Section IV.2.2) and $g^* : F^* \to E^*$ is the linear mapping adjoint to g (see Section IV.1.3).

Lemma II.3.14 *Let the pair* $(\beta, y^*) \in \mathbb{R} \times F^*$ *be given. Then*

$$(\beta, y^*) \in S^* \text{ (see (II.3.11))} \Leftrightarrow y^* \in S(Y^*, g^*, f), \quad \textit{and} \quad \beta \leqslant 0.$$

Proof

1. Let $(\beta, y^*) \in S^*$. Then

$$f(x) - y^*(g(x)) \geqslant \beta - y^*(y) \qquad \text{for all } x \in X \text{ and } y \in Y. \tag{II.3.22}$$

If there were a $y \in Y$ with $y^*(y) < 0$, then choose $x \in X$ and fix it and $\lambda > 0$ so large that

$$-y^*(\lambda y) = -\lambda y^*(y) > f(x) - y^*(g(x)) - \beta.$$

Since $\lambda y \in Y$ that would contradict (II.3.22). This proves $y^*(y) \geqslant 0$ for all $y \in Y$, i.e. $y^* \in Y^*$. Analogously one shows

$$f(x) - y^*(g(x)) \geqslant 0 \qquad \text{for all } x \in X \Leftrightarrow g^*(y^*) \leqslant f.$$

For $x = \Theta_E$ and $y = \Theta_F$ it follows finally that $\beta \leqslant 0$ which proves the implication '$\Rightarrow$'.

2. Conversely, let $y^* \in S(Y^*, g^*, f)$ and $\beta \leqslant 0$. Then it follows for all $x \in X$ and $y \in Y$ that

$$f(x) - y^*(g(x)) \geqslant 0 \geqslant \beta - y^*(y)$$

i.e. (II.3.22) is satisfied which means $(\beta, y^*) \in S^*$.

Therefore, maximizing $\beta + y^*(b)$ for $(\beta, b) \in S^*$ is obviously equivalent to maximizing $y^*(b)$ for $y^* \in S(Y^*, g^*, f)$. This proves the equivalence of the two dual problems in Sections I.3.1 and I.3.2. The linear optimization theory developed in Sections I.3 and I.4 is thus contained in the convex theory.

II.3.5 Bibliographical remarks

There is an extensive literature dealing with the theory of finite dimensional convex optimization problems. We shall not go into this here since we are concerned primarily with infinite dimensional problems. This literature has been treated in text books, of which we name a few, namely Karlin [59], Rockafellar [69], and Stoer and Witzgall [70].

The theory of convex optimization in partially ordered vector spaces parallels to a large extent the theory in finite dimensional spaces since the fundamental separation theorems for convex sets are valid in general. In the course of the last two decades, various duality theories have been developed which in part lead to equivalent statements.

First, there is a well developed theory, which begins with a paper of Fenchel [49] (see also Fenchel [53]) on conjugate, convex functionals. It leads to very symmetric duality and existence theorems and is connected with the names Brøndsted [64], Dieter [66], Moreau [62], and Rockafellar [67]. Numerous papers of Moreau and Rockafellar have been quoted in a large survey article of Ioffe and Tikhomirov [68], which concerns this theory and applies it to variational, control and approximation problems. For these, a very geometrical theory is developed. Chapter 3 of the book of Göpfert [73] is also devoted to the Fenchel duality idea. It also uses the book of Luenberger [69] and contributions of Joly and Laurent [71] and Mosco [71]. A very complete presentation of this circle of ideas with an extensive bibliography is to be found in Sander [73].

Another access to a duality theory of convex optimization problems consists in formulating the problem posed as a min–sup problem as in Section II.4.1 and then setting up the dual problem as a max–inf problem as in Section II.4.2. This is how Gol'stein [67] proceeds. Analogous to Duffin's idea of subconsistency, Gol'stein introduces generalized consistent elements and defines a subvalue for the convex optimization problem. A major result of his paper is the statement that the subvalue of the problem coincides with the extremal value of the dual problem when the set of generalized consistent solutions is not empty, which corresponds to Theorem II.3.10.

A generalization of this statement was also derived by Ioffe and Tikhomirov [68].

The approach to duality chosen here is a more geometrical formulation of that just mentioned. It goes back to a paper of van Slyke and Wets [68] and has the advantage of easy applicability. They do not operate with the concept of subconsistency (see Section II.3.1) but immediately apply the idea of normality and obtain as a main result the corollary to Theorem II.3.10. These ideas were also partially followed by Luenberger [69], who for example proves Theorem II.4.9 in §8.6 of his book.

Rubinshtein [70] gives an approach to duality theorems which is based on an abstract set theory principle whose proof has no need for separation theorems for convex sets. Among numerous applications he treats a generalized problem of best approximation.

II.4 MIN–SUP, MAX–INF AND SADDLE POINT THEOREMS

II.4.1 The optimization problem as min–sup problem

We next consider more generally than in Section II.3 the following situation. Suppose X is a non-empty subset of a linear vector space E. Further suppose F is a partially ordered normed vector space with Y as its positive cone, $f : X \to \mathbb{R}$ is a given functional and $g : X \to F$ is a given mapping. We set

$$S = \{x \in X : g(x) \in Y\} \qquad \text{(II.4.1)}$$

and consider problem (P): we seek an $\hat{x} \in S$ satisfying

$$f(\hat{x}) \leqslant f(x) \qquad \text{for all } x \in S. \tag{II.4.2}$$

Let Y^* be the cone in F^* adjoint to the cone Y in F (see Section IV.2.2):

$$Y^* = \{y^* \in F^* : y^*(y) \geqslant 0 \qquad \text{for all } y \in Y\}.$$

A central role will be played in the sequel by the so-called Lagrange functional $\Phi : X \times Y^* \to \mathbb{R}$, which is defined by

$$\Phi(x, y^*) = f(x) - y^*(f(x)) \qquad \text{for } x \in X, y^* \in Y^*. \tag{II.4.3}$$

Lemma II.4.1 *Let the positive cone Y of F be closed. Then the following statements hold:*
(a) If $x \in S$, then (even without Y being closed) it follows that

$$f(x) = \Phi(x, \Theta_{F^*}) = \max_{y^* \in Y^*} \Phi(x, y^*). \tag{II.4.4}$$

(b) If for a given $x \in X$ there is a $y_x^* \in Y^*$ satisfying

$$\Phi(x, y_x^*) = \max_{y^* \in Y^*} \Phi(x, y^*) \tag{II.4.5}$$

then $x \in S$ and the following two equivalent statements hold:

$$f(x) = \Phi(x, y_x^*) \qquad \text{and} \qquad y_x^*(g(x)) = 0. \tag{II.4.6}$$

(c) The set S in (II.4.1) is empty if and only if for every $x \in X$

$$\sup_{y^* \in Y^*} \Phi(x, y^*) = +\infty. \tag{II.4.7}$$

Proof (*a*) Suppose $x \in S$. Then $g(x) \in Y$ and

$$y^*(g(x)) \geqslant 0 \qquad \text{for all } y^* \in Y^*$$

which immediately implies (II.4.4).
(*b*) From (II.4.5) follows

$$y_x^*(g(x)) \leqslant y^*(g(x)) \qquad \text{for all } y^* \in Y^* \tag{II.4.8}$$

which is possible only if

$$y^*(g(x)) \geqslant 0 \qquad \text{for all } y^* \in Y^*.$$

By Theorem IV.2.3 this implies $g(x) \in Y$, i.e. $x \in S$. Thus $y_x^*(g(x)) \geqslant 0$ and by (II.4.8), $y_x^*(g(x)) \leqslant 0$, which implies (II.4.6).
(*c*) If S is empty, then for each $x \in X$ we have $g(x) \notin Y$, which again by Theorem IV.2.3 implies the existence of a $y_x^* \in Y^*$ with $y_x^*(g(x)) < 0$. From this, however, follows (II.4.7), since λy_x^* belongs to Y^* for all $\lambda > 0$. If S is not empty, then for every $x \in S$, we have by (*a*)

$$\sup_{y^* \in Y^*} \Phi(x, y^*) = f(x) < +\infty$$

i.e. (II.4.7) is violated for at least one $x \in X$.

On the basis of Lemma II.4.1, problem (P) is obviously equivalent to the following problem.

Problem (P)* We seek a pair $(\hat{x}, \hat{y}^*) \in X \times Y^*$ satisfying

$$\Phi(\hat{x}, \hat{y}^*) = \min_{x \in X} \sup_{y^* \in Y^*} \Phi(x, y^*). \tag{II.4.9}$$

If $\hat{x} \in S$ is a solution of problem (P), then $(\hat{x}, \Theta_{F^*}) \in X \times Y^*$ is a solution of problem (P*). If $(\hat{x}, \hat{y}^*) \in X \times Y^*$ is a solution of problem (P*), then $\hat{x}$ solves problem (P) and

$$f(\hat{x}) = \Phi(\hat{x}, \hat{y}^*) \qquad \text{i.e.} \qquad \hat{y}^*(g(\hat{x})) = 0.$$

If one defines the extremal value of problem (P) by

$$v(P) = \begin{cases} \inf_{x \in S} f(x) & \text{if } S \neq \phi \\ +\infty & \text{otherwise} \end{cases} \tag{II.4.10}$$

then $v(P)$ is also given by

$$v(P) = v(P^*) := \inf_{x \in X} \sup_{y^* \in Y^*} \Phi(x, y^*). \tag{II.4.11}$$

II.4.2 The dual problem as a max–inf problem

By analogy to (II.3.11) we again define

$$S^* = \{(\beta, y^*) \in \mathbb{R} \times F^* : f(x) - y^*(g(x)) \geqslant \beta - y^*(y) \quad \text{for all } x \in X, y \in Y\} \tag{II.4.12}$$

and consider as dual problem the problem (D): we seek a pair $(\hat{\beta}, \hat{y}^*) \in S^*$ satisfying

$$\hat{\beta} \geqslant \beta \qquad \text{for all } (\beta, y^*) \in S^*. \tag{II.4.13}$$

Lemma II.4.2
(a) $(\beta, y^*) \in S^*$ if and only if

$$y^* \in Y^* \qquad \textit{and} \qquad \beta \leqslant \inf_{x \in X} \Phi(x, y^*) \tag{II.4.14}$$

where Φ is defined by (II.4.3).
(b) S^* is empty if and only if for every $y^* \in Y^*$

$$\inf_{x \in X} \Phi(x, y^*) = -\infty. \tag{II.4.15}$$

Proof (a) Suppose $(\beta, y^*) \in S^*$. Then it follows as in the proof of Lemma II.3.14 that $y^* \in Y^*$ which implies $\beta \leqslant \inf_{x \in X} \Phi(x, y^*)$ and hence (II.4.14) holds.

Conversely if (II.4.14) is satisfied, then it follows for every $x \in X$ and $y \in Y$ that

$$f(x) - y^*(g(x)) \geqslant \inf_{x \in X} \Phi(x, y^*) \geqslant \beta - y^*(y)$$

that is

$$(\beta, y^*) \in S^*.$$

(*b*) If S^* is not empty, then by (*a*) there is a $y^* \in Y^*$ satisfying (II.4.14), i.e. (II.4.15) is violated for at least one $y^* \in Y^*$. If conversely there is a $y^* \in Y^*$ such that (II.4.15) is violated, then it follows that $(\inf_{x \in X} \Phi(x, y^*), y^*) \in S^*$, i.e. S^* is not empty.

Consequently the dual problem (D) is equivalent to problem (D*): we seek a $\hat{y}^* \in Y^*$ such that

$$\inf_{x \in X} \Phi(x, \hat{y}^*) \geqslant \inf_{x \in X} \Phi(x, y^*) \qquad \text{for all } y^* \in Y^*. \tag{II.4.16}$$

If one defines the extremal value of the problem (D) by

$$v^*(D) = \begin{cases} \sup_{(\beta, y^*) \in S^*} \beta & \text{if } S^* \neq \phi \\ -\infty & \text{otherwise} \end{cases} \tag{II.4.17}$$

then $v^*(\mathrm{D})$ is also given by

$$v^*(D) = v^*(D^*) := \sup_{y^* \in Y^*} \inf_{x \in X} \Phi(x, y^*). \tag{II.4.18}$$

If $\hat{y}^* \in Y^*$ is a solution of problem (D*), then $(\inf_{x \in X} \Phi(x, \hat{y}^*), \hat{y}^* \in S^*$ is a solution of problem (D). Conversely, if $(\hat{\beta}, \hat{y}^*) \in S^*$ is a solution of problem (D), then we must necessarily have $\hat{y}^* \in Y^*$ as well as $\hat{\beta} = \inf_{x \in X} \Phi(x, \hat{y}^*)$, and $\hat{y}^*$ solves the problem (D*).

We now have the following theorem concerning the simultaneous solvability of these two problems.

Theorem II.4.3 *If the positive cone Y of F is closed, then the following two statements are equivalent:*
(*a*) *Problem (P) and problem (D) are solvable and the extremal values of both problems are equal.*
(*b*) $\min_{x \in X} \sup_{y^* \in Y^*} \Phi(x, y^*) = \max_{y^* \in Y^*} \inf_{x \in X} \Phi(x, y^*).$ (II.4.19)

Proof
1. Suppose (*a*) is satisfied. If $\hat{x} \in S$ is a solution of problem (P), then by Section II.4.1, $(\hat{x}, \Theta_{F^*})$ is a solution of problem (P*) and

$$f(\hat{x}) = \Phi(\hat{x}, \Theta_{F^*}) = \min_{x \in X} \sup_{y^* \in Y^*} \Phi(x, y^*).$$

If $(\hat{\beta}, \hat{y}^*) \in S^*$ is a solution of problem (D), then on the basis of the above considerations $\hat{y}^* \in Y^*$ is a solution of problem (D*) and

$$\hat{\beta} = \inf_{x \in X} \Phi(x, \hat{y}^*) = \max_{y^* \in Y^*} \inf_{x \in X} \Phi(x, y^*).$$

Since $\hat{\beta} = f(\hat{x})$ we therefore have (II.4.19).
2. Suppose (b) holds. Choose $\hat{x} \in X$ and $\hat{y}^* \in Y^*$ such that

$$\sup_{y^* \in Y^*} \Phi(\hat{x}, y^*) = \min_{x \in X} \sup_{y^* \in Y^*} \Phi(x, y^*)$$

and

$$\inf_{x \in X} \Phi(x, \hat{y}^*) = \max_{y^* \in Y^*} \inf_{x \in X} \Phi(x, y^*)$$

respectively. Then, by Section II.4.1, $\hat{x}$ is a solution of problem (P) with $f(\hat{x}) = \sup_{y^* \in Y^*} \Phi(\hat{x}, y^*)$ and, by Section II.4.2, $\hat{y}^*$ is a solution of problem (D) with extremal value $\inf_{x \in X} \Phi(x, \hat{y}^*)$ and because of (II.4.19) the extremal values of both problems are equal.

II.4.3 The equivalence of the min–sup = max–inf statement with a saddle point statement

Definition A point $(\hat{x}, \hat{y}^*) \in X \times Y^*$ is called a saddle point of the Lagrange functional Φ in (II.4.3) if

$$\Phi(\hat{x}, y^*) \leqslant \Phi(\hat{x}, \hat{y}^*) \leqslant \Phi(x, \hat{y}^*) \tag{II.4.20}$$

holds for all $x \in X$ and $y^* \in Y^*$.

Theorem II.4.4 A point $(\hat{x}, \hat{y}^*) \in X \times Y^*$ *is a saddle point of* Φ *if and only if*

$$\begin{aligned} \sup_{y^* \in Y^*} \Phi(\hat{x}, y^*) &= \min_{x \in X} \sup_{Y^* \in Y^*} \Phi(x, y^*) \\ &= \max_{y^* \in Y^*} \inf_{x \in X} \Phi(x, y^*) = \inf_{x \in X} \Phi(x, \hat{y}^*). \end{aligned} \tag{II.4.27}$$

Proof
1. If (II.4.21) is true, then

$$\Phi(\hat{x}, \hat{y}^*) \leqslant \sup_{y^* \in Y^*} \Phi(\hat{x}, y^*) = \inf_{x \in X} \Phi(x, \hat{y}^*) \leqslant \Phi(\hat{x}, \hat{y}^*)$$

which implies equality and in turn (II.4.20), i.e. $(\hat{x}, \hat{y}^*)$ is a saddle point of Φ.
2. Let $(\hat{x}, \hat{y}^*) \in X \times Y^*$ be a saddle point of Φ. Then

$$\sup_{y^* \in Y^*} \Phi(\hat{x}, y^*) = \Phi(\hat{x}, \hat{y}^*) = \inf_{x \in X} \Phi(x, \hat{y}^*). \tag{II.4.22}$$

For every $\tilde{x} \in X$ and $y^* \in Y^*$ we have obviously

$$\inf_{x \in X} \Phi(x, y^*) \leqslant \Phi(\tilde{x}, y^*)$$

and hence

$$\sup_{y^* \in Y^*} \inf_{x \in X} \Phi(x, y^*) \leqslant \sup_{y^* \in Y^*} \Phi(\tilde{x}, y^*)$$

which implies

$$\sup_{y^* \in Y^*} \inf_{x \in X} \Phi(x, y^*) \leqslant \inf_{x \in X} \sup_{y^* \in Y^*} \Phi(x, y^*). \tag{II.4.23}$$

The statement (II.4.21) then follows from (II.4.22) and (II.4.23).

Remark The proof of Theorem II.4.4 shows that Φ could in general be any functional in any two variables and the theorem would also be true.

By combining Theorems II.4.3 and II.4.4 we obtain the next theorem.

Theorem II.4.5 *If the positive cone Y of F is closed, then a point $(\hat{x}, \hat{y}^*) \in X \times Y^*$ is a saddle point of the Lagrange functional Φ in (II.4.3) if and only if $\hat{x} \in S$, $\hat{x}$ is a solution of problem (P), and $(f(\hat{x}), \hat{y}^*)$ is a solution of problem (D).*

II.4.4 The generalized theorem of Kuhn and Tucker

For the sequel we now assume that the set $X \subseteq E$ is not empty and is convex, $f: X \to \mathbb{R}$ is a convex functional (see Section II.2.1) and $g : X \to F$ is a concave mapping (see Section II.2.2). Then we have the following generalized theorem from Kuhn and Tucker [51].

Theorem of Kuhn and Tucker *Suppose that the interior $\mathring{Y}$ of the positive cone Y of F is not empty and for every $y^* \geqslant \Theta_{F^*}$ with $y^* \neq \Theta_{F^*}$ there is an $x_* \in X$ with*

$$y^*(g(x_*)) > 0. \tag{II.4.24}$$

If $\hat{x} \in X$ is a solution of problem (P), then there is a $\hat{y}^* \in Y^*$ such that $(\hat{x}, \hat{y}^*)$ is a saddle point of the Lagrange functional Φ of (II.4.3).

Proof We define (see (II.3.3*b*))

$$\begin{aligned} K &= \bigcup_{x \in X} \{(\alpha, z) \in \mathbb{R} \times F : f(x) \leqslant \alpha, g(x) \geqslant z\} \\ &= \bigcup_{x \in X} \{(f(x) + r, g(x) - y) : r \geqslant 0, y \in Y\}. \end{aligned}$$

By Lemma II.2.10 the set K is convex and contains the open convex subset

$$K_0 = \bigcup_{x \in X} \{(f(x) + r, g(x) - y) : r > 0, y \in \mathring{Y}\}.$$

Furthermore $K \subseteq \bar{K}_0$, for if $(f(x) + r, g(x) - y) \in K$ is given, then we choose an $r_0 > 0$ and a $y_0 \in \mathring{Y}$. Then for all $\lambda \in [0, 1]$

$$r_\lambda = \lambda r + (1 - \lambda) r_0 > 0 \qquad \text{and} \qquad y_\lambda = \lambda y + (1 - \lambda) y_0 \in \mathring{Y}$$

(by Lemma II.2.3) and thus

$$\begin{aligned} &(f(x) + r_\lambda g(x) - y_\lambda) \in K_0 \\ &(f(x) + r, g(x) - y) = \lim_{\lambda \to 1} (f(x) + r_\lambda, g(x) - y_\lambda). \end{aligned}$$

Obviously $(f(\hat{x}), \Theta_F) \notin K_0$, for otherwise there would be an $x \in X$, an $r > 0$ and a

$y \in \overset{\circ}{Y}$ with

$$f(\hat{x}) = f(x) + r \qquad \text{for } g(x) = y \in Y$$

which would contradict the optimality of $\hat{x}$.

From the separation theorem 1 in Section IV.3.2 (for $A = \{(f(\hat{x}), \Theta_F)\}$ and $B = \overset{\circ}{B} = K_0$) used together with the representation formula (IV.1.18), we conclude the existence of $(\lambda, y^*) \in \mathbb{R} \times F^*$ and $\rho \in \mathbb{R}$ such that

$$\lambda f(\hat{x}) \leqslant \rho < \lambda\alpha + y^*(z) \qquad \text{for all } (\alpha, z) \in K_0 \tag{II.4.25a}$$

which because of $K \subseteq \bar{K}_0$ implies

$$\lambda f(\hat{x}) \leqslant \rho \leqslant \lambda[f(x) + r] + y^*(g(x) - y) \qquad \text{for all } r \geqslant 0, x \in X \text{ and } y \in Y. \tag{II.4.25b}$$

This implies $\hat{y}^* = -y^* \in Y^*$ and $\lambda \geqslant 0$. If $\lambda = 0$, then it follows from (II.4.25b) that

$$\hat{y}^*(g(x)) \leqslant 0 \qquad \text{for all } x \in X$$

and because of (II.4.25a), $\hat{y}^* \neq \Theta_{F^*}$. This is a contradiction to the assumption (II.4.24). Thus $\lambda > 0$ and without loss of generality $\lambda = 1$. From (II.4.25b) follows furthermore

$$f(\hat{x}) \leqslant f(x) - \hat{y}^*(g(x)) = \Phi(x, \hat{y}^*) \qquad \text{for all } x \in X$$

and for $x = \hat{x}$ follows $\hat{y}^*(g(\hat{x})) \leqslant 0$, whence $\hat{y}^*(g(\hat{x})) = 0$ since $\hat{y}^* \in Y^*$ and $g(\hat{x}) \in Y$. Thus the right-hand inequality of (II.4.21) is demonstrated. The left-hand inequality is trivial.

The requirement (II.4.24) is implied by the generalized Slater condition for $b = \Theta_F$ (see Section II.3.3).

Lemma II.4.6 *If the interior $\overset{\circ}{Y}$ of Y is not empty and if there is an $x_0 \in X$ with $g(x_0) \in \overset{\circ}{Y}$, then for every $y^* \in Y^*$ with $y^* \neq \Theta_{F^*}$, we have $y^*(g(x_0)) > 0$ (so (II.4.24) is satisfied for $x_* = x_0$).*

Proof Suppose $y^* \in Y^*$ is given with $y^*(g(x_0)) \leqslant 0$. Then it follows from Theorem IV.1.5 that $y^* = \Theta_{F^*}$.

II.4.5 Existence theorems for the dual problem

By combining Theorem II.4.5, the Theorem of Kuhn and Tucker and Lemma II.4.6, we obtain the following theorem.

Theorem II.4.7 *Suppose problem (P) satisfies the generalized Slater condition for $b = \Theta_F$ (in Section II.3.3) and suppose Y is closed. If $\hat{x} \in S$ is a solution of problem (P), then there is a $\hat{y} \in Y^*$ such that $(f(\hat{x}), \hat{y}^*)$ solves the dual problem (D) (and obviously the extremal values of the two problems coincide).*

Theorem II.4.7 supplements the following statement, which Lemma II.3.11 and the corollary to Theorem II.3.10 yield: if problem (P) satisfies the generalized

Slater condition for $b = \Theta_F$ and its extremal value is finite, then the dual problem (D) is also consistent and its extremal value coincides with that of problem (P).

This statement can be sharpened still further, which is done in the next theorem and this is likewise a generalization of Theorem I.4.14.

Theorem II.4.8 *Suppose that problem (P) satisfies the generalized Slater condition for* $b = \Theta_F$ *(in Section II.3.3). If its extremal value* $v(P)$ *is finite, then the dual problem (D) is solvable and its extremal value* $v^*(D)$ *coincides with* $v(P)$.

Proof We define two sets K and K_0 as in the proof of the theorem of Kuhn and Tucker. Then it follows that $(v(P), \Theta_F) \notin K_0$, for otherwise we would have $v(P) = f(x) + r$ for $r > 0$ and some $x \in X$ with $g(x) \in \mathring{Y} \subseteq Y$ in contradiction to the definition of $v(P)$. By applying the separation theorem 1 in Section IV.3.2 (for $A = \{(v(P), \Theta_F)\}$ and $B = \mathring{B} = K_0$ and the representation formula (IV.1.18), we again conclude the existence of $(\lambda, y^*) \in \mathbb{R} \times F^*$ and $\rho \in \mathbb{R}$ with

$$\lambda v(P) \leqslant \rho < \lambda\alpha + y^*(z) \qquad \text{for all } (\alpha, z) \in K_0. \tag{II.4.25c}$$

Because of $K \subseteq \bar{K}_0$ (see the proof of the Theorem of Kuhn and Tucker) this implies

$$\lambda v(P) \leqslant \rho \leqslant \lambda\ [f(x) + r]\ + y^*(g(x) - y) \qquad \text{for all } x \in X, r \geqslant 0 \text{ and } y \in Y \tag{II.4.25d}$$

and further $\lambda \geqslant 0$ as well as $\hat{y}^* = -y^* \in Y^*$. If λ were 0, then

$$\hat{y}^*(g(x)) \leqslant 0 \qquad \text{for all } x \in X.$$

Because of the generalized Slater condition there is an $x_0 \in X$ with $g(x_0) \in \mathring{Y}$, which by Theorem IV.1.5 implies $\hat{y}^* = \Theta_{F^*}$. However, because of (II.4.25c), that is not possible and hence $\lambda > 0$. Suppose without loss of generality that $\lambda = 1$. Then (II.4.25d) yields

$$v(P) - \hat{y}^*(y) \leqslant f(x) - \hat{y}^*(g(x)) \qquad \text{for all } x \in X, y \in Y$$

i.e. $(v(P), \hat{y}^*) \in S^*$ (see (II. 3.11)) $\Rightarrow v(P) \leqslant v^*(D)$. From Lemma II.3.8 it follows that $v^*(D) \leqslant v(P) \Rightarrow v(P) = v^*(D)$, and consequently $(v(P), \hat{y}^*) \in S^*$ is a solution of problem (D).

By Section II.4.2, Theorem II.4.8 may be formulated as follows.

Theorem II.4.9 *Suppose that problem (P) satisfies the generalized Slater condition in Section II.3.3 for* $b = \Theta_F$. *If its extremal value* $v(P)$, *(II.4.10) or also (II.4.11), is finite, then (see (II.4.18))*

$$v(P) = v^*(D) = v^*(D^*) = \max_{y^* \in Y^*} \inf_{x \in X} \Phi(x, y^*). \tag{II.4.26}$$

II.4.6 Bibliographical remarks

There is a vast literature dealing with min–sup, max–inf, and saddle point statements. We shall therefore give only a few representative contributions dealing with finite dimensional optimization problems. First there is the paper of Slater [50] in which the condition named after him concerning the validity of the saddle point statement appears for the first time. Other results in this direction were obtained for example by Stoer [63, 64] and Uzawa [58]. Complete presentations can be found in Stoer and Witzgall [70] and Karlin [59].

Generalizations to infinite dimensional spaces were already undertaken in the thirties and forties in connection with the famous saddle point theorem of von Neumann [28]. A compilation of the literature revolving around this set of ideas up to 1958 can be found for example in Sion [58]. The generalized Kuhn–Tucker Theorem proven in Section II.4.4 was first established in an even more general form by Hurwicz [58]. The same saddle point statement as in Section II.4.4 was also proven by Neustadt [70] with the inclusion of a finite number of affine linear side conditions. Neustadt also gives a differential form of the saddle point theorem, together with an application to a control problem. After weakening the Slater condition, Norris [67] derived another saddle point theorem and also applied this result to a control problem.

II.5 APPLICATION TO APPROXIMATION PROBLEMS

II.5.1 Existence theorems in the case of convex approximation problems in normed vector spaces

Let E be a normed vector space, X a non-empty convex subset of E and $z \in E$ a fixed element. The distance of z from X is then given by

$$\rho(z, X) = \inf_{x \in X} \| x - z \| \ (\geqslant 0) \tag{II.5.1}$$

and $\rho(z, X) = 0$ if and only if z belongs to the closure $\bar{X}$ of X. We then seek an $\hat{x} \in X$ with

$$\| \hat{x} - z \| = \rho(z, X). \tag{II.5.2}$$

Define the functional $f : E \to \mathbb{R}$ by

$$f(y) = \| y - z \| \qquad \text{for } y \in E. \tag{II.5.3}$$

Then since

$$| f(y_1) - f(y_2) | \leqslant \| y_1 - y_2 \| \qquad \text{for all } y_1, y_2 \in E$$

f is continuous and because of

$$\begin{aligned} f(\lambda y_1 + (1 - \lambda) y_2) &= \| \lambda y_1 + (1 - \lambda) y_2 - z \| \\ &= \| \lambda (y_1 - z) + (1 - \lambda)(y_2 - z) \| \leqslant \lambda \| y_1 - z \| \\ &\quad + (1 - \lambda) \| y_2 - z \| \\ &= \lambda f(y_1) + (1 - \lambda) f(y_2) \qquad \text{for all } \lambda \in [0, 1], y_1, y_2 \in E \end{aligned}$$

f is convex on X.

If one fixes an $x_0 \in X$ and defines

$$X_0 = \{x \in X : f(x) \leqslant f(x_0)\} \tag{II.5.4}$$

then X_0 is a non-empty, convex subset of X (Proof = Exercise), and since

$$X_0 \subseteq \{x \in X : \| x \| \leqslant f(x_0) + \| z \| \}$$

X_0 is bounded in E. If X is closed, then X_0 is also closed (Proof = Exercise) and the existence theorem in Section II.2.1 yields the following result.

Theorem II.5.1 *If E is a reflexive Banach space and X is a non-empty, closed convex subset of E, then for every $z \in E$ there is an $\hat{x} \in X$ satisfying (II.5.2).*

Proof If X_0 is defined by (II.5.4), then obviously $\rho(z, X_0) = \rho(z, X)$ where ρ is defined by (II.5.1). Seeking an $\hat{x} \in X$ is equivalent to seeking an $\hat{x} \in X_0$ satisfying

$$f(\hat{x}) \leqslant f(x) \qquad \text{for all } x \in X_0$$

where f is defined by (II.5.3). As remarked above, f is continuous and convex on X (and hence also on X_0). Furthermore, as was also remarked above, X_0 is not empty, convex, closed and bounded so that the assertion follows immediately from the existence theorem in Section II.2.1.

Remark If one fixes any $x_0 \in X$, then obviously

$$\rho(z, X) = \rho(z - x_0, X - x_0) \qquad \text{with } X - x_0 = \{x - x_0 : x \in X\}.$$

Moreover $\| x^* - (z - x_0) \| = \rho(z - x_0, X - x_0)$ holds for $x^* \in X - x$ if and only if (II.5.2) is satisfied for $\hat{x} = x^* + x_0$. If $z \in E$ has been chosen and is fixed, then in place of E, one could consider the linear subspace F of E which is spanned by $z - x_0$ and $V = L(X - x_0)$, where V is the linear subspace of E generated by $X - x_0$, i.e. it consists of all elements of the form

$$v = \sum_{i \in I} \lambda_i (x_i - x_0) \qquad \lambda_i \in \mathbb{R}, x_i \in X \text{ for all } i \in I$$

where I is an arbitrary finite index set.

If $V = L(X - x_0)$ is reflexive, then the linear subspace F of E spanned by V and $z - x_0$ is also reflexive (Proof = Exercise), and Theorem II.5.1 yields

Theorem II.5.2 *If X is a non-empty, closed, convex subset of a normed vector space E such that for some $x_0 \in X$ the linear subspace of E generated by $X - x_0$ is reflexive (for example is finite dimensional), then for each $z \in E$ there is an $\hat{x} \in X$ satisfying (II.5.2).*

II.5.2 Uniform approximation of functions

II.5.2.1 The general convex case

We again take up the approximation problem in Section I.2.1 in a somewhat more general form. Therefore, let M be a compact metric space and $E = C(M)$ be the vector space of continuous real-valued functions on M equipped with the maximum norm (I.1.2). Further, let X be a non-empty, convex subset of $C(M)$ and $z \in C(M)$ be a given function. We again seek an $\hat{x} \in X$ satisfying (II.5.2), where $\| \ \|$ denotes the maximum norm (I.2.1) in $E = C(M)$.

Since E is not reflexive, Theorem II.5.1 cannot be applied. One the basis of the remarks at the end of Theorem II.5.1, we shall assume without loss of generality that X contains the null function $\Theta_E \equiv 0$. If X is closed and the linear subspace $V = L(X)$ generated by X is reflexive then, for each function $z \in C(M)$, Theorem II.5.2 assures the existence of an $\hat{x} \in X$ satisfying (II.5.2). Now let W be the linear subspace of $C(M)$ generated by the function $e \equiv 1$ and $z \in C(M)$, and let $F = W \times W$. If one defines

$$g(v, \gamma) = \begin{pmatrix} g_1(v, \gamma) \\ g_2(v, \gamma) \end{pmatrix}$$

with

$$g_1(v, \gamma)(t) = v(t) + \gamma - z(t)$$

and

$$g_2(v, \gamma)(t) = -v(t) + \gamma + z(t) \qquad \text{for } t \in M$$

then $g : V \times \mathbb{R} \to F$ is an affine linear mapping and it is concave (see Section II.2.2) provided one defines the positive cone Y in F appropriately, for example by $Y = Y_W \times Y_W$ with

$$Y_W = \{w \in W : w(t) \geqslant 0 \qquad \text{for all } t \in M\}.$$

If, finally, one defines another linear (and hence convex) functional (see Section II.2.1) $f : V \times \mathbb{R} \to \mathbb{R}$ by $f(v, \gamma) = \gamma$, $(v, \gamma) \in V \times \mathbb{R}$, then seeking an $\hat{x} \in X$ satisfying (II.5.2) is equivalent to the problem (P) of minimizing the functional $f(x, \gamma) = \gamma$ subject to the side conditions

$$g(x, \gamma) \in Y \qquad \text{for } (x, \gamma) \in X \times \mathbb{R}$$

(see in this respect Section I.2.1.1). If one defines

$$S = \{(x, \gamma) \in X \times \mathbb{R} : g(x, \gamma) \in Y\}$$

then

$$\rho(z, X) = \inf_{(x, \gamma) \in S} f(x, \gamma) \geqslant 0.$$

The positive cone Y of F has a non-empty interior $\mathring{Y}$, which is given by $\mathring{Y} = \mathring{Y}_W \times \mathring{Y}_W$, where

$$\mathring{Y}_W = \{w \in W : w(t) > 0 \qquad \text{for all } t \in M\}$$

(Proof = Exercise) since $e \in Y_W$ is not empty.

If one chooses $x_0 \in X$ arbitrarily and $\gamma_0 > \| x_0 - z \|$, then $g(x_0, \gamma_0) \in \mathring{Y}$, i.e. the generalized Slater condition in Section II.3.3 with $b = \Theta_F$ is satisfied. Since the extremal value of problem (P) is bounded from below by zero, Theorem II.4.8 is applicable. After taking into account the representation formula (IV.1.18), Theorem

II.4.8 yields the existence of two continuous linear functionals $\hat{y}_1^*, \hat{y}_2^* \in W^*$ (the topological dual space of W, see Section IV.1.2) with

$$\gamma - \hat{y}_1^*(g_1(x,\gamma)) - \hat{y}_2^*(g_2(x,\gamma)) \geqslant \rho(z,X) - \hat{y}_1^*(y_1) - \hat{y}_2^*(y_2)$$

for all $\gamma \in \mathbb{R}$, $x \in X$ and $y_1, y_2 \in Y_W$

which is equivalent to

$$\gamma - (\hat{y}_1^* - \hat{y}_2^*)(x) - \gamma(\hat{y}_1^* + \hat{y}_2^*)(e) + (\hat{y}_1^* - \hat{y}_2^*)(z) \geqslant \rho(z,X) - \hat{y}_1^*(y_1) - \hat{y}_2^*(y_2)$$

for all $\gamma \in \mathbb{R}$, $x \in X$ and $y_1, y_2 \in Y_W$. (II.5.5)

Assertion The statement (II.5.5) is equivalent to

$$\hat{y}_1^*(e) + \hat{y}_2^*(e) = 1 \tag{II.5.6a}$$

$$(\hat{y}_1^* - \hat{y}_2^*)(z - x) \geqslant \rho(z,X) \qquad \text{for all } x \in X \tag{II.5.6b}$$

and

$$\hat{y}_1^*(y) \geqslant 0 \text{ as well as } \hat{y}_2^*(y) \geqslant 0 \qquad \text{for all } y \in Y_W. \tag{II.5.6c}$$

Problem II.5.1 Prove this assertion.

If $w \in W$ is given, then for all $t \in M$

$$-\|w\|e \leqslant w(t) \leqslant \|w\|e$$

and hence for $i = 1, 2$

$$-\|w\|\hat{y}_i^*(e) \leqslant \hat{y}_i^*(w) \leqslant \|w\|\hat{y}_i^*(e) \Rightarrow |\hat{y}_i^*(w)| \leqslant \hat{y}_i^*(e)\|w\|$$

which implies $\hat{y}_i^*(e) \geqslant \|\hat{y}_i^*\|_W$. On the other hand, since $e \in W$

$$\hat{y}_i^*(e) \leqslant \|\hat{y}_i^*\|_W = \sup_{\substack{w \in W \\ \|w\| = 1}} |\hat{y}_i^*(w)|$$

and hence $\hat{y}_i^*(e) = \|\hat{y}_i^*\|_W$ for $i = 1, 2$ (see also Section IV.2.3). Setting $\hat{y}^* = \hat{y}_1^* - \hat{y}_2^*$, then

$$\|\hat{y}^*\|_W \leqslant \|\hat{y}_1^*\|_W + \|\hat{y}_2^*\|_W = \hat{y}_1^*(e) + \hat{y}_2^*(e) = 1 \tag{II.5.7}$$

and

$$\rho(z,X) \leqslant \inf_{x \in X} \hat{y}^*(z - x). \tag{II.5.8}$$

Conversely, choosing $y^* \in W^*$ with $\|y^*\|_W \leqslant 1$ arbitrarily, then for all $x \in X$

$$y^*(z - x) \leqslant \|y^*\|_W \|z - x\| \leqslant \|z - x\|$$

whence

$$\inf_{x \in X} y^*(z - x) \leqslant \rho(z,X)$$

follows, which together with (II.5.7) and (II.5.8) furnishes the following theorem.

Theorem II.5.3 *If one sets*

$$B_{W^*} = \{y^* \in W^* : \| y^* \|_W \leq 1\}$$

then the distance (II.5.1) of z from X is

$$\rho(z, X) = \max_{y^* \in B_{W^*}} \inf_{x \in X} y^*(z - x). \tag{II.5.9}$$

From Theorem IV.1.4 it follows that

$$\| z - x \| = \max_{y^* \in B_{W^*}} y^*(z - x).$$

(This statement in fact is a trivial consequence of the definition of the maximum norm.) It yields

$$\rho(z, X) = \inf_{x \in X} \max_{y^* \in B_{W^*}} y^*(z - x). \tag{II.5.10}$$

II.5.2.2 The general linear case

Theorem II.5.4 *If X is a linear subspace of $E = C(M)$, then for every $y^* \in W^*$ one has the equivalence*

$$\inf_{x \in X} y^*(z - x) > -\infty \Leftrightarrow y^*(x) = 0 \qquad \text{for all } x \in X.$$

In both cases

$$y^*(z) = \inf_{x \in X} y^*(z - x) \tag{II.5.11}$$

which (by Theorem II.5.3) implies

$$\rho(z, X) = \max_{y^* \in B_{W^*} \cap X^{\perp}_{W^*}} y^*(z) \tag{II.5.12}$$

where

$$X^{\perp}_{W^*} = \{y^* \in W^* : y^*(x) = 0 \qquad \text{for all } x \in X\}.$$

Proof If $y^* \in X^{\perp}_{W^*}$, then it follows that $\inf_{x \in X} y^*(z - x) = y^*(z) > -\infty$.
Conversely, if $\inf_{x \in X} y^*(z - x) = \alpha > -\infty$, then

$$-y^*(x) \geq \alpha - y^*(z) \qquad \text{for all } x \in X$$

which implies $y^* \in X^{\perp}_{W^*}$.

If X is a linear subspace of $E = C(M)$, then the problem dual to problem (P) is equivalent with the problem (D) of maximizing the linear form $y^*(z)$ subject to the side conditions

$$y^* \in W^*, \| y^* \|_W \leq 1, y^*(x) = 0 \qquad \text{for all } x \in X.$$

This problem is always solvable and its extremal value equals $\rho(z, X)$. In addition the following holds.

Theorem II.5.5 *Suppose X is a linear subspace of $E = C(M)$ and $\rho(z, X) > 0$. Then (II.5.2) is satisfied for $\hat{x} \in X$ if and only if a $\hat{y}^* \in W^*$ exists with*

$$\| \hat{y}^* \|_W = \sup_{\substack{w \in W \\ \| W \| = 1}} \hat{y}^*(w) = 1 \tag{II.5.13}$$

$$\hat{y}^*(x) = 0 \qquad \text{for all } x \in X \tag{II.5.14}$$

and

$$\hat{y}^*(z - \hat{x}) = \| z - \hat{x} \|. \tag{II.5.15}$$

Proof

1. If $\| \hat{x} - z \| = \rho(z, X)$ for some $\hat{x} \in X$, then by Theorem II.5.4 there is a $\hat{y}^* \in W^*$ satisfying $\| \hat{y}^* \|_W \leqslant 1$, (II.5.14), and (II.5.15), which also implies (II.5.13).
2. If conversely for a given $\hat{x} \in X$, there is a $\hat{y}^* \in W^*$ satisfying (II.5.13), (II.5.14), (II.5.15), then for all $x \in X$ it follows that

$$\| z - \hat{x} \| = \hat{y}^*(z - \hat{x}) = \hat{y}^*(z - x) \leqslant \| z - x \|$$

that is

$$\| \hat{x} - z \| = \rho(z, X).$$

If X is an n-dimensional linear subspace of $C(M)$, then Theorem II.5.5 can be sharpened still further to Theorem I.5.6 in the case that $\rho(z, X) > 0$, if one uses Theorem IV.2.8 and the corresponding corollary.

For this purpose we choose in particular for V the linear subspace of $C(M)$ which is spanned by X and the function $z \in C(M)$ which is to be approximated. Then $r = \dim V \leqslant (n + 1)$, and the linear subspace W of $C(M)$ appearing in Theorem II.5.5 is spanned by V and $e \equiv 1$. Every element $y \in W$ therefore also belongs to V. Now Theorem II.5.5 and Theorem IV.2.8 together with its corollary, plus an application of problem I.5.3*b*, yield the Theorem I.5.6.

II.5.3 An approximation problem with a mixed norm

We again take up the approximation problem (II.1.34′), (II.1.43) in a somewhat more general form. Let E be the vector space $C(M)$ of the continuous real functions h on $M = [a, b] \times [a, b], a < b$, equipped with the norm

$$\| h \| = \max_{t \in [a, b]} \int_a^b | h(t, s) | \, ds. \tag{II.5.16}$$

Let X be a non-empty convex subset of $E = C(M)$ and $z \in E$ be some function. We again seek an $\hat{x} \in X$ satisfying (II.5.2). For the question of the existence of $\hat{x}$, one can again apply Theorem II.5.2. Further, we shall give necessary and sufficient conditions for $\hat{x}$. For this purpose we reformulate the approximation problem as a problem of convex optimization. Let $F = C[a, b]$, equipped with the maximum norm and the natural partial ordering $y_1 \geqslant y_2 \Leftrightarrow y_1(t) \geqslant y_2(t)$ for all $t \in [a, b]$, $y_1, y_2 \in F$.

If one defines a mapping $g : X \times \mathbb{R} \to F$ by

$$g(x,\gamma)(t) = \gamma - \int_a^b |x(t,s) - z(t,s)|\,ds \qquad \text{for } t \in [a,b]$$

then g is concave on $X \in \mathbb{R}$ (see Section II.2.2; Proof = Exercise). Seeking an $\hat{x} \in X$ satisfying (II.5.2) (with the norm (II.5.16)) is equivalent to the problem (P) of minimizing the linear functional $f(x,\gamma) = \gamma$ subject to the side conditions

$$g(x,\gamma) \in Y \qquad \text{for } (x,\gamma) \in X \times \mathbb{R}.$$

Here

$$Y = \{y \in F = C[a,b] : y(t) \geqslant 0 \text{ for all } t \in [a,b]\}$$

is the positive cone of F. The interior of Y is given by

$$\mathring{Y} = \{y \in F : y(t) > 0 \text{ for all } t \in [a,b]\}.$$

If one chooses $x_0 \in X$ arbitrarily and $\gamma_0 > \| x_0 - z \|$, then $g(x_0,\gamma_0) \in \mathring{Y}$, i.e. the Slater condition of Section II.3.3 is satisfied with $b = \Theta_F$.

Since besides

$$\rho(z,X) = \inf_{(x,\gamma) \in S} f(x,\gamma) \geqslant 0$$

with

$$S = \{(x,\gamma) \in X \times \mathbb{R} : g(x,\gamma) \in Y\}$$

the dual problem is solvable by Theorem II.4.8 and the extreme values coincide.

From this follows the existence of a $\hat{y}^* \in F^*$ with

$$\gamma - \hat{y}^*(g(x,\gamma)) \geqslant \rho(z,X) - \hat{y}^*(y) \qquad \text{for all } (x,\gamma) \in X \times \mathbb{R} \text{ and } y \in Y. \tag{II.5.17}$$

This statement is equivalent to

$$\hat{y}^*(y) \geqslant 0 \qquad \text{for all } y \in Y, \text{ i.e., } \hat{y}^* \geqslant \Theta_{F^*} \tag{II.5.18}$$

$$\hat{y}^*(e) = \| \hat{y}^* \| = 1 \tag{II.5.19}$$

where $e \equiv 1$ (see Lemma IV.2.4) and

$$\inf_{x \in X} \hat{y}^*\left(\int_a^b |x(\cdot,s) - z(\cdot,s)|\,ds \right) \geqslant \rho(z,X). \tag{II.5.20}$$

By Section IV.2.3 there is for $\hat{y}^*$ a monotonic function $\hat{g}$ on $[a,b]$ such that

$$\hat{y}^*(h) = \int_a^b h(t)\,d\hat{g}(t)$$

for all $h \in F$ where the integral is a Riemann–Stieltjes integral. By (II.5.19) we have further that

$$\int_a^b d\hat{g}(t) = \hat{g}(b) - \hat{g}(a) = 1$$

and (II.5.20) reads

$$\inf_{x \in X} \int_a^b \int_a^b |x(t,s) - z(t,s)| \, ds \, d\hat{g}(t) \geqslant \rho(z, X). \tag{II.5.21}$$

Conversely, if g is a monotonic function on $[a, b]$ with $g(b) - g(a) = 1$, then it follows for all $x \in X$, that

$$\int_a^b \int_a^b |x(t,s) - z(t,s)| \, ds \, dg(t) \leqslant \| x - z \| (g(b) - g(a)) = \| x - z \|$$

and hence

$$\inf_{x \in X} \int_a^b \int_a^b |x(t,s) - z(t,s)| \, ds \, dg(t) \leqslant \rho(z, X). \tag{II.5.22}$$

With this we have the next theorem.

Theorem II.5.6 *If B is the set of monotonic functions g on $[a, b]$ with $g(b) - g(a) = 1$, then*

$$\rho(z, X) = \max_{g \in B} \inf_{x \in X} \int_a^b \int_a^b |x(t,s) - z(t,s)| \, ds \, dg(t). \tag{II.5.23}$$

From this theorem one easily obtains the next theorem.

Theorem II.5.7 *Equation (II.5.2) is satisfied for $\hat{x} \in X$ if and only if there is a $\hat{g} \in B$ satisfying*

$$\| \hat{x} - z \| = \int_a^b \int_a^b |\hat{x}(t,s) - z(t,s)| \, ds \, d\hat{g}(t)$$

$$= \max_{g \in B} \inf_{x \in X} \int_a^b \int_a^b |x(t,s) - z(t,s)| \, ds \, dg(t). \tag{II.5.24}$$

Proof

1. That (II.5.24) implies (II.5.2) follows directly from (II.5.23).
2. If conversely (II.5.2) is satisfied, then (II.5.23) yields the existence of a function $\hat{g} \in B$ with

$$\| \hat{x} - z \| = \inf_{x \in X} \int_a^b \int_a^b |x(t,s) - z(t,s)| \, ds \, d\hat{g}(t)$$

$$\leqslant \int_a^b \int_a^b |\hat{x}(t,s) - z(t,s)| \, ds \, dg(t) \leqslant \| \hat{x} - z \|$$

which implies

$$\| \hat{x} - z \| = \int_a^b \int_a^b |\hat{x}(t,s) - z(t,s)| \, ds \, d\hat{g}(t)$$

$$= \inf_{x \in X} \int_a^b \int_a^b |x(t,s) - z(t,s)| \, ds \, d\hat{g}(t)$$

and together with (II.5.23) and (II.5.2) yields the assertion.

II.5.4 Calculation of the minimal deviation for a convex approximation problem

We again take up the convex approximation problem in Section II.5.1; however, we shall use the notation of Section II.1.1 in order to be able to connect up with the general convex optimization problem in Section II.3.1 without having to introduce any other notation. We consider, therefore, a normed vector space Z over $\mathbb{R}$, a non-empty, convex subset V of Z and an element $z \in Z$. We seek a $\hat{v} \in V$ satisfying (II.1.1). Our goal now consists in calculating the minimal deviation $\rho(z, V)$ given by (II.1.2) independently of the solvability of this problem. For this purpose, we shall reformulate the approximation problem (II.1.1) as an equivalent convex optimization problem and apply Theorem II.4.9 to it.

By Section II.1.1 the problem (II.1.1) is equivalent to that of minimizing the linear functional $f(v, \gamma) = \gamma$ subject to the side conditions (II.1.9). (We saw this before for a linear subspace V of Z. However, the derivation is valid for any non-empty subset V of Z.) We now define $E = F = Z \times \mathbb{R}$ equipped with the norm $\|(y, \gamma)\| = \|y\| + |\gamma|$, $(y, \gamma) \in F$. Furthermore, we introduce in F a partial ordering which is induced by the closed convex cone

$$Y = \{(y, \gamma) \in F : L(y) + \gamma \geqslant 0 \qquad \text{for all } L \in B^*\}. \tag{II.5.25}$$

Here B^* again denotes the unit ball (II.1.6) of the dual space Z^* of Z. If one now defines $X = \{z - V\} \times \mathbb{R}$ and $g : E \to F$ is the identity mapping, then X is a convex subset of E and g is a concave mapping, and the side conditions (II.1.10) go over into the conditions

$$(y, \gamma) \in X \qquad \text{for } g(y, \gamma) \in Y. \tag{II.5.26}$$

With this the approximation problem (II.1.1) is equivalent to the problem of minimizing the functional $f(y, \gamma) = \gamma$ subject to the side conditions (II.5.26). We are faced, therefore, with a convex optimization problem as in Section II.3.1 (with $b = \Theta_F$) (see also Section II.4.1). In order to be able to apply Theorem II.4.9 to this problem, we need the cone Y^* in F^* which is adjoint to the one in Y (see Section IV.2.2).

Lemma II.5.8 *The cone Y^* which is adjoint to Y of (II.5.25) is given by*

$$K(B^* \times \{1\}) = \{\lambda(L, 1) : L \in B^*, \lambda \geqslant 0\}. \tag{II.5.27}$$

Proof Let $y^* = \lambda(L, 1) \in K(B^* \times \{1\})$ be given. Then for every $(y, \gamma) \in Y$ follows

$$y^*(y, \gamma) = \lambda L(y) + \lambda\gamma = \lambda(L(y) + \gamma) \geqslant 0 \qquad \text{i.e.} \qquad y^* \in Y^*.$$

Now, conversely, let $y^* = (\hat{L}, \hat{\lambda}) \in Y^*$ be given. Then by definition

$$\hat{L}(y) + \hat{\lambda}\gamma \geqslant 0 \qquad \text{for all } (y, \gamma) \in Y. \tag{II.5.28}$$

If one equips Z^* with the weak* topology, then F^{**} consists precisely of all linear forms of the form

$$y^{**}(L, \lambda) = L(y) + \lambda\gamma \qquad \text{for } L \in Z^*, \lambda \in \mathbb{R}$$

where $(y, \gamma) \in F$ is an arbitrary but fixed element, and B^* is compact in Z^*. Thus

$B^* \times \{1\}$ is a convex, compact subset of F^* and $\Theta_{F^*} \notin B \times \{1\}$. By Theorem IV.2.1[1] $K(B^* \times \{1\})$ is thus closed. Now we assume that $(\hat{L}, \hat{\lambda}) \notin K(B^* \times \{1\})$. Then by Theorem IV.2.3[1] there is a $(y, \gamma) \in F$ with

$$\lambda L(y) + \lambda\gamma \geqslant 0 \qquad \text{for all } L \in B^* \quad \text{and} \quad \lambda \geqslant 0 \Rightarrow (y, \gamma) \in Y$$

and $\hat{L}(y) + \hat{\lambda}\gamma < 0$, a contradiction to (II.5.28). Consequently $(\hat{L}, \hat{\lambda}) \in K(B^* \times \{1\})$.

Next we prove the following lemma.

Lemma II.5.9 The cone Y defined by (II.5.25) has a non-empty interior $\mathring{Y}$.

Proof If one chooses $\hat{y} \in Z$ arbitrarily and $\hat{\gamma} = \| \hat{y} \| + \epsilon$ for some $\epsilon > 0$, then for every $L \in B^*$ follows

$$L(\hat{y}) + \hat{\gamma} \geqslant -\| \hat{y} \| + \| \hat{y} \| + \epsilon = \epsilon > 0 \qquad \text{i.e.} \qquad (\hat{y}, \hat{\gamma}) \in Y$$

and for every $(y, \gamma) \in Z$ with $\| y - \hat{y} \| \leqslant \epsilon/4$, $|\gamma - \hat{\gamma}| \leqslant \epsilon/4$ follows

$$\begin{aligned} L(y) + \gamma &= L(\hat{y}) + \hat{\gamma} + L(y - \hat{y}) + \gamma - \hat{\gamma} \\ &\geqslant \epsilon - \| y - \hat{y} \| - |\gamma - \hat{\gamma}| \geqslant \epsilon/2 > 0 \qquad \text{i.e.} \qquad (y, \gamma) \in Y \end{aligned}$$

which implies $(\hat{y}, \hat{\gamma}) \in \mathring{Y}$.

In the same way one sees that a pair $(y, \gamma) \in X$ exists with $(y, \gamma) \in \mathring{Y}$. Therefore the generalized Slater condition of Section II.3.3 (for $b = \Theta_F$) is satisfied. Since the extremal value of the problem subject to the side conditions (II.5.26) is equal to the minimal deviation $\rho(z, V)$ (II.1.2) and hence is bounded from below by zero, one can apply Theorem II.4.9 and with the aid of Lemma II.5.8 one obtains the statement

$$\rho(z, V) = \max_{\substack{(\lambda L, \lambda) \\ L \in B^*, \lambda \geqslant 0}} \inf_{v \in V, \gamma \in \mathbb{R}} \{\gamma - [\lambda L(z - v) + \lambda\gamma]\}.$$

In forming the maximum, only pairs such as $(\lambda L, \lambda) \in K(B^* \times \{1\})$ come into question such that

$$\inf_{v \in V, \gamma \in \mathbb{R}} \{\gamma - [\lambda L(z - v) + \lambda\gamma]\} = \inf_{v \in V, \gamma \in \mathbb{R}} \{\gamma(1 - \lambda) + \lambda L(v - z)\} > -\infty$$

which can obviously only occur for $\lambda = 1$. With that we obtain

$$\rho(z, V) = \max_{L \in B^*} \inf_{v \in V} L(v - z) = \max_{L \in B^*} \inf_{v \in V} L(z - v) \qquad \text{(II.5.29}a\text{)}$$

since L belongs to B^*, so does $-L$. This can be rewritten as

$$\rho(z, V) = \max_{L \in B^*} \{L(z) - \sup_{v \in V} L(v)\}. \qquad \text{(II.5.29}b\text{)}$$

This statement can be proven directly much more briefly (for this see Köthe [66], page 348). In Holmes [72] it was derived using conjugate functionals.

Problem II.5.2 In analogy to Theorems II.5.4 and II.5.5, prove for the case of a

[1] Both theorems hold also for $E = Z^*$ equipped with the weak* topology.

linear subspace V of Z the following two statements: (*a*)

$$\rho(z, V) = \max_{L \in B^* \cap V^\perp} L(z)$$

where

$$V^\perp = \{L \in Z^* : L(v) = 0 \qquad \text{for all } v \in V\}$$

and (*b*) $\hat{v} \in V$ is a best approximation to $z \notin \overline{V}$ in V if and only if there is an $\hat{L} \in V^\perp$ with $\| \hat{L} \| = 1$ and

$$\hat{L}(z - \hat{v}) = \| z - \hat{v} \|.$$

If V is a linear subspace of Z, and if one chooses as the positive cone in E the linear subspace $V \times \mathbb{R}$ and defines the linear mapping $A : E \to F$ by $A(y, \gamma) = (-y, \gamma)$ then the above optimization problem can also be formulated as a linear optimization problem (Section I.3), which consists of minimizing the linear form $c(y, \gamma) = \gamma$ subject to the side conditions

$$(y, \gamma) \in V \times \mathbb{R} \qquad \text{for } A(y, \gamma) + (z, 0) \in Y$$

where Y is the positive cone in $F = E$ defined by (II.5.25).

In conclusion we shall show how one can obtain the formula (I.5.52) in Section I.5.5 from (II.5.29*b*). We consider for this a slight generalization of the situation as it occurs there: let W be another normed vector space (whose norm we also denote by $\| \cdot \|$) and $S : W \to Z$ a given linear mapping and also U a non-empty convex subset of W. The convex subset V in Z we take to be the image $S(U)$ of U. Then it follows from (5.29*b*) that

$$\rho(z, S(U)) = \max_{L \in B^*} \{L(z) - \sup_{u \in U} L(S(u))\}. \qquad \text{(II.5.30)}$$

This formula is obviously a generalization of (I.5.52) in Section I.5.5. There $W = C[0, T]$, $Z = C[-1, +1]$, equipped with the maximum norm $\| \cdot \|_\infty$, $U = \{u \in W : \| u \|_\infty \leqslant 1\}$ and the mapping S is defined by (I.1.8) with $S = B$.

II.6 CONVEX OPTIMIZATION PROBLEMS IN FUNCTION SPACES

II.6.1 Posing the problem and characterizing the optimality

We first start from a very general situation and consider a linear vector space E, a non-empty subset X of E, a functional $f : X \to \mathbb{R}$ and a mapping $g : X \to C(T)$, where $C(T)$ is the vector space of continuous, real-valued functions defined on a compact Hausdorff space T. We imagine $C(T)$ to be equipped with the maximum norm (I.2.1) and partially ordered in the natural way (see Section IV.1.1). We denote the positive cone of $C(T)$ by Y. We assume that the set

$$S = \{x \in X : g(x) \in Y\} \qquad \text{(II.6.1)}$$

is not empty. We seek an $\hat{x} \in S$ such that

$$f(\hat{x}) \leqslant f(x) \qquad \text{for all } x \in S. \tag{II.6.2}$$

We want to give necessary and sufficient conditions for such optimal elements $\hat{x} \in S$. For this purpose, we associate with every $\hat{x} \in X$ the value

$$\delta(x) = \inf_{t \in T} g(x, t) \tag{II.6.3}$$

and the non-empty set

$$I(x) = \{t \in T : g(x, t) = \delta(x)\} \tag{II.6.4}$$

Then we have the following theorem.

Theorem II.6.1 *An element $\hat{x} \in S$ is optimal, i.e. (II.6.2) holds if for each $x \in X$ the following implication is true:*

$$g(x, t) \geqslant 0 \qquad \text{for all } t \in I(\hat{x}) \Rightarrow f(\hat{x}) \leqslant f(x) \tag{II.6.5}$$

i.e. if $\hat{x}$ minimizes the functional f on the set

$$S(\hat{x}) = \{x \in X : g(x, t) \geqslant 0 \qquad \text{for all } t \in I(\hat{x})\}. \tag{II.6.6}$$

The proof of Theorem II.6.1 follows immediately from the fact that $S \subseteq S(\hat{x})$ and from the equivalence of the implication (II.6.5) with the statement $f(\hat{x}) \leqslant f(x)$ for all $x \in S(\hat{x})$.

The question arises under what requirements the validity of the implication (II.6.5) for all $x \in X$ is also necessary for the optimality of $\hat{x} \in S$. For the investigation of this question we need the following definitions.

Definition (a) The set X is called star-shaped with respect to $\hat{x} \in X$ if

$$\lambda \in [0, 1], x \in X \Rightarrow \lambda x + (1 - \lambda)\hat{x} \in X.$$

Definition (b) The functional $f : X \to \mathbb{R}$, and the mapping $g : X \to C(T)$, respectively, is called convex, or concave, with respect to $\hat{x} \in X$, if X is star-shaped with respect to $\hat{x}$ and if for all $x \in X$ and $\lambda \in [0, 1]$ one has

$$f(\lambda x + (1 - \lambda)\hat{x}) \leqslant \lambda f(x) + (1 - \lambda) f(\hat{x})$$

or

$$g(\lambda x + (1 - \lambda)\hat{x}) \geqslant \lambda g(x) + (1 - \lambda) g(\hat{x}) \tag{II.6.7}$$

respectively. Here (II.6.7) is equivalent to

$$g(\lambda x + (1 - \lambda)\hat{x}, t) \geqslant \lambda g(x, t) + (1 - \lambda) g(\hat{x}, t) \qquad \text{for all } t \in T$$

which is equivalent to the concavity with respect to $\hat{x}$ of all functionals $g(\cdot, t) : X \to \mathbb{R}, t \in T$.

If T is a finite set (and hence a compact Hausdorff space, if one equips it with the discrete topology), then the next theorem holds.

Theorem II.6.2 *Suppose $\hat{x} \in S$ is optimal, i.e. suppose (II.6.2) holds. If T is finite,*

X star-shaped, f convex and g concave with respect to $\hat{x}$, then the implication (II.6.5) holds for all $x \in X$.

Proof We assume that there is an $x^* \in X$ such that

$$g(x^*, t) \geqslant 0 \qquad \text{for all } t \in I(\hat{x}) \quad \text{and} \quad f(x^*) < f(\hat{x}). \tag{II.6.8}$$

Then we define

$$B = \{t \in T : g(x^*, t) - g(\hat{x}, t) < 0\}$$

and set

$$\hat{\lambda} = \begin{cases} \min\limits_{t \in B} \dfrac{g(\hat{x}, t)}{g(\hat{x}, t) - g(x^*, t)} & \text{if } B \text{ is non-empty} \\ 1 & \text{if } B \text{ is empty.} \end{cases}$$

Surely $\hat{\lambda} > 0$, for in the case that B is not empty, we have $B \cap I(\hat{x}) = \phi$ on the basis of the assumption (II.6.8), which implies $g(\hat{x}, t) > 0$ for all $t \in B$. For $\lambda = \min(\hat{\lambda}, 1)$ we have then $\lambda \in (0, 1)$, and because of the fact that X is star-shaped with respect to $\hat{x}$, follows $x_\lambda = \lambda x^* + (1 - \lambda)\hat{x} \in X$. From the concavity of g with respect to $\hat{x}$ follows further with the definition of λ

$$\begin{aligned} g(x_\lambda, t) &\geqslant \lambda g(x^*, t) + (1 - \lambda) g(\hat{x}, t) \\ &= g(\hat{x}, t) + \lambda[g(x^*, t) - g(\hat{x}, t)] \geqslant 0 \qquad \text{for all } t \in T \end{aligned}$$

i.e. $x_\lambda \in S$. From the convexity of f with respect to $\hat{x}$ and (II.6.8) follows finally because of $\lambda > 0$

$$f(x_\lambda) \leqslant \lambda f(x^*) + (1 - \lambda) f(\hat{x}) = f(\hat{x}) + \lambda\{f(x^*) - f(\hat{x})\} < f(\hat{x})$$

a contradiction to the optimality of $\hat{x}$. With that the assumption (II.6.8) is false.

If one assumes that T is not finite, then only the following weaker theorem can be proven.

Theorem II.6.3 *Under the same assumptions as in Theorem II.6.2, the optimality of $\hat{x} \in S$ for all $x \in X$ implies*

$$g(x, t) > 0 \quad \text{for all } t \in I(\hat{x}) \Rightarrow f(\hat{x}) \leqslant f(x) \tag{II.6.9}$$

(which is obviously a consequence of the implication (II.6.5)).

Proof We assume there is an $x^* \in X$ with

$$g(x^*, t) > 0 \qquad \text{for all } t \in I(\hat{x}) \quad \text{and} \quad f(\hat{x}) > f(x^*). \tag{II.6.10}$$

If one defines

$$\delta = \min_{t \in I(\hat{x})} g(x^*, t)$$

then $\delta > 0$, and the set

$$\tilde{I} = \{t \in T : g(x^*, t) > \tfrac{1}{2}\delta\}$$

is open and contains $I(\hat{x})$. If $\tilde{I} = T$, then $x^* \in S$ and the assumption (II.6.10) is a contradiction to the optimality of $\hat{x}$.

If $\tilde{I} \neq T$, then the complement B of $\tilde{I}$ is a non-empty, closed subset of T and

$$\mu_1 = \min_{t \in B} g(\hat{x}, t) > 0.$$

If now

$$g(x^*, t) \geqslant g(\hat{x}, t) \qquad \text{for all } t \in T \tag{II.6.11}$$

then again $x^* \in S$ and the assumption (II.6.10) is a contradiction to the optimality of $\hat{x}$. If (II.6.11) is not satisfied, then

$$\mu_2 = \min_{t \in T} \; [g(x^*, t) - g(\hat{x}, t)] < 0.$$

If one chooses $\lambda := \min(1, \hat{\lambda})$ with $\hat{\lambda} = \mu_1/(-\mu_2)$ then

$$x_\lambda = \lambda x^* + (1 - \lambda)\hat{x} \in X.$$

and

$$g(x_\lambda, t) \geqslant \lambda g(x^*, t) + (1 - \lambda) g(\hat{x}, t)$$

$$= g(\hat{x}, t) + \lambda [g(x^*, t) - g(\hat{x}, t)] \begin{cases} > \lambda\delta/2 & \text{for all } t \in \tilde{I} \\ \geqslant \mu_1 + \lambda\mu_2 \geqslant 0 & \text{for all } t \in B \end{cases}$$

whence $x_\lambda \in S$. Furthermore, because $\lambda \in (0, 1]$, we have

$$f(x_\lambda) \leqslant \lambda f(x^*) + (1 - \lambda) f(\hat{x}) = f(\hat{x}) + \lambda [f(x^*) - f(\hat{x})] < f(\hat{x})$$

which contradicts the optimality of $\hat{x}$. Thus, the assumption (II.6.10) is false.

The question now arises, under what assumptions the implication (II.6.5) (for all $x \in X$) follows from the implication (II.6.9). The next theorem gives some information concerning this.

Theorem II.6.4 *Let E be a normed vector space and $f: X \to \mathbb{R}$ a continuous functional. If for a given $\hat{x} \in S$ the set*

$$S_0(\hat{x}) = \{x \in X : g(x, t) > 0 \qquad \text{for all } t \in I(\hat{x})\} \tag{II.6.12}$$

is not empty and if for $S(\hat{x})$ defined by (II.6.6) the statement

$$S(\hat{x}) \subseteq \overline{S_0(\hat{x})} = \text{closure of } S_0(\hat{x})$$

holds, then the implication (II.6.5) (for all $x \in X$) is a consequence of the implication (II.6.9), i.e. the two implications are equivalent.

Proof If, for some $x \in X$,

$$g(x, t) \geqslant 0 \qquad \text{for all } t \in I(\hat{x})$$

then $x \in S(\hat{x})$ and there is a sequence $\{x_k\}$ of points $x_k \in S_0(\hat{x})$ with $x = \lim\limits_{k \to \infty} x_k$.

From (II.6.9) it then follows that $f(\hat{x}) \leqslant f(x_k)$ for all k. From this $f(\hat{x}) \leqslant f(x)$ because of the continuity of f, which proves the implication (II.6.5).

Lemma II.6.5 *If E is a normed vector space, X a non-empty subset of E, g concave on X (see Section II.2.2), and the set*

$$S_0 = \{x \in X : g(x, t) > 0 \qquad \text{for all } t \in T\} \tag{II.6.13}$$

is not empty, then for every $\hat{x} \in S$, the set $S_0(\hat{x})$ defined by (II.6.12) is not empty, and

$$S(\hat{x}) \subseteq \overline{S_0(\hat{x})}$$

where $S(\hat{x})$ is defined by (II.6.6).

Proof By assumption there is an $x_0 \in X$ with

$$g(x_0, t) > 0 \qquad \text{for all } t \in T$$

whence $x_0 \in S_0(\hat{x})$ for all $\hat{x} \in S$. Now let an $x \in S(\hat{x})$ be given. Then for all $k \geqslant 1$ we have

$$x_k = \frac{1}{k} x_0 + \left(1 - \frac{1}{k}\right) x \in X$$

and

$$g(x_k, t) \geqslant \frac{1}{k} g(x_0, t) + \left(1 - \frac{1}{k}\right) g(x, t) > 0 \qquad \text{for all } t \in I(\hat{x})$$

i.e. $x_k \in S_0(\hat{x})$. Furthermore, $x = \lim_{k \to \infty} x_k$, which implies $x \in \overline{S_0(\hat{x})}$.

Summarizing, we have from Theorems II.6.1, II.6.3, II.6.4, and Lemma II.6.5 the following theorem.

Theorem II.6.6 *Let E be a normed vector space, X be non-empty, convex subset of E, f continuous and convex on X, g concave on X, and the set S_0 defined by (II.6.13) be non-empty. Then we have the assertion that: an element $\hat{x} \in S$ is optimal if and only if for all $x \in X$ the implication (II.6.5) holds (which is equivalent to the implication (II.6.9). If T is finite, then the assumption $S_0 \neq \phi$ is superfluous by Theorem II.6.2.*

This theorem is actually interesting only in the case that

$$I(\hat{x}) = \{t \in T : g(\hat{x}, t) = 0\}$$

i.e. $\delta(\hat{x}) = 0$ where δ is given by (II.6.3). If $\delta(\hat{x}) > 0$, then we have in fact the next theorem.

Theorem II.6.7 *Suppose $\hat{x} \in S$ is optimal and $\delta(\hat{x}) > 0$. If X is star-shaped, f convex and g concave with respect to $\hat{x}$, then there follows*

$$f(\hat{x}) \leqslant f(x) \qquad \textit{for all } x \in X$$

i.e. $\hat{x}$ is in fact a minimal point of f on the set X.

Proof We assume there were an $x^* \in X$ with $f(x^*) < f(\hat{x})$. If $g(x^*, t) \geqslant g(\hat{x}, t)$ for all $t \in T$, then $x^* \in S$, in contradiction to the optimality of $\hat{x}$. Therefore we have

$$\mu = \min_{t \in T} [g(x^*, t) - g(\hat{x}, t)] < 0.$$

If one chooses $\lambda = \min(1, \delta(\hat{x})/(-\mu))$, then $\lambda \in (0, 1]$ and

$$x_\lambda = \lambda x^* + (1 - \lambda)\hat{x} \in X$$

and

$$g(x_\lambda, t) \geqslant g(\hat{x}, t) + \lambda [g(x^*, t) - g(\hat{x}, t)]$$
$$\geqslant \delta(\hat{x}) + \lambda\mu \geqslant 0 \qquad \text{for all } t \in T, \quad \text{i.e. } x_\lambda \in S.$$

Finally $f(x_\lambda) \leqslant f(\hat{x}) + \lambda(f(x^*) - f(\hat{x})) < f(\hat{x})$, a contradiction to the optimality of $\hat{x}$. Thus the assumption is false.

The converse of Theorem II.6.7 obviously holds without the assumptions made.

II.6.2 A mixed linear convex problem

Let $E = \mathbb{R}^n$, equipped with any norm, and X be a non-empty, convex subset of E. Further let T be a compact Hausdorff space, let $v : T \to \mathbb{R}^n$ be a continuous mapping, let $\alpha : T \to \mathbb{R}$ be a continuous functional and let $c \in \mathbb{R}^n$ be a given vector. If, for every $x \in \mathbb{R}^n$ and every $t \in T$, one defines

$$g(x, t) = \langle v(t), x \rangle - \alpha(t) \tag{II.6.14}$$

where $\langle \cdot, \cdot \rangle$ denotes the ordinary inner product in $\mathbb{R}^n$, then $g : E \to C(T)$ is an affine linear, and hence concave, mapping.

Problem (P) Minimize the continuous linear functional $f(x) = \langle c, x \rangle$ subject to the side conditions

$$x \in X \quad \text{and} \quad g(x, t) \geqslant 0 \qquad \text{for all } t \in T.$$

For every $\hat{x} \in X$, we define

$$T(X, \hat{x}) = \overline{\bigcup_{\lambda > 0} \{\lambda(x - \hat{x}) : x \in X\}}. \tag{II.6.15}$$

Assertion $T(X, \hat{x})$ is a closed, convex cone in $E = \mathbb{R}^n$ (Proof = Exercise, see Figure II.6.1).

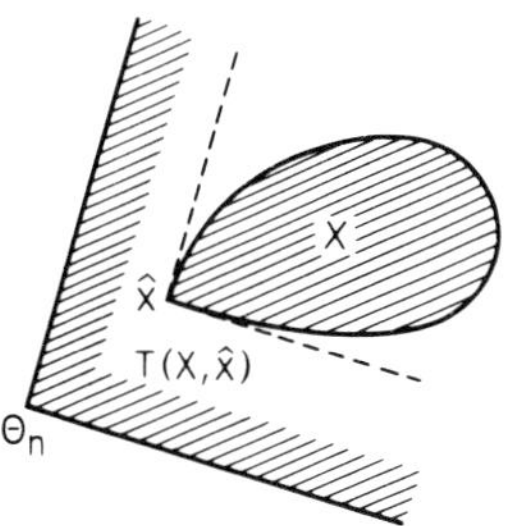

Figure II.6.1 The cone $T(X, \hat{x})$

Theorem II.6.8 *An element $\hat{x} \in S$, with S defined according to (II.6.1), where $\delta(\hat{x}) > 0$ and δ is defined according to (II.6.3) is optimal if and only if*

$$c \in T(X, \hat{x})^{\circ} \tag{II.6.16}$$

where $T(X, \hat{x})$ is defined according to (II.6.12), and where $T(X, x)^{\circ}$ is the convex cone defined by (IV.2.14).

Proof By Theorem II.6.7, $\hat{x}$ is optimal if and only if

$$\langle c, x - \hat{x} \rangle \geqslant 0 \qquad \text{for all } x \in X.$$

This statement is, however, equivalent to

$$\langle c, h \rangle \geqslant 0 \qquad \text{for all } h \in T(X, \hat{x})$$

that is

$$c \in T(X, \hat{x})^{\circ}$$

(Proof = Exercise).

Next we consider an $\hat{x} \in S$ with $\delta(\hat{x}) = 0$. If the set S_0 defined by (II.6.13) is not empty or if T is a finite set (equipped with the discrete topology), then by Theorem II.6.6 the element $\hat{x}$ is optimal if and only if

$$\langle c, x - \hat{x} \rangle \geqslant 0 \qquad \text{for all } x \in X$$

with

$$\langle v(t), x \rangle - \alpha(t) \geqslant 0 \qquad \text{for all } t \in I(\hat{x}) \tag{II.6.17}$$

where

$$I(\hat{x}) = \{t \in T : \langle v(t), \hat{x} \rangle - \alpha(t) = 0\} \neq \phi.$$

If one defines

$$L(E, \hat{x}) = \{h \in \mathbb{R}^n : \langle v(t), h \rangle \geqslant 0 \qquad \text{for all } t \in I(\hat{x})\} \tag{II.6.18}$$

then $L(E, \hat{x})$ is obviously a closed convex cone in $E = \mathbb{R}^n$, and we have the following theorem.

Theorem II.6.9 *If the set S_0 defined by (II.6.13) is not empty, then an $\hat{x} \in S$ with $\delta(\hat{x}) = 0$ (see (II.6.3)) is optimal if and only if*

$$c \in L(E, \hat{x})^{\circ} + T(X, \hat{x})^{\circ}. \tag{II.6.19}$$

Proof

1. Suppose (II.6.19) holds. Then with the aid of (IV.2.10′) and (IV.2.8′) it follows that

$$c \in (L(E, \hat{x}) \cap T(X, \hat{x}))^{\circ} \subseteq (L(E, \hat{x}) \cap (X - \hat{x}))^{\circ}$$

which is equivalent to the statement (II.6.17) and hence to the optimality of $\hat{x}$.

2. Suppose $\hat{x}$ is optimal. Since S_0 is not empty there is an $x^* \in S \subseteq X$ with

$$\langle v(t), x^* - \hat{x} \rangle > 0 \qquad \text{for all } t \in I(\hat{x}). \tag{II.6.20}$$

Now let $h \in L(E, \hat{x}) \cap T(X, \hat{x})$ be given. If one defines for every $k \geqslant 1$

$$h_k = \frac{1}{k}(x^* - \hat{x}) + \left(1 - \frac{1}{k}\right) h$$

then $h_k \in T(X, \hat{x})$ for all k and

$$\langle v(t), h_k \rangle > 0 \qquad \text{for all } t \in I(\hat{x}).$$

Now for every k, there is a sequence $\{x_j^k\}$, $x_j^k \in X$ and a sequence $\{\lambda_j^k\}$ of positive numbers λ_j^k with $h_k = \lim_{j \to \infty} \lambda_j^k (x_j^k - \hat{x})$. It follows that for every k there exists a j_k with

$$\langle v(t), x_j^k - \hat{x} \rangle \geqslant 0 \qquad \text{for all } t \in I(\hat{x}) \text{ and all } j \geqslant j_k.$$

Because of the optimality of $\hat{x}$ the statement (II.6.17) holds, which implies $\langle c, x_j^k - \hat{x} \rangle \geqslant 0$ for all $j \geqslant j_k$ and further $\langle c, h_k \rangle \geqslant 0$ for all k. Since $h = \lim_{k \to \infty} h_k$, it also follows that $\langle c, h \rangle \geqslant 0$. Thus, making use of (IV.2.12′) and (IV.2.9′), we have

$$\begin{aligned} c \in (L(E, \hat{x}) \cap T(X, \hat{x}))^\circ &= [(L(E, \hat{x})^\circ)^\circ \cap (T(X, \hat{x})^\circ)^\circ]^\circ \\ &= [(L(E, \hat{x})^\circ + T(X, \hat{x})^\circ)^\circ]^\circ \end{aligned} \tag{II.6.21}$$

Next, we show that

$$(-L(E, \hat{x})^\circ) \cap T(X, \hat{x})^\circ = \{\Theta_n\}. \tag{II.6.22}$$

If one defines

$$Q_{\hat{x}} = \{v(t) \in \mathbb{R}^n : t \in I(\hat{x})\} \text{ and } K(Q_{\hat{x}}) = \{\lambda \cdot h : h \in H(Q_{\hat{x}}), \lambda \geqslant 0\} \tag{II.6.23}$$

where $H(Q_{\hat{x}})$ is the convex hull of $Q_{\hat{x}}$, then $Q_{\hat{x}}$ is a non-empty, compact subset of $\mathbb{R}^n$ and $H(Q_{\hat{x}})$ is also compact by Theorem IV.3.2. Furthermore, $\Theta_n \notin H(Q_{\hat{x}})$, for otherwise there would be numbers $\lambda_1 \geqslant 0, \ldots, \lambda_m \geqslant 0$ with $\Sigma_{i=1}^m \lambda_i = 1$ and points $t_1, \ldots, t_m \in I(\hat{x})$ with

$$\sum_{i=1}^m \lambda_i v(t_i) = \Theta_n.$$

For $x^* \in S$ in (II.6.20) it follows then that

$$\sum_{i=1}^m \lambda_i \langle v(t_i), x^* - \hat{x} \rangle = 0 \quad \text{with} \quad \langle v(t_i), x^* - \hat{x} \rangle > 0 \qquad \text{for all } i$$

a contradiction to $\lambda_i \geqslant 0$ for all i and $\Sigma_{i=1}^n \lambda_i = 1$.

By Theorem IV.2.1 with $Q = H(Q_{\hat{x}})$ and $K = \{\Theta_n\}$ the convex cone $K(Q_{\hat{x}})$ is closed. Furthermore

$$L(E, \hat{x}) = \{h \in \mathbb{R}^n : \langle h, k \rangle \geqslant 0 \qquad \text{for all } k \in K(Q_{\hat{x}})\}$$

and consequently $L(E,\hat{x}) = K(Q_{\hat{x}})^\circ$, whence from (IV.2.12′) it follows that $L(E,x)^\circ = K(Q_{\hat{x}})$. Now let $y \in (-L(E,\hat{x})^\circ) \cap T(X,\hat{x})^\circ$ be given. Then it follows that $-y = \sum_{i=1}^m \lambda_i v(t_i)$ with certain points $t_1, \ldots, t_m \in I(\hat{x})$ and numbers $\lambda_1 \geqslant 0, \ldots, \lambda_m \geqslant 0$, and

$$\langle y, h \rangle \geqslant 0 \qquad \text{for all } h \in T(X,\hat{x})$$

from which in particular we conclude $\langle y, x^* - x \rangle \geqslant 0$ with x^* defined by (II.6.20). Hence

$$\sum_{i=1}^m \lambda_i \langle v(t_i), x^* - \hat{x} \rangle = \langle -y, x^* - \hat{x} \rangle \leqslant 0 \text{ and } \langle v(t_i), x^* - \hat{x} \rangle > 0 \qquad \text{for all } i$$

which implies $\lambda_i = 0$ for all i and consequently $y = \Theta_n$, which concludes the proof of (II.6.22). The assertion (II.6.19) follows therefore from (II.6.21), (II.6.22) with the help of Theorem IV.2.1 and (IV.2.12′).

From the proof of Theorem II.6.9 we obtain the following corollary.

Corollary *Under the assumptions of Theorem II.6.9 we have, for an optimal* $\hat{x}$,

$$c \in K(Q_{\hat{x}}) + T(X,\hat{x})^\circ \tag{II.6.24}$$

with $T(X,\hat{x})$ *defined by (II.6.15) and* $K(Q_{\hat{x}})$ *defined by (II.6.23).*

Remarks If one considers instead of problem (P) the following somewhat more general problem of minimizing the continuous linear functional $c = c(x)$ subject to the side conditions

$$x \in X \quad \text{and} \quad a(x,t) - \alpha(t) \geqslant 0 \qquad \text{for all } t \in T$$

where X is a non-empty convex subset of a normed vector space E, $a : E \to C(T)$ is a continuous linear mapping and $\alpha \in C(T)$ is a given function, then defining $T(X,x)$ according to (II.6.15),

$$L(E,x) = \begin{cases} \bigcap_{t \in I(x)} \{h \in E : a(h,t) \geqslant 0\} & \text{if } \delta(x) = 0 \\ E & \text{if } \delta(x) > 0 \end{cases} \tag{II.6.25}$$

with

$$\delta(x) = \min_{t \in T} [a(x,t) - \alpha(t)]$$

and

$$S_0 = \{x \in X : a(x,t) - \alpha(t) > 0 \text{ for all } t \in T\} \tag{II.6.26}$$

the following theorem may be proven.

Theorem II.6.10 *If* S_0 *is not empty, then an element*

$$\hat{x} \in S = \{x \in X : a(x,t) - \alpha(t) \geqslant 0 \text{ for all } t \in T\}$$

is optimal if and only if

$$c \in (L(E,\hat{x}) \cap T(X,\hat{x}))^*. \tag{II.6.27}$$

Here we denote by K^* the cone which is adjoint to K (see (IV.2.6)), where K is a given convex cone in E. We shall not give the proof here, but will prove the next lemma instead.

Lemma II.6.11 *If E is a finite dimensional normed vector space and if*

$$(-L(E,\hat{x})^*) \cap T(X,\hat{x})^* = \{\Theta_{E^*}\} \tag{II.6.28}$$

with $L(E,\hat{x})$ given by (II.6.25) and $T(X,\hat{x})$ by (II.6.15), then there follows

$$(L(E,\hat{x}) \cap T(X,\hat{x}))^* = L(E,\hat{x})^* + T(X,\hat{x})^*. \tag{II.6.29}$$

Proof By (IV.2.10) we have

$$L(E,\hat{x})^* + T(X,\hat{x})^* \subseteq (L(E,\hat{x}) \cap T(X,\hat{x}))^*.$$

By Section IV.1.2, example 1, E^* is also a finite dimensional normed vector space with the norm (IV.1.9). For every convex cone K in E the adjoint cone K^* in E^* is closed (Proof = Exercise). Thus $-L(E,\hat{x})^*$ and $T(X,\hat{x})^*$ are closed in E^* and by Theorem IV.2.2, $L(E,\hat{x})^* + T(X,\hat{x})^*$ is a convex closed cone in E^*.

Now let $x^* \in (L(E,\hat{x}) \cap T(X,\hat{x}))^*$, but $x^* \notin L(E,\hat{x})^* + T(X,\hat{x})^*$. Since E is reflexive, there is by the separation theorem 2 in Section IV.3.2 an $x \in E$ with

$$h^*(x) < x^*(x) \qquad \text{for all } h^* \in L(E,\hat{x})^* + T(X,\hat{x})^* \tag{II.6.30}$$

which implies $h^*(x) \leqslant 0$ for all $h^* \in L(E,\hat{x})^* + T(X,\hat{x})^*$. This is, in turn, equivalent to

$$h^*(x) \leqslant 0 \qquad \text{for all } h^* \in L(E,\hat{x})^*$$

and

$$h^*(x) \leqslant 0 \qquad \text{for all } h^* \in T(X,x)^*$$

i.e. $-x \in L(E,\hat{x}) \cap T(X,\hat{x})$ by Theorem IV.2.3, which implies $x^*(x) \leqslant 0$. This is, however, a contradiction to (II.6.30) for $h^* = \Theta_{E^*}$, whence $x^*(x) > 0$. Thus we have

$$(L(E,\hat{x}) \cap T(X,\hat{x}))^* \subseteq L(E,\hat{x})^* + T(X,\hat{x})^*.$$

Problem II.6.1 Prove (II.6.28) under the assumption that the set S_0 defined by (II.6.26) is not empty.

II.6.3 Applications

II.6.3.1 Uniform linear approximation with interpolation

We consider as in Section I.2.1.2 the uniform linear approximation problem on a compact metric space M with interpolation side conditions, which, as was shown there is equivalent to the following problem.

Optimization problem Minimize γ subject to the side conditions

$$\sum_{j=1}^{n} v_j(t)y_j + \gamma - f(t) \geqslant 0$$

$$\text{for all } t \in M, \quad \sum_{j=1}^{n} v_j(t_i)y_j - f(t_i) = 0,$$

$$\text{for } i = 1, \ldots r(<n).$$

$$-\sum_{j=1}^{n} v_j(t)y_j + \gamma + f(t) \geqslant 0$$

We define $T = M \times \{1, 2\}$, equip $\{1, 2\}$ with the discrete topology and T with the product topology. Then T is a compact Hausdorff space. Further we define

$$v(t, 1) = (v_1(t), \ldots, v_n(t), 1)^{\mathrm{T}}$$
$$v(t, 2) = (-v_1(t), \ldots, -v_n(t), 1)^{\mathrm{T}}$$
$$\alpha(t, 1) = f(t), \quad \alpha(t, 2) = -f(t) \qquad \text{for all } t \in M$$

as well as

$$\hat{v}(t_i) = (v_1(t_i), \ldots, v_n(t_i), 0)^{\mathrm{T}} \qquad \text{for } i = 1, \ldots, r$$

and

$$X = \{x = (y^{\mathrm{T}}, \gamma)^{\mathrm{T}} \in \mathbb{R}^n \times \mathbb{R} : \langle \hat{v}(t_i), x \rangle - f(t_i) = 0 \qquad \text{for all } i = 1, \ldots, r\} \tag{II.6.31}$$

where $\langle \mathrm{x}, \mathrm{x} \rangle$ denotes the scalar product in $\mathbb{R}^{n+1}$. If one defines finally $c = (\Theta_n^T, 1)^T$, Θ_n = null vector in $\mathbb{R}^n$, then the above problem goes over into the problem of minimizing the continuous linear form $\langle c, x \rangle$ subject to the side conditions

$$x \in X \text{ and } \begin{cases} \langle v(t, 1), x \rangle - \alpha(t, 1) \geqslant 0 \\ \langle v(t, 2), x \rangle - \alpha(t, 2) \geqslant 0 \end{cases} \qquad \text{for all } t \in M. \tag{II.6.32}$$

This is, however, precisely the problem (P) as treated in Section II.6.2.

The set X defined by (II.6.31) is in fact even a linear manifold in $\mathbb{R}^{n+1}$ (see Section IV.3.2) and hence convex. If X is not empty, then it is possible to satisfy the side conditions (II.6.32) strictly, i.e. the set S_0 defined by (II.6.13) is not empty.

Lemma II.6.12 *For every $\hat{x} \in X$ the closed convex cone $T(X, \hat{x})$ defined by (II.6.15) is equal to the linear subspace of $\mathbb{R}^{n+1}$ generated by $X - \hat{x}$, which is obviously equal to $X - \hat{x}$ itself, i.e.*

$$T(X, \hat{x}) = X - \hat{x} = \{x \in \mathbb{R}^{n+1} : \langle \hat{v}(t_i), x \rangle = 0 \text{ for all } i = 1, \ldots, r\}. \tag{II.6.33}$$

Proof $X - \hat{x}$ is closed since it is the intersection of r hyperplanes

$$H_i = \{x \in \mathbb{R}^{n+1} : \langle \hat{v}(t_i), x \rangle = 0\} \qquad \text{for } i = 1, \ldots, r$$

which are closed (why?). From this we obtain immediately $T(X, \hat{x}) \subseteq X - \hat{x}$. The inclusion $X - \hat{x} \subseteq T(X, \hat{x})$ is clear.

This lemma yields first by Sections II.5.1 and II.5.2 the existence of an optimal $\hat{x}$.

Corollary $T(X, \hat{x})^*$ is equal to the linear subspace $L(\hat{v}(t_1), \ldots, \hat{v}(t_r))$ of $\mathbb{R}^{n+1}$ generated by $\{\hat{v}(t_i)\}_{i=1,\ldots,r}$.

Problem II.6.2 Prove this corollary.

If one defines for an optimal $\hat{x}$

$$I(\hat{x}) = I_1(\hat{x}) \cup I_2(\hat{x})$$

with

$$I_1(\hat{x}) = \{(t, 1) \in T : \langle v(t, 1), \hat{x} \rangle - f(t) = 0\}$$

and

$$I_2(\hat{x}) = \{(t, 2) \in T : \langle v(t, 2), \hat{x} \rangle + f(t) = 0\}$$

then $I(\hat{x})$ is not empty (i.e. $\delta(\hat{x}) = 0$ where δ is defined by (II.6.3); Proof = Exercise).
From the corollary to Theorem II.6.9 therefore follows

$$c \in K(Q_{\hat{x}}) + L(\hat{v}(t_1), \ldots, \hat{v}(t_r)) \tag{II.6.34}$$

with

$$Q_{\hat{x}} = \{v(t, 1) : (t, 1) \in I_1(\hat{x})\} \cup \{v(t, 2) : (t, 2) \in I_2(\hat{x})\}$$

and $K(Q_{\hat{x}})$ defined according to (II.6.23). We define further

$$E_1(\hat{x}) = \{t \in M : (t, 1) \in I_1(\hat{x})\} = \{t \in M : \sum_{j=1}^{n} v_j(t)\hat{y}_j - f(t) = -\hat{\gamma}\} \tag{II.6.35a}$$

and

$$E_2(\hat{x}) = \{t \in M : (t, 2) \in I_2(\hat{x})\} = \{t \in M : \sum_{j=1}^{n} v_j(t)\hat{y}_j - f(t) = +\hat{\gamma}\} \tag{II.6.35b}$$

(where $\hat{x} = (\hat{y}^{\mathrm{T}}, \hat{\gamma})$ and $\hat{\gamma} = \| \sum_{j=1}^{n} v_j(\cdot)\hat{y}_j - f \|_\infty = \rho_\infty(f, V_0)$ with V_0 by (I.2.6)).
For the sequel we assume that $\hat{\gamma} = \rho_\infty(f, V_0) > 0$. Then the intersection $E_1(\hat{x}) \cap E_2(\hat{x})$ is empty and (II.6.34) yields the existence of two finite subsets $\hat{E}_1$ of $E_1(\hat{x})$, and $\hat{E}_2$ of $E_2(\hat{x})$, respectively, of which at least one is not empty, and numbers $\lambda_t^1 \geqslant 0, t \in \hat{E}_1, \lambda_t^2 \geqslant 0, t \in \hat{E}_2$, as well as additional numbers $\mu_1, \ldots, \mu_r \in \mathbb{R}$ such that

$$c = \sum_{t \in \hat{E}_1} \lambda_t^1 v(t, 1) + \sum_{t \in \hat{E}_2} \lambda_t^2 v(t, 2) + \sum_{i=1}^{r} \mu_i \hat{v}(t_i) \tag{II.6.36}$$

whereby $\hat{E}_1 \cap \hat{E}_2$ is empty.
In summary we obtain the theorem below.

Theorem II.6.13 *If $\hat{x} = (\hat{y}^{\mathrm{T}}, \hat{\gamma})^{\mathrm{T}}$ is optimal, i.e. if $\sum_{j=1}^{n} \hat{v}_j v_j$ is a solution of the approximation problem subject to the interpolation side conditions and if further*

$$\hat{\gamma} = \left\| \sum_{j=1}^{n} \hat{y}_j v_j - f \right\|_\infty = \rho_\infty(f, V_0) > 0$$

then there is a non-empty subset $\hat{E}$ *of*

$$E(\hat{x}) = \left\{ t \in M : \left| \sum_{j=1}^{n} \hat{y}_j v_j(t) - f(t) \right| = \hat{\gamma} \right\} \tag{II.6.37}$$

and numbers $c_t \in \mathbb{R}$, $t \in \hat{E}$ *as well as numbers* $\mu_1, \ldots, \mu_r \in \mathbb{R}$ *with*

$$\sum_{t \in \hat{E}} |c_t| = 1 \tag{II.6.38}$$

$$\sum_{t \in \hat{E}} c_t v_j(t) + \sum_{i=1}^{r} \mu_i v_j(t_i) = 0 \qquad j = 1, \ldots, n \tag{II.6.39}$$

and

$$\operatorname{sgn} c_t = - \operatorname{sgn} \left\{ \sum_{j=1}^{n} \hat{y}_j v_j(t) - f(t) \right\} \tag{II.6.40}$$

provided $c_t \neq 0$.

For the proof one has only to define

$$c_t = \begin{cases} \lambda_t^1 & \text{for } t \in \hat{E}_1 \\ -\lambda_t^2 & \text{for } t \in \hat{E}_2 \end{cases}$$

and $\hat{E} = \hat{E}_1 \cup \hat{E}_2$.

Conversely, if for a given $\hat{x} = (\hat{y}^{\mathrm{T}}, \hat{\gamma})^{\mathrm{T}}$ *satisfying (II.6.32) and*

$$\hat{\gamma} = \left\| \sum_{j=1}^{n} \hat{y}_j v_j - f \right\|_{\infty}$$

there is a finite subset $\hat{E}$ *of* $E(\hat{x})$ *defined by (II.6.37) and numbers* $c_t \in \mathbb{R}$, $t \in \hat{E}$ *as well as* $\mu_1, \ldots, \mu_r \in \mathbb{R}$ *defined by (II.6.38), (II.6.39), and (II.6.40), then* $\hat{x}$ *is optimal.*

For the proof we choose any $x = (y^{\mathrm{T}}, \gamma)^{\mathrm{T}}$ satisfying (II.6.32). Then

$$\begin{aligned}
\hat{\gamma} &= \sum_{\substack{t \in \hat{E} \\ c_t \neq 0}} |c_t| \left| \sum_{j=1}^{n} \hat{y}_j v_j(t) - f(t) \right| \\
&= \sum_{\substack{t \in \hat{E} \\ c_t \neq 0}} c_t \left\{ f(t) - \sum_{j=1}^{n} \hat{y}_j v_j(t) \right\} = \sum_{\substack{t \in \hat{E} \\ c_t \neq 0}} c_t f(t) + \sum_{i=1}^{r} \mu_i \sum_{j=1}^{n} \hat{y}_j v_j(t_i) \\
&= \sum_{\substack{t \in \hat{E} \\ c_t \neq 0}} c_t f(t) + \sum_{i=1}^{r} \mu_i f(t_i) = \sum_{\substack{t \in \hat{E} \\ c_t \neq 0}} c_t f(t) + \sum_{i=1}^{r} \mu_i \sum_{j=1}^{n} y_j v_j(t_i) \\
&= \sum_{\substack{t \in \hat{E} \\ c_t \neq 0}} c_t \left\{ f(t) - \sum_{j=1}^{n} y_j v_j(t) \right\} \leqslant \left\| \sum_{j=1}^{n} y_j v_j - f \right\|_{\infty} \leqslant \gamma
\end{aligned}$$

whence follows the assertion.

II.6.3.2 A semi-infinite problem in the control of air pollution

We again take up the problem of Section I.1.3. This consists in minimizing the linear functional

$$c(x_1, \ldots, x_n) = \sum_{j=1}^{n} c_j x_j$$

subject to the side conditions

$$\sum_{j=1}^{n} u_j(s)x_j \geq \sum_{j=0}^{n} u_j(s) - \varphi(s) \qquad s \in S, 0 \leq x_j \leq 1 \text{ for } j = 1, \ldots, n. \tag{II.6.41}$$

Here S is a given planar region and $u_0, \ldots, u_n$, and φ are real-valued functions defined on S which describe the yearly average of certain air pollutant sources, and a certain prescribed standard, respectively. The variables x_j are reduction factors for the controllable sources and the functional c describes the costs arising in the reduction. We assume without loss of generality that all $c_j > 0$. We assume that S is closed and bounded and the functions $u_0, \ldots, u_n, \varphi$ are continuous on S.

One then defines

$$v(s) = (u_1(s), \ldots, u_n(s))^{\mathrm{T}} \qquad \alpha(s) = \sum_{j=0}^{n} u_j(s) - \varphi(s) \qquad s \in S$$
$$g(x, s) = \langle v(s), x \rangle - \alpha(s) \qquad x \in \mathbb{R}^n \tag{II.6.42}$$

where $\langle x, x \rangle$ denotes the scalar product in $\mathbb{R}^n$. Let $X = \{x \in \mathbb{R}^n : 0 \leq x_j \leq 1$ for $j = 1, \ldots, n\}$, then taking $T = S$ we are faced precisely with a problem (P) as described in Section II.6.2. We assume that

$$u_0(s) \leq \varphi(s) \qquad \text{for all } s \in S \tag{II.6.43}$$

i.e. the pollution originating from an uncontrollable source lies in the whole region S under the prescribed standard (which is sensible, since by reducing the other sources one could not achieve or get under the standard at all points). The the set $S(X, g)$ is closed and bounded so that the linear, and hence continuous, functional c assumes its minimum on $S(X, g)$. There is, consequently, an $\hat{x} \in S(X, g)$ with

$$\langle c, \hat{x} \rangle \leq \langle c, x \rangle \qquad \text{for all } x \in S(X, g). \tag{II.6.44}$$

If one defines $\delta(\hat{x})$ according to (II.6.3) with $T = S$ and g according to (II.6.42), then there are two possible cases.

(*a*) $\delta(\hat{x}) > 0$. By Theorem II.6.8, that is precisely the case if $c \in T(X, \hat{x})^{\circ}$ with $T(X, \hat{x})$ given by (II.6.15) and $T(X, \hat{x})^{\circ}$ by (IV.2.14). In the case at hand we have

$$T(X, \hat{x}) = \{x \in \mathbb{R}^n : x_j \in \mathbb{R} \text{ for } \hat{x}_j \in (0, 1), x_j \geq 0 \text{ for } \hat{x}_j = 0, x_j \leq 0 \text{ for } \hat{x}_j = 1\}$$

and

$$T(X, \hat{x})^{\circ} = \{x \in \mathbb{R}^n : x_j = 0 \text{ for } \hat{x}_j \in (0, 1), x_j \geq 0 \text{ for } \hat{x}_j = 0, x_j \leq 0 \text{ for } \hat{x}_j = 1\}. \tag{II.6.45}$$

Since all $c_j > 0$ by assumption, $c \in T(X, x)^\circ$ is possible only in the case when $\hat{x} = \Theta_n$, which implies

$$\sum_{j=0}^{n} u_j(s) \leqslant \varphi(s) \qquad \text{for all } s \in S \tag{II.6.46}$$

so that no reduction is necessary in order not to exceed the standard and consequently no costs arise. Conversely (II.6.46) is naturally also sufficient in order that $\hat{x} = \Theta_n$ belongs to $S(X, g)$ and is optimal, i.e. (II.6.44) is fulfilled. We exclude this trivial case from now on and assume that for at least one $s \in S$

$$\sum_{j=0}^{n} u_j(s) > \varphi(s) \tag{II.6.47}$$

For each $\hat{x} \in S(X, g)$ satisfying (II.6.44) we are faced necessarily with the following case.

(b) $\delta(\hat{x}) = 0$. In this case we suppose further that

$$u_0(s) < \varphi(s) \qquad \text{for all } s \in S \tag{II.6.48}$$

i.e. the pollution arising from the non-controllable sources lies strictly under the prescribed standard in all points of the region S. Then it follows that

$$g(\Theta_n, s) > 0 \qquad \text{for all } s \in S$$

with g given by (II.6.42), so that the set S defined by (II.6.13) in the present case is not empty and one can apply the corollary to Theorem II.6.9. This leads to our next theorem.

Theorem II.6.14 *Under the assumptions (II.6.47) and (II.6.48), $\hat{x} \in S(X, g)$ is optimal i.e. satisfies (II.6.44) if and only if the (non-empty) set*

$$I(\hat{x}) = \left\{ s \in S : \sum_{j=1}^{n} u_j(s)\, \hat{x}_j = \sum_{j=0}^{n} u_j(s) - \varphi(s) \right\}$$

contains finitely many points s_i, $i \in I$, and there exist numbers $\lambda_i \geqslant 0$ such that

$$c_j \begin{cases} = \sum\limits_{i \in I} \lambda_i u_j(s_i) & \text{for } \hat{x}_j \in (0, 1) \\ \leqslant \sum\limits_{i \in I} \lambda_i u_j(s_i) & \text{for } \hat{x}_j = 1 \\ \geqslant \sum\limits_{i \in I} \lambda_i u_j(s_i) & \text{for } \hat{x}_j = 0. \end{cases} \tag{II.6.49}$$

Problem II.6.3a Verify that (II.6.49) is precisely the condition (II.6.24) of the corollary to Theorem II.6.9.

Problem II.6.3b Prove directly the sufficiency of the condition (II.6.49) for the optimality of $\hat{x} \in S(X, g)$.

SOLUTIONS AND HINTS TO THE PROBLEMS

Problem II.1.1a At first we observe that (II.1.42) is an immediate consequence of (II.1.31) and the orthonormality of the φ_j.

If $x_{\lambda,y}$ is defined by (II.1.35) with $\rho_j(\lambda)$ by (II.1.41), then it follows that

$$\int_a^b K(\lambda, t, s) x_{\lambda,y}(s)\, ds = \sum_{j=1}^{n} [\lambda_j \varphi_j(t) b_j(y) + \lambda_j \rho_j(\lambda) \varphi_j(t)]$$

$$= \sum_{j=1}^{n} \rho_j(\lambda)\varphi_j(t) = x_{\lambda,y}(t) - y(t).$$

Problem II.1.1b The statement (II.1.42) implies

$$|\lambda_j|\,|\varphi_j(t)| \leqslant \int_a^b |K(\lambda, t, s)|\, ds\, \|\varphi_j\|_\infty \qquad \text{for all } t \in [a, b]$$

hence,

$$|\lambda_j|\,\|\varphi_j\| \leqslant \max_{t\in[a,b]} \int_a^b |K(\lambda, t, s)|\, ds\, \|\varphi_j\|_\infty \Rightarrow |\lambda_j| < 1 \Rightarrow \lambda_j \neq 1$$

since $\|\varphi_j\|_\infty > 0$ for all $j = 1, \ldots, n$.

Problem II.2.1a By Lemma IV.3.6 the implication

$$(x_k, x \in X, x_k \rightharpoonup x) \Rightarrow x_k \to x$$

holds and the continuity of f on X implies

$$f(x) = \lim_{k\to\infty} f(x_k) = \liminf_{k\to\infty} f(x_k).$$

Problem II.2.1b By Theorem IV.3.8 the compactness of X implies that X is weakly sequentially compact. The assertion then follows from (*a*) and Theorem II.2.4.

Problem II.2.2 Put $E = F = G = \mathbb{R}$, $X = [-\frac{1}{2}\pi, \frac{1}{2}\pi]$, $g(x) = -\cos x$, and $h(y) = y^2$. Then $g : X \to F$ is convex, $h : F \to G$ is convex but not monotonic and $(h \circ g)(x) = (\cos x)^2$, $x \in X$, is concave.

Problem II.2.3 By Section IV.1.2, example 1 and problem IV.1.2 there exists, for each $y^* \in F^*$, exactly one $y \in F$ such that

$$y^*(f) = \sum_{i=1}^{m} y_i f_i \qquad \text{for all } f \in F. \tag{12}$$

Conversely, if y^* is defined by (12) for some $y \in F$, then $y^* \in F^*$. Further,

$$y^* \geqslant \Theta_{F^*} \Leftrightarrow y_i \geqslant 0 \qquad \text{for all } i = 1, \ldots, m$$

and $g(x) = (g_1(x), \ldots, g_m(x))^T$, $x \in X$, is convex if and only if all functionals $g_i : X \to \mathbb{R}$ are convex. Therefore Lemma II.2.9 reads in this case:

$$\left.\begin{array}{l} X \text{ convex subset of } E, y_1 \geqslant 0, \ldots, y_m \geqslant 0, \\ g_1, \ldots, g_m : X \to \mathbb{R} \text{ convex} \end{array}\right\} \Rightarrow$$

$$f(x) = \sum_{j=1}^{m} y_j g_j(x),\ x \in X \text{ convex on } X.$$

Problem II.2.4a (1) Let $g : X \to F$ be convex, $(x_1, y_1), (x_2, y_2) \in E_g$ and $\lambda \in [0, 1]$. Then we have

$$\lambda x_1 + (1 - \lambda)x_2 \in X$$

and

$$g(\lambda x_1 + (1 - \lambda)x_2) \leqslant \lambda g(x_1) + (1 - \lambda)g(x_2) \leqslant \lambda y_1 + (1 - \lambda)y_2$$

that is

$$\lambda(x_1, y_1) + (1 - \lambda)(x_2, y_2) = (\lambda x_1 + (1 - \lambda)x_2, \lambda y_1 + (1 - \lambda)y_2) \in E_g.$$

The proof of the implication

$$g : X \to F \text{ concave} \Rightarrow H_g \text{ concave}$$

is analogous.

(2) Assume E_g to be convex. Consider $x_1, x_2 \in X$ and $\lambda \in [0, 1]$. Then $(x_1, g(x_1)), (x_2, g(_2)) \in E_g$ implies, by the convexity of E_g, that

$$\lambda(x_1, g(x_1)) + (1 - \lambda)(x_2, g(x_2)) \in E_g$$

which is equivalent to

$$(\lambda x_1 + (1 - \lambda)x_2, \lambda g(x_1) + (1 - \lambda)g(x_2)) \in E_g$$

and, by the definition of E_g, to

$$g(\lambda x_1 + (1 - \lambda)x_2) \leqslant \lambda g(x_1) + (1 - \lambda)g(x_2).$$

The proof of the implication

$$H_g \text{ concave} \Rightarrow g : X \to F \text{ concave}$$

is analogous.

Problem II.2.4b Let $\{f_i\}_{i \in I}$ be a family of convex functionals on the convex set X. Consider $x_1, x_2 \in X$ and $\lambda \in [0, 1]$. Then

$$f_i(\lambda x_1 + (1 - \lambda)x_2) \leqslant \lambda f_i(x_1) + (1 - \lambda)f_i(x_2) \leqslant \lambda \sup_{i \in I} f_i(x_1) + (1 - \lambda) \sup_{i \in I} f_i(x_2)$$

for all $i \in I$ which implies

$$\sup_{i \in I} f_i(\lambda x_1 + (1 - \lambda)x_2) \leqslant \lambda \sup_{i \in I} f_i(x_1) + (1 - \lambda) \sup_{i \in I} f_i(x_2).$$

The second part of the assertion is proved similarly.

Problem II.2.5 Let $u_1, u_2 \in \Omega$ and $\lambda \in [0, 1]$ be given. Then $\lambda u_1 + (1-\lambda)u_2 \in U$ since U is a linear space and

$$\| \lambda u_1(t) + (1-\lambda)u_2(t) \| \leqslant \lambda \| u_1(t) \| + (1-\lambda) \| u_2(t) \| \leqslant \lambda\gamma + (1-\lambda)\gamma = \gamma$$

for almost all $t \in [0, 1]$, hence $\lambda_1 u_1 + (1-\lambda)u_2 \in \Omega$, i.e. Ω is convex. Further, $u \in \Omega$ implies

$$\| u \|_{L_2 [0,1]^r} = \left(\sum_{k=1}^{r} \int_0^1 u_k(t)^2 \, dt \right)^{1/2} \leqslant \sqrt{r}\ \gamma$$

i.e. Ω is bounded. In order to prove the closedness of Ω we refer to Theorem IV.3.7 and prove the weak closedness of Ω. For that purpose we put

$$H_i = \{x \in \mathbb{R}^r : \langle e_i, x \rangle \leqslant \gamma\}$$

where e_i denotes the unit vector with +1 or −1 as ith component. Then

$$\Omega = \bigcap_i \Omega_i \quad \text{with} \quad \Omega_i = \{u \in U : u(t) \in H_i \text{ for almost all } t \in [0, 1]\}.$$

We have to show that each Ω_i is weakly closed. Then Ω is weakly closed as an intersection of weakly closed sets. Consider $u^k \rightharpoonup u$, $u^k \in \Omega_i$ for all k. Then $u \in U$ since U is finite dimensional.

Assumption

$$\langle e_i, u(t) \rangle > \gamma \qquad \text{for all } t \in M \subseteq [0, 1] \quad \text{with} \quad \mu(M) > 0.$$

If we define

$$v(t) = \begin{cases} e_i & \text{for } t \in M \\ \Theta_r & \text{for } t \notin M \end{cases}$$

then $v \in L_2 [0, 1]^r$ and

$$\int_0^1 \langle v(t), u(t) \rangle \, dt > \gamma\mu(M) \geqslant \int_0^1 \langle v(t), u^k(t) \rangle \, dt \qquad \text{for all } k$$

which contradicts $u^k \rightharpoonup u$. Hence

$$\langle e_i, u(t) \rangle \leqslant \gamma \text{ for almost all } t \in [0, 1], \text{ i.e. } u \in \Omega_i.$$

Problem II.5.1 The implication (II.5.6*a*–*c*) ⇒ (II.5.5) is clear. Assume (II.5.5), and choose $x = \Theta_E$, $y_1 = y_2 = \Theta_W$. Then

$$\gamma[1 - (\hat{y}_1^* + \hat{y}_2^*)(e)] \geqslant \rho(z, X) + (y_1^* - y_2^*)(z) \qquad \text{for all } \gamma \in \mathbb{R}$$

which is possible only if (II.5.6*a*) holds. Hence (II.5.5) can be rewritten in the form

$$(\hat{y}_1^* - \hat{y}_2^*)(z - x) \geqslant \rho(z, X) - \hat{y}_1^*(y_1) - \hat{y}_2^*(y_2)$$
$$\text{for all } x \in X, y_1, y_2 \in Y_W. \qquad (13)$$

If $\hat{y}_1^*(\hat{y}_1) < 0$ for some $\hat{y}_1 \in Y_W$, then (13) would be violated for $x = \Theta_E, y_2 = \Theta_W$ and $y_1 = \lambda\hat{y}_1$ with $\lambda > 0$ sufficiently large. Therefore $\hat{y}_1^*(y_1) \geqslant 0$ for all $y_1 \in Y_W$. Similarly the second assertion of (II.5.6*c*) is proved. Finally, we conclude from (13) that

$$(\hat{y}_1^* - \hat{y}_2^*)(z - x) \geqslant \rho(z, X) \qquad \text{for all } x \in X$$

on putting $y_1 = y_2 = \Theta_W$.

Problem II.5.2a The statement

$$L(z) - \sup_{v \in V} L(v) \leqslant \rho(z, V) \qquad \text{for } L \in B^* \tag{14}$$

is equivalent with

$$L(v) = 0 \qquad \text{for all } v \in V. \tag{15}$$

For, if (15) holds for $L \in B^*$, then

$$L(z) = L(z - v) \leqslant \| z - v \| \qquad \text{for all } v \in V$$

which implies $L(z) \leqslant \rho(z, V)$ and hence (14). If (15) were false, then

$$L(\hat{v}) \neq 0 \qquad \text{for some } \hat{v} \in V$$

and we put $v = -\lambda \operatorname{sgn} L(\hat{v})$. Then, for sufficiently large $\lambda > 0$, we have

$$L(z) - L(v) = L(z) + \lambda \mid L(\hat{v}) \mid > \rho(z, V)$$

i.e. (14) is violated. From (II.5.29*b*) we therefore conclude

$$\rho(z, V) = \max_{L \in B^* \cap V^\perp} L(z)$$

where

$$V^\perp = \{L \in Z^* : L(v) = 0 \text{ for all } v \in V\}.$$

Problem II.2.5b Let $\hat{v} \in V$ be a best approximation of $z \in V$. Choose $\hat{L} \in V^\perp$ such that $\| \hat{L} \| \leqslant 1$ and $\hat{L}(z) = \rho(z, V)$ which implies $L(z - \hat{v}) = \| z - \hat{v} \|$ and hence $\| \hat{L} \| = 1$. If conversely $\hat{v} \in V$ is given such that there exists an $\hat{L} \in V^\perp$ with $\|\hat{L} \| = 1$ and $\hat{L}(z - \hat{v}) = \| z - \hat{v} \|$, then, for each $v \in V$, we have

$$\| z - \hat{v} \| = \hat{L}(z - \hat{v}) = L(z - v) \leqslant \| z - v \|.$$

Problem II.6.1 Let $x^* \in (-L(E, \hat{x}))^* \cap T(X, \hat{x})^*$ be given. Then we distinguish two cases.

(1) $\delta(\hat{x}) > 0$. Then $L(E, \hat{x}) = E$ and $x^*(h) \leqslant 0$ for all $h \in E$ which implies $x^* = \Theta_{E^*}$.

(2) $\delta(\hat{x}) = 0$. Choose $\bar{x} \in S_0$. Then

$$a(\bar{x} - \hat{x}, t) = a(\bar{x}, t) - \alpha(t) > 0 \qquad \text{for all } t \in I(\hat{x})$$

which implies

$$\bar{x} - \hat{x} \in \overset{\circ}{\widehat{L(E, \hat{x})}} \qquad \text{and} \qquad x^*(\bar{x} - \hat{x}) \leqslant 0.$$

Because of $\bar{x} - \hat{x} \in T(X, \hat{x})$ it follows that $x^*(\bar{x} - \hat{x}) \geqslant 0$, hence $x^*(\bar{x} - \hat{x}) = 0$. Therefore Theorem IV.1.5 is applicable to $-x^*$ and yields the result $x^* = \Theta_{E^*}$.

Problem II.6.2 The cone $T(X, \hat{x})^*$ can be identified with the set of all $h \in \mathbb{R}^{n+1}$ such that

$$\langle h, x \rangle \geqslant 0 \quad \text{for all } x \in \mathbb{R}^{n+1} \text{ with } \langle \hat{v}(t_i), x \rangle = 0 \quad i = 1, \ldots, r.$$

If we define $K = L[\hat{v}(t_1), \ldots, \hat{v}(t_r)]$ then by problem IV.2.2 it follows that $T(X, \hat{x})^* = K^{**} = K$ since K is closed and $E = \mathbb{R}^{n+1}$ is reflexive. (The isomorphism Φ in (IV.2.12) can be taken as the identity map.)

Problem II.6.3a The proof consists of pure verification.

Problem II.6.3b For each $x \in S(X, g)$ it follows by (II.6.49) that

$$\sum_{j=1}^{n} c_j(\hat{x}_j - x_j) \leqslant \sum_{j=1}^{n} \left(\sum_{i \in I} \lambda_i u_j(s_i) \right) (\hat{x}_j - x_j)$$

$$\sum_{i \in I} \lambda_i \sum_{j=1}^{n} u_j(s_i)(\hat{x}_j - x_j)$$

$$\sum_{i \in I} \lambda_i \left(\sum_{j=1}^{n} u_j(s_i) - \varphi(s_i) - \sum_{j=1}^{n} u_j(s_i) x_j \right) \leqslant 0.$$

CHAPTER III

Nonlinear Problems

III.1 SOME EXAMPLES OF NONLINEAR APPROXIMATION AND OPTIMIZATION PROBLEMS

III.1.1 Nonlinear approximation in normed vector spaces

III.1.1.1 General remarks

In Section I.1.1 we considered the general linear approximation problem in a normed vector space E, which consists of minimizing the convex functional $f(v) = \| v - x \|$ on a linear subspace V of E. Here $x \in E$ is an arbitrary but fixed element and $\| \quad \|$ denotes the norm in E. One can also say that x is to be approximated by an element from V as well as possible in the sense of the norm of E. If one now replaces V by an arbitrary non-empty subset of E, which we again denote by V, then we are faced with a general nonlinear approximation problem in E, which consists of minimizing a convex functional on a non-empty subset of E. We shall take this standpoint in Section III.2.2 in order to derive a general necessary condition for minimal points of f, i.e. for best approximations of x in V.

As in Section II.1.1, one can also consider the equivalent problem of minimizing the functional $f(v, \gamma) = \gamma$ subject to the side conditions

$$(v, \gamma) \in V \times \mathbb{R}$$

$$\varphi_L(v, \gamma) = L(v) - \gamma - L(x) \leqslant 0 \qquad \text{for all } L \in B^*. \tag{III.1.1}$$

Here B^* denotes the unit sphere (II.1.6) of the topological dual space E^* of E (see Section IV.1.2).

As we shall see in Section III.2.2 in connection with the theorems of Section II.6.1 this standpoint leads to refined necessary conditions for best approximations of x in V.

The question of the existence of best approximations can be answered positively only in a very restricted way. It leads to considerable difficulties even in special cases of the general nonlinear approximation problem. One such important special case will be treated in the following section.

III.1.1.2 Uniform approximation of functions

As in Section II.5.2 (see also Section I.2.1) we choose for E the vector space $C(M)$ of continuous real-valued functions on a compact metric space M and equip $C(M)$ with the maximum norm (I.2.1). The element x is then a continuous real-valued function on M, which is to be approximated as well as possible in the sense of the maximum norm of $E = C(M)$ by a function v from a given family of functions V in $C(M)$. Under this formulation, all problems of the uniform approximation of functions may be subsumed, even those in which explicit side conditions in the form of equations (see, for example, Sections I.2.1.2, or II.6.3.1) or in the form of inequalities (see Section III.1.2) arise, which then must be taken into the definition of V. This is, however, not always useful. In contrast, finer statements about the properties of best approximations are encountered when, as in Sections I.2.1.1, or II.1.1, the equivalent problem of minimizing the functional $f(v, \gamma) = \gamma$ is considered subject to the side conditions

$$(v, \gamma) \in V \times \mathbb{R} \qquad \text{(III.1.2a)}$$

$$\left.\begin{aligned} \varphi_t^1(v, \gamma) &= v(t) - \gamma - x(t) \leqslant 0 \\ \varphi_t^2(v, \gamma) &= -v(t) - \gamma + x(t) \leqslant 0 \end{aligned}\right\} \quad \text{for all } t \in M. \qquad \text{(III.1.2b)}$$

If necessary, explicit side conditions in V are specified.

Instead of (III.1.2b), under certain circumstances it is also appropriate to consider the equivalent side conditions

$$[v(t) - x(t)]^2 - \gamma^2 \leqslant 0 \qquad \text{for all } t \in M \qquad \text{(III.1.3)}$$

and minimize $\lambda = \gamma^2$.

III.1.1.3 General rational approximation

In Collatz and Krabs [73] numerous problems of nonlinear uniform approximation are treated. By way of illustration, we shall pick the case of general rational approximation. For this purpose we consider two linear subspaces U and W of $C(M)$ of dimensions $(r + 1)$ and $(s + 1)$, $(r, s \geqslant 0)$, which are spanned by the functions $u_0, \ldots, u_r \in C(M)$ and $w_0, \ldots, w_s \in C(M)$ and assume that the convex cone

$$W^+ = \{w \in W : w(t) > 0 \text{ for all } t \in M\} \qquad \text{(III.1.4)}$$

is not empty. Then we define

$$V = \{v = u/w : u \in U \text{ and } w \in W^+\}. \qquad \text{(III.1.5)}$$

The general rational approximation problem consists therefore in approximating a function $x \in C(M)$ as well as possible in the sense of the maximum norm by quotients $u/w \in V$. It is equivalent to the problem of minimizing the functional $f(a_0, \ldots, a_r, b_0, \ldots, b_s, \gamma) = \gamma$ subject to the side conditions

$$\begin{aligned} \sum_{j=0}^{r} u_j(t)a_j - (\gamma + x(t)) \sum_{k=0}^{s} w_k(t)b_k &\leqslant 0 \\ -\sum_{j=0}^{r} u_j(t)a_j - (\gamma - x(t)) \sum_{k=0}^{s} w_k(t)b_k &\leqslant 0 \qquad \text{for all } t \in M \qquad \text{(III.1.6)} \\ -\sum_{k=0}^{s} w_k(t)b_k &< 0 \end{aligned}$$

for $a_0, \ldots, a_r, b_0, \ldots, b_s, \gamma \in \mathbb{R}$. This is a nonlinear optimization problem.

The existence question cannot be answered positively, in this general case (see, for example, Collatz and Krabs [73] or also Cheney [66]). The existence is, however, assured in the case of ordinary rational approximation, where M is a finite, closed real interval and U and W consist of all polynomials of degree less than or equal to r and s respectively. If one requires, in addition, that the numerator and demoninator of the quotients in V have no common factors, then there is precisely one best approximation of x in V.

We shall go into a characterization of best approximations in Section III.2.2.3.

III.1.2 One- and two-sided approximation in nonlinear boundary value problems

III.1.2.1 The general case

We begin with a nonlinear boundary value problem of the form

$$y'' + H(x, y, y') = 0 \qquad \text{on } (a, b) \tag{III.1.7}$$

$$y(a) = \gamma \qquad y(b) = \gamma_2. \tag{III.1.8}$$

Here $H : [a, b] \times D_1 \times D_2 \to \mathbb{R}$ is a continuous function in all variables and D_1, D_2 are convex subsets of $\mathbb{R}$. We seek a function $y \in C^2[a, b]$ which satisfies (III.1.7), (III.1.8). The existence and uniqueness of such a solution of the boundary value problem is not always assured. Independently of that, under certain circumstances, approximate solutions can be given which enclose a solution of the boundary value problem, if one exists.

The search for optimal approximate solutions with this property leads to one-two-sided approximation problems with side conditions similar to some already considered in Sections I.2.4.2, and I.3.3.2 in the case of linear boundary value problems.

We assume that the partial derivatives $\partial H(x, y, z)/\partial y$ and $\partial H(x, y, z)/\partial z$ exist and are continuous for all $(x, y, z) \in G = [a, b] \times D_1 \times D_2$. Finally suppose

$$\partial H/\partial y(x, y, z) \leqslant 0 \qquad \text{for all } (x, y, z) \in G. \tag{III.1.9}$$

Then the following monotonicity assertion is valid.

Theorem III.1.1 *Under the above assumptions suppose that two functions $u, v \in C^2 [a, b]$ are given with*

$$u(x), u'(x)) \qquad (v(x), v'(x)) \in D_1 \times D_2 \qquad \textit{for all } x \in [a, b] \tag{III.1.10}$$

$$u'' + H(x, u, u') \leqslant 0 \leqslant v'' + H(x, v, v') \qquad \textit{on } (a, b) \tag{III.1.11}$$

and

$$u(a) \geqslant \gamma_1 \geqslant v(a) \qquad u(b) \geqslant \gamma_2 \geqslant v(b). \tag{III.1.12}$$

Then it follows that

$$u \geqslant v \qquad \textit{on } [a, b]. \tag{III.1.13}$$

Remark Inequality (III.1.13) holds also under the assumptions

$$u'' + H(x, u, u') \leqslant v'' + H(x, v, v') \qquad \text{on } (a, b)$$

$$u(a) \geqslant v(a) \qquad u(b) \geqslant v(b).$$

This theorem is an immediate consequence of Theorem 21 in Chapter I in Protter and Weinberger [67].

In particular, if we are faced with a linear boundary value problem with

$$H(x, y, y') = -f(x)y' - g(x)y - h(x) \tag{III.1.14}$$

$f, g, h \in C[a, b]$, $D_1 = D_2 = R$, then (III.1.9) means

$$g(x) \geqslant 0 \qquad \text{for all } x \in [a, b]$$

and is sufficient for the monotonic type of boundary value problem (III.1.7), (III.1.8) (see Section I.2.4.2), i.e. it is sufficient for the implication

$$\left.\begin{array}{ll} w \in C^2[a, b] & \\ -w'' + f \cdot w' + gw \geqslant 0 & \text{on } (a, b) \\ w(a) \geqslant 0 \qquad w(b) \geqslant 0 & \end{array}\right\} \Rightarrow w \geqslant 0 \qquad \text{on } [a, b].$$

The monotonic type assures the unique solvability of the boundary value problem, so that in the linear case Theorem III.1.1 always leads to an inclusion relation

$$u \geqslant y \geqslant v \qquad \text{on } [a, b] \tag{III.1.15}$$

whereby $y \in C^2[a, b]$ is the unique solution of the linear boundary value problem (III.1.7), (III.1.8) with $H(x, y, y')$ given by (III.1.14).

One can now try to select the functions $u, v \in C^2[a, b]$ in Theorem III.1.1 from two classes of functions of $C^2[a, b]$ so that the conditions (III.1.10), (III.1.11), (III.1.12) are satisfied and

$$\| u - v \| = \max_{x \in [a, b]} \{u(x) - v(x)\} \tag{III.1.16}$$

is made as small as possible. Then one obtains as good an inclusion of a solution $y \in C^2[a, b]$ as possible for the boundary value problem (III.1.7), (III.1.8) with $(y(x), y'(x)) \in D_1 \times D_2$ for all $x \in [a, b]$, as long as such a solution exists. This is, by the way, uniquely determined as a consequence of Theorem III.1.1 (Proof = Exercise).

In order to give two such function classes, we imagine $u_0, v_0 \in C^2[a, b]$ to be chosen such that

$$u_0(a) = v_0(a) = \gamma_1 \qquad \text{and} \qquad u_0(b) = v_0(b) = \gamma_2 \tag{III.1.17}$$

and also functions $u_1, \ldots, u_r, v_1, \ldots, v_s \in C^2[a, b]$ chosen so that

$$u_j(a) = u_j(b) = 0 \qquad \text{for } j = 1, \ldots, r \tag{III.1.18a}$$

and

$$v_k(a) = v_k(b) = 0 \qquad \text{for } k = 1, \ldots, s. \tag{III.1.18b}$$

Thus we form the two classes of all functions

$$u(\alpha) = u_0 + \sum_{j=1}^{r} \alpha_j u_j \qquad \text{and} \qquad v(\beta) = v_0 + \sum_{k=1}^{s} \beta_k v_k \tag{III.1.19}$$

where

$$\alpha = (\alpha_1, \ldots, \alpha_r)^T \in \mathbb{R}^r \qquad \text{and} \qquad \beta = (\beta_1, \ldots, \beta_s)^T \in \mathbb{R}^s$$

are chosen arbitrarily at first. Then (III.1.17), (III.1.18) yield

$$u(\alpha, a) = \gamma_1 \qquad u(\alpha, b) = \gamma_2 \qquad \text{for all } \alpha \in \mathbb{R}^r$$

and

$$v(\beta, a) = \gamma_1 \qquad v(\beta, b) = \gamma_2 \qquad \text{for all } \beta \in \mathbb{R}^s$$

so that the conditions (III.1.12) of Theorem III.1.1 are satisfied.

In order to come to an optimal inclusion (III.1.15) for a solution y of the boundary value problem (III.1.7), (III.1.8) with $(y(x), y'(x)) \in D_1 \times D_2$ for all $x \in [a, b]$ with $u = u(\alpha)$ and $v = v(\beta)$ given by (III.1.19), one must minimize the functional

$$f(\alpha, \beta) = \| u(\alpha) - v(\beta) \| = \max_{x \in [a,b]} \{u(\alpha, x) - v(\beta, x)\}$$

subject to the side conditions $\alpha \in \mathbb{R}^r, \beta \in \mathbb{R}^s$

$$(u(\alpha, x), u'(\alpha, x)) \qquad (v(\beta, x), v'(\beta, x)) \in D_1 \times D_2 \qquad \text{for all } x \in | a, b | \tag{III.1.10$'$}$$

$$u''(\alpha) + H(x, u(\alpha), u'(\alpha)) \leqslant 0 \qquad \text{for all } x \in (a, b) \tag{III.1.11$'a$}$$

and

$$-v''(\alpha) - H(x, v(\beta), v'(\beta)) \leqslant 0 \qquad \text{for } x \in (a, b). \tag{III.1.11$'b$}$$

Here it is a question of dealing with a linear approximation problem (see Section I.2.1) subject to nonlinear side conditions. On the other hand the set

$$\{(\alpha, \beta) \in \mathbb{R}^{r+s} : (u(\alpha, x), u'(\alpha, x)) \text{ and } (v(\beta, x), v'(\beta, x)) \in D_1 \times D_2 \quad \text{for all } x \in [a, b]\} \tag{III.1.20}$$

so convex is that the actual nonlinearity in the side conditions is to be found in (III.1.11$'$).

The above approximation problem is equivalent to the nonlinear optimization problem of minimizing the functional $f(\alpha, \beta, \gamma) = \gamma$ subject to the side conditions $\alpha \in \mathbb{R}^r, \beta \in \mathbb{R}^s$, (III.1.10$'$), (III.1.11$'$), $\gamma \in \mathbb{R}$, and

$$u(\alpha, x) - v(\beta, x) - \gamma \leqslant 0 \qquad \text{for all } x \in [a, b].$$

In the case of a linear boundary value problem with $H(x, y, y')$ given by (III.1.14), the side conditions (III.1.10$'$) are fulfilled naturally because $D_1 = D_2 = \mathbb{R}$, and the side conditions (III.1.11$'$) are affine linear. Hence altogether we are faced with a problem of semi-infinite linear optimization (see Section I.3.2). In the procedure so far, a solution of the boundary value problem (III.1.7), (III.1.8) was approximated from two sides. As in Sections I.2.4.2, and I.3.3.2, one naturally could also try a one-sided approximation in the linear case. This leads to the following simpler problem, namely, to minimize the functional

$$f(\alpha) = \max_{x \in [a,b]} |u''(\alpha, x) + H(x, u(\alpha, x), u'(\alpha, x))|$$

subject to the side conditions

$$(u(\alpha, x), u'(\alpha, x)) \in D_1 \times D_2 \qquad \text{for all } x \in [a, b]. \qquad \text{(III.1.10'')}$$

$$u''(\alpha) + H(x, u(\alpha), u'(\alpha)) \begin{cases} \leqslant 0 \\ \geqslant 0 \end{cases} \qquad \text{on } (a, b). \qquad \text{(III.1.11'')}$$

Then for the solution $y \in C^2[a, b]$ of the boundary value problem (III.1.7), (III.1.8) with $(y(x), y'(x)) \in D_1 \times D_2$ for all $x \in [a, b]$ (when it exists) (III.1.10″) and (III.1.11″) yield the estimate

$$u(\alpha, x) \begin{Bmatrix} \geqslant \\ \leqslant \end{Bmatrix} y(x) \qquad \text{for all } x \in [a, b].$$

Instead of the error $\| u(\alpha) - y \|$, the defect $\| u''(\alpha) + H(\cdot, u(\alpha), u'(\alpha)) \|$ is minimized. This procedure will be justified by the considerations in Section III.1.3.

III.1.2.2 An example

Consider the nonlinear boundary value problem

$$y'' - 6xy^2 = 0 \qquad \text{on } (0, 1) \qquad \text{(III.1.7')}$$

$$y(0) = y(1) = 1. \qquad \text{(III.1.8')}$$

In this case, $H(x, y, y') = -6xy$ and also

$$H_y(x, y, y') = -12xy^2 \qquad H_{y'}(x, y, y') = 0 \qquad \text{for all } x \in [0, 1], y, y' \in \mathbb{R}.$$

Furthermore, one has

$$H_y(x, y, y') \leqslant 0 \qquad \text{for all } x \in [0, 1] \text{ and } y \geqslant 0.$$

Therefore one chooses $D_1 = \{y \in \mathbb{R} : y \geqslant 0\}$ and $D_2 = \mathbb{R}$, so that (III.1.9) is satisfied and Theorem III.1.1 can be applied. We now choose $r = s = 1$ and set $u_0 = v_0 = 1$, $u_1(x) = x - x^4$, $v_1(x) = x - x^3$. Then the boundary conditions (III.1.8′) are satisfied for all functions

$$u(\alpha) = u_0 + \alpha u_1 \qquad v(\beta) = v_0 + \beta v_1 \qquad \alpha, \beta \in \mathbb{R} \qquad \text{(III.1.19')}$$

The boundary value problem (III.1.7′), (III.1.8′) has a non-negative solution $y \in C^2[0, 1]$. With the aid of the functions (III.1.19′) we arrive at an optimal inclusion of y if we minimize the functional

$$f(\alpha, \beta) = \max_{x \in [0,1]} \{\alpha(x - x^4) - \beta(x - x^3)\}$$

subject to the side conditions

$$\left.\begin{aligned} u(\alpha, x) &= 1 + \alpha(x - x^4) \geqslant 0 \\ v(\beta, x) &= 1 + \beta(x - x^3) \geqslant 0 \end{aligned}\right\} \quad \text{for all } x \in [0, 1] \qquad \text{(III.1.10}''')$$

$$u''(\alpha, x) - 6xu(\alpha, x)^2 = -12\alpha x^2 - 6x[1 + \alpha(x - x^4)]^2 \leqslant 0 \qquad \text{(III.1.11}'''a)$$

$$-v''(\beta, x) + 6xv(\beta, x)^2 = 6\beta x + 6x[1 + \beta(x - x^3)]^2 \leqslant 0 \qquad \text{for all } x \in (0, 1). \qquad \text{(III.1.11}'''b)$$

If one chooses $\alpha = -0.43$ and $\beta = -1$, then the side conditions (III.1.10″) and (III.1.11″) are satisfied and we conclude that $f(\alpha, \beta) \lesssim 0.187$.

III.1.3 Defect estimates for nonlinear boundary value problems

We again consider the nonlinear boundary value problem (III.1.7), (III.1.8), although the following considerations may be extended to more general boundary value problems. The linear boundary value problem associated with (III.1.7), (III.1.8) is

$$-y'' = r \qquad \text{on } (a, b) \qquad \text{(III.1.21)}$$

$$y(a) = \gamma_1 \qquad y(b) = \gamma_2. \qquad \text{(III.1.8)}$$

Here $r \in C[a, b]$ is a given function. The boundary value problem (III.1.21), (III.1.8) is uniquely solvable in the form

$$y(x) = g(x) + \int_a^b G(x, \xi) r(\xi)\, d\xi$$

with

$$g(x) = \frac{\gamma_1 b - \gamma_2 a}{b - a} + \frac{\gamma_2 - \gamma_1}{b - a} x \qquad \text{(III.1.22)}$$

and

$$G(x, \xi) = \begin{cases} \dfrac{(b - x)(\xi - a)}{b - a} & \text{for } a \leqslant x \leqslant b \\[2ex] \dfrac{(b - \xi)(x - a)}{b - a} & \text{for } a \leqslant x \leqslant \xi \leqslant b. \end{cases} \qquad \text{(III.1.23)}$$

From this we conlcude that $y \in C^2[a, b]$ is a solution to the nonlinear boundary value problem (III.1.7), (III.1.8) if and only if

$$y(x) = g(x) + \int_a^b G(x, \xi) H(\xi, y(\xi), y'(\xi))\, d\xi \qquad \text{(III.1.24)}$$

for $x \in [a, b]$ with g given by (III.1.22) and G by (III.1.23) (Proof = Exercise). Furthermore, for all $x \in [a, b]$, one has

$$y'(x) = g'(x) + \int_a^x G_x(x, \xi) H(\xi, y(\xi), y'(\xi))\, d\xi + \int_x^b G_x(x, \xi) H(\xi, y(\xi), y'(\xi))\, d\xi. \qquad \text{(III.1.25)}$$

From now on we assume the existence of a unique solution $y \in C^2[a, b]$ of the boundary value problem (III.1.7), (III.1.8) with $(y(x), y'(x)) \in D_1 \times D_2$, where again D_1 and D_2 are two suitable convex subsets of $\mathbb{R}$. Further, we assume the existence of two non-negative continuous functions L_1, L_2 on $[a, b]$ such that, for all $(x, y, z), (x, \hat{y}, \hat{z}) \in [a, b] \times D_1 \times D_2$,

$$|H(x, y, z) - H(x, \hat{y}, \hat{z})| \leqslant L_1(x) |y - \hat{y}| + L_2(x) |z - \hat{z}|. \qquad \text{(III.1.26)}$$

Suppose now a function $v \in C^2[a, b]$ is given which satisfies the boundary conditions (III.1.8). Substituting v into the differential equation (III.1.7), we obtain the defect

$$h_v(x) = -v''(x) - H(x, v(x), v'(x)) \qquad \text{for } x \in [a, b]. \qquad \text{(III.1.27)}$$

We shall assume further that

$$(v(x), v'(x)) \in D_1 \times D_2 \qquad \text{for all } x \in [a, b]. \qquad \text{(III.1.28)}$$

On the basis of the above assumptions, under certain circumstances it is possible to estimate the error $\| y - v \|$ with the aid of the defect. For this we remark first that by analogy with (III.1.24), we have the integral equation for v

$$v(x) = g(x) + \int_a^b G(x, \xi)[H(\xi, v(\xi), v'(\xi)) + h_v(\xi)] \, d\xi \qquad \text{for } x \in [a, b] \qquad \text{(III.1.29)}$$

with g given by (III.1.22) and G by (III.1.23). Furthermore one has

$$v'(x) = g'(x) + \int_a^x G_x(x, \xi)[H(\xi, v(\xi), v'(\xi)) + h_v(\xi)] \, d\xi$$
$$+ \int_a^b G_x(x, \xi)[H(\xi, v(\xi), v'(\xi)) + h_v(\xi)] \, d\xi. \qquad \text{(III.1.30)}$$

Subtraction of (III.1.29) from (III.1.24) and (III.1.30) from (III.1.25) yields

$$y(x) - v(x) = -\int_a^b G(x, \xi) h_v(\xi) \, d\xi$$
$$+ \int_a^b G(x, \xi)[H(\xi, y(\xi), y'(\xi)) - H(\xi, v(\xi), v'(\xi)] \, d\xi$$

and

$$y'(x) - v'(x) = -\int_a^x G_x(x, \xi) h_v(\xi) \, d\xi - \int_a^b G_x(x, \xi) h_v(\xi)] \, d\xi$$
$$+ \int_a^x G_x(x, \xi)[H(\xi, y(\xi), y'(\xi)) - H(\xi, v(\xi), v'(\xi))] \, d\xi$$
$$+ \int_x^b G_x(x, \xi)[H(\xi, y(\xi), y'(\xi)) - H(\xi, v(\xi), v'(\xi))] \, d\xi.$$

Taking (III.1.26) into account we conclude that

$$\| y - v \| \leqslant \max_{x \in [a,b]} \int_a^b G(x, \xi)\, d\xi \, \| h_v \|$$

$$+ \max_{x \in [a,b]} \int_a^b G(x, \xi)[L_1(\xi) + L_2(\xi)]\, d\xi \max \; \| y - v \|, \| y' - v' \|$$

$$\| y' - v' \| \leqslant \max_{x \in [a,b]} \left(\int_a^x | G_x(x, \xi) |\, d\xi + \int_x^b | G_x(x, \xi) |\, d\xi] \, \| h_v \| \right)$$

$$+ \max_{x \in [a,b]} \left(\int_a^x | G_x(x, \xi) | [L_1(\xi) + L_2(\xi)]\, d\xi \right.$$

$$\left. + \int_x^b | G_x(x, \xi) | [L_1(\xi) + L_2(\xi)]\, d\xi \right) \max \{ \| y - v \|, \| y' - v' \| \}.$$

Here $\| \;\; \|$ denotes the maximum norm in $C[a, b]$.

Now

$$\max_{x \in [a,b]} \int_a^b G(x, \xi)\, d\xi = \tfrac{1}{8}(b - a)^2$$

and

$$\max_{x \in [a,b]} \left(\int_a^x | G_x(x, \xi) |\, d\xi + \int_x^b | G_x(x, \xi) |\, d\xi \right) = \tfrac{1}{2}(b - a).$$

We now make the assumptions

$$\max_{x \in [a,b]} \int_a^b G(x, \xi)[L_1(\xi) + L_2(\xi)]\, d\xi \leqslant \alpha < 1 \qquad \text{(III.1.31}a\text{)}$$

and

$$\max_{x \in [a,b]} \left(\int_a^x | G_x(x, \xi) | [L_1(\xi) + L_2(\xi)]\, d\xi \right.$$
$$\left. + \int_x^b | G_x(x, \xi) | [L_1(\xi) + L_2(\xi)]\, d\xi \right) \leqslant \beta < 1. \qquad \text{(III.1.31}b\text{)}$$

Let $c = \max \{ (b - a)^2/8, (b - a)/2 \}$. Then

$$\max\{ \| y - v \|, \| y' - v' \| \} \leqslant c \| h_v \| + \max \{ \alpha, \beta \} \max \{ \| y - v \|, \| y' - v' \| \}$$

whence finally

$$\max\{ \| y - v \|, \| y' - v' \| \} \leqslant \frac{c}{1 - \max\{\alpha, \beta\}} \| h_v \|. \qquad \text{(III.1.32)}$$

If H does not depend explicitly on z, then one may choose $L_2 \equiv 0$ and one does not have to take into account the derivatives. This leads to the estimate

$$\| y - v \| \leqslant \frac{1}{8} \frac{(b - a)^2}{1 - \alpha} \| h_v \| \qquad \text{(III.1.33)}$$

provided the assumption (III.1.31*a*) is satisfied. The two estimates (III.1.32), (III.1.33) suggest letting the approximation v run through a class of functions in $C^2[a, b]$ satisfying (III.1.28) (in this regard, see the remark at the end of Section III.1.2.1). For example, one could again choose v_0, and $v_1, \ldots, v_s \in C^2[a, b]$ satisfying (III.1.17), and (III.1.18*b*), respectively. The class V consists then of all functions $v = v(\beta)$ given by (III.1.19) satisfying (III.1.28), and we obtain the non-linear approximation problem of minimizing the functional

$$f(\beta) = \| v''(\beta) + H(\cdot, v(\beta), v'(\beta)) \|$$

subject to the side conditions

$$\left.\begin{array}{l} v(\beta, x) \in D_1 \\ v'(\beta, x) \in D_2 \end{array}\right\} \text{ for all } x \in [a, b]. \qquad \begin{array}{r} \text{(III.1.28}a\text{)} \\ \text{(III.1.28}b\text{)} \end{array}$$

If H does not depend explicitly on z, then one dispenses with the side condition (III.1.28*b*). As an example, we again consider the boundary value problem (III.1.7′), (III.1.8′). This has precisely one non-negative solution $y \in C^2[0, 1]$.

On the basis of the considerations in Sections III.1.2.1 and III.1.2.2 we have

$$y(x) \leqslant 1 - 0.43(x - x^4) \leqslant 1 \qquad \text{for all } x \in [0, 1]$$

so that one can choose $D_1 = [0, 1]$. Then we find

$$| H(x, y) - H(x, \hat{y}) | \leqslant 12x | y - \hat{y} | \qquad \text{for all } x \in [0, 1] \text{ and } y, \hat{y} \in D_1$$

i.e. (III.1.26) is satisfied with $L_1(x) = 12x$ and $L_2 = 0$. Further, by (III.1.23)

$$G(x, \xi) = \begin{cases} (1 - x)\xi & \text{for } 0 \leqslant x \leqslant 1 \\ (1 - \xi)x & \text{for } 0 \leqslant \xi \leqslant 1 \end{cases}$$

and

$$\max_{x \in [0,1]} \int_0^1 G(x, \xi) L_1(\xi) \, d\xi = \tfrac{4}{9}\sqrt{3} < 0.8$$

so that (III.1.31*a*) is satisfied with $\alpha = 0.8$. For each $v \in C^2[0, 1]$ with $v(0) = v(1) = 1$ and $0 \leqslant v(x) \leqslant 1$ for all $x \in [0, 1]$, (III.1.33) yields the error estimate

$$\| y - v \| \leqslant \tfrac{5}{8} \max_{x \in [0,1]} | v''(x) - 6xv(x)^2 |. \qquad \text{(III.1.33′)}$$

If v runs through functions of the form, say,

$$v(\beta, x) = 1 + \beta_1(x - x^3) + \beta_2(x - x^4)$$

for $x \in [0, 1]$, $\beta_1, \beta_2 \in \mathbb{R}$, then we obtain the approximation problem of minimizing the functional

$$\max_{x \in [0,1]} | 6\beta_1 x + 12\beta_2 x^2 + 6x[1 + \beta_1(x - x^3) + \beta_2(x - x^4)]^2 |$$

subject to the side conditions

$$1 \geqslant 1 + \beta_1(x - x^3) + \beta_2(x - x^4) \geqslant 0 \qquad \text{for all } x \in [0, 1]. \qquad \text{(III.1.28′}a\text{)}$$

III.2 MINIMIZING CONVEX FUNCTIONALS ON ARBITRARY SETS

In Section III.1.1 we have already indicated that the nonlinear approximation in a normed vector space consists of minimizing a convex functional on an arbitrary subset of the space. In the sequel, this problem will be investigated more precisely with the goal of giving necessary conditions for minimal points, which for the approximation problem then lead to the generalized Kolmogoroff condition (see Section III.2.2.3). As a decisive aid we need

III.2.1 Tangent cones in normed vector spaces

Let E be a normed vector space (over the real or complex numbers), X a non-empty subset of E, and x any point of X. A vector $h \in E$ is called a tangent vector to X at x if there is a sequence $\{x_k\}$ of elements $x_k \in X$ and a sequence $\{\lambda_k\}$ of positive real numbers λ_k with

$$\lim_{k\to\infty} x_k = x \tag{III.2.1a}$$

$$\lim_{k\to\infty} \lambda_k(x_k - x) = h. \tag{III.2.1b}$$

Let $T(X, x)$ be the set of all tangent vectors to X at x. Since $T(X, x)$ certainly contains the null vector Θ_E of E, $T(X, x)$ is not empty and moreover has the property

$$h \in T(X, x), \lambda \geqslant 0 \Rightarrow \lambda h \in T(X, x)$$

i.e. $T(X, x)$ is a cone with vertex Θ_E (see Section IV.1.1). We call $T(X, x)$ the tangent cone to X at x. The tangent cone to a set at a point is in general not convex (see Figure III.2.1). However, the following lemma does hold.

Lemma III.2.1 *The tangent cone $T(X, x)$ to X at x ($\in X$) is closed.*

Proof Let $\{h^m\}$ be a sequence in $T(X, x)$ with $\lim_{m\to\infty} h^m = h$. Then for every m there is a sequence $\{x_k^m\}$ in X and a sequence $\{\lambda_k^m\}$ of positive real numbers with

$$\lim_{k\to\infty} x_k^m = x \qquad \text{and} \qquad \lim_{k\to\infty} \lambda_k^m(x_k^m - x) = h^m.$$

In particular, for each $m \geqslant 1$ there is a $k(m)$ with

$$\| x_k^m - x \| \leqslant 1/m \qquad \text{and} \qquad \| \lambda_k^m(x_k^m - x) - h^m \| \leqslant 1/m$$

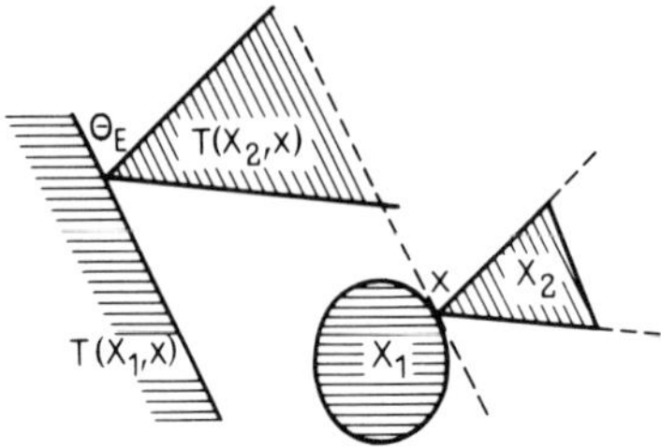

Figure III.2.1 Example of a non-convex tangent cone

$$X = X_1 \cup X_2$$

$$T(X, x) = T(X_1, x) \cup T(X_2, x)$$

for all $k \geqslant k(m)$. Therefore, if one defines $y_m = x^m_{k(m)}$ and $\mu_m = \lambda^m_{k(m)}$, then $\{y_m\}$ is a sequence in X and $\{\mu_m\}$ is a sequence of positive real numbers with

$$\| y_m - x \| \leqslant 1/m \qquad \text{and} \qquad \| \mu_m(y_m - x) - h \| \leqslant 1/m + \| h^m - h \|$$

whence

$$\lim_{m\to\infty} y_m = x \qquad \text{and} \qquad \lim_{m\to\infty} \mu_m(y_m - x) = h$$

follows, i.e. $h \in T(X, x)$.

Examples of tangent cones

1. Let X be a non-empty open subset of E and let $x \in X$ be arbitrary. If one fixes $h \in E$ arbitrarily, then, for a suitable $\lambda_h > 0$, certainly $x + \lambda_h h \in X$ for all $\lambda \in [0, \lambda_h]$. If one chooses k so large that $1/k \leqslant \lambda_h$, and defines $x_k = x + (1/k)h$, then $x_k \in X$ for all $k \geqslant 1/\lambda_h$, $\lim_{k\to\infty} x_k = x$ and $k(x_k - x) = h$ for all k, whence $\lim_{k\to\infty} k(x_k - x) = h$, i.e. $h \in T(X, x)$.

Therefore

$$T(X, x) = E \qquad \text{for all } x \in X \text{ (open).} \tag{III.2.2}$$

2. Let X be a non-empty convex subset of E and let $x \in X$ be arbitrary. If one chooses $y \in X$ arbitrarily, then

$$x + \frac{1}{k}(y - x) = \left(1 - \frac{1}{k}\right)x + \frac{1}{k}y \in X \qquad \text{for all } k \geqslant 1.$$

If one defines $x_k = x + (1/k)(y - x)$, then $\lim_{k\to\infty} x_k = x$ and $k(x_k - x) = y - x$, whence $\lim_{k\to\infty} k(x_k - x) = y - x$, i.e. $y - x \in T(X, x)$. Since $T(X, x)$ is a cone, we have $\lambda(y - x) \in T(X, x)$ for all $\lambda \geqslant 0$. From this follows

$$\bigcup_{\lambda\geqslant 0} \{\lambda(y - x) : y \in X\} \subseteq T(X, x)$$

and further

$$\overline{\bigcup_{\lambda\geqslant 0} \{\lambda(y - x) : y \in X\}} \subseteq T(X, x)$$

since $T(X, x)$ is closed by Lemma III.2.1.

Now let $h \in T(X, x)$ be given. Then there is a sequence $\{x_k\}$ in X and a sequence $\{\lambda_k\}$ of positive numbers satisfying (III.2.1*a*), (III.2.1*b*), which implies

$$h \in \overline{\bigcup_{\lambda\geqslant 0} \{\lambda(y - x) : y \in X\}}.$$

Therefore one has

$$T(X, x) = \overline{\bigcup_{\lambda\geqslant 0} \{\lambda(y - x) : y \in X\}} \qquad \text{for all } x \in X \text{ (convex).} \tag{III.2.3}$$

Remark The closed convex cone $T(X, \hat{x})$ defined by (II.6.15) is therefore the tangent cone to the convex set X at $\hat{x}$. In the case of a convex set, the tangent cone at each of its points is therefore also convex.

3. In particular, let X be a linear manifold in E (see Section IV.3.2), i.e. $X = x + V$, where V is a linear subspace of E and is independent of the choice of $x \in X$. Then, in particular, (III.2.3) implies

$$T(X, x) = \bar{V} \qquad \text{for all } x \in X. \tag{III.2.4}$$

4. *Problem III.2.1* Prove the following statements:

(*a*) If $x \in X \subseteq Y$, or $x \in X \cap Y$, respectively, then

$$T(X, x) \subseteq T(Y, x) \tag{III.2.5}$$

or

$$T(X \cap Y, x) \subseteq T(X, x) \cap T(Y, x) \tag{III.2.6}$$

respectively.

(*b*) If E, F are normed vector spaces and X, Y are non-empty subspaces of E, F, respectively, then for every $x \in X$ and $y \in Y$

$$T(X \times Y, (x, y)) \subseteq T(X, x) \times T(Y, y) \tag{III.2.7}$$

if one introduces say the norm $\| (x, y) \| = \| x \| + \| y \|$, $(x, y) \in E \times F$. If one of the two sets X or Y is open, then the equality sign holds in (III.2.7).

III.2.2 Necessary conditions for minimal points of convex functionals on arbitrary sets

III.2.2.1 A general theorem

Again let a normed vector space E (over $\mathbb{R}$ or $\mathbb{C}$) be given, let a non-empty subset X of E be given and $f: E \to \mathbb{R}$ (see Section II.2.1) be a given continuous linear functional. Then we have the following theorem.

Theorem III.2.2 *If $\hat{x} \in X$ is a minimal point of f on X, i.e. if*

$$f(\hat{x}) \leqslant f(x) \qquad \textit{for all } x \in \bar{X} \tag{III.2.8}$$

and if f is convex on E with respect to $\hat{x}$ (see Section II.6.1), then

$$f(\hat{x}) \leqslant f(\hat{x} + h) \qquad \textit{for all } h \in T(X, \hat{x}) \tag{III.2.9}$$

whereby $T(X, \hat{x})$ is the tangent cone to X at $\hat{x}$.

Proof Let us assume there is an $h \in T(X, \hat{x})$ with

$$f(\hat{x}) - f(\hat{x} + h) > \delta > 0.$$

By the definition of h there is a sequence $\{x_k\}$ of elements $x_k \in X$ and a sequence $\{\lambda_k\}$ of positive numbers satisfying (III.2.1*a*), (III.2.1*b*). We set $h_k = \lambda_k(x_k - \hat{x})$ for every k.

From (III.2.1*b*) and the continuity of f it follows then that

$$| f(\hat{x} + h_k) - f(\hat{x} + h) | \leqslant \delta \qquad \text{for all sufficiently large } k$$

Furthermore

$$\frac{1}{\lambda_k} h_k = x_k - \hat{x} \qquad \lim_{k\to\infty} \| h_k - h \| = 0 \Rightarrow \lim_{k\to\infty} \| h_k \| = \| h \|.$$

From (III.2.1a) as well as $\| h \| > 0$ it follows that $\lim_{k\to\infty}(1/\lambda_k) = 0$. Hence $\eta_k = 1/\lambda_k \in (0, 1)$ for all sufficiently large k and

$$f(x_k) = f((1 - \eta_k)\hat{x} + \eta_k(\hat{x} + h_k)) \leqslant (1 - \eta_k)f(\hat{x}) + \eta_k f(\hat{x} + h_k)$$
$$\leqslant (1 - \eta_k)f(\hat{x}) + \eta_k [f(\hat{x} + h) + \delta] < (1 - \eta_k)f(\hat{x}) + \eta_k f(\hat{x}) = f(\hat{x})$$

for sufficiently large k which contradicts (III.2.8).

If X is convex, then it follows from (III.2.3) that $x - \hat{x} \in T(X, \hat{x})$ for all $x \in X$, so that (III.2.9) again implies (III.2.8) and, hence, both statements are equivalent for continuous convex functionals $f: E \to \mathbb{R}$.

III.2.2.2 Application to approximation in normed spaces

We again take as the basis for our discussion the situation of Section III.1.1. Therefore let E be a normed vector space over $\mathbb{R}$, V a non-empty subset of E, and fix an arbitrary point $x \in E$. We wish to minimize on V the functional

$$f(y) = \| y - x \| \qquad \text{for } y \in V.$$

Now, f is continuous and convex so that Theorem III.2.2 can be immediately applied.

If $\hat{v} \in V$ is a best approximation of x in V, i.e. if

$$\| \hat{v} - x \| \leqslant \| v - x \| \qquad \text{for all } v \in V \tag{III.2.10}$$

then it follows that

$$\| \hat{v} - x \| \leqslant \| \hat{v} + h - x \| \qquad \text{for all } h \in T(V, \hat{v}) \tag{III.2.11}$$

where $T(V, \hat{v})$ denotes the tangent cone to V at $\hat{v}$ (see also Figure III.2.2). On the basis of the remark at the end of Theorem III.2.2, the two statements (III.2.10) and (III.2.11) are equivalent if V is a convex subset of E.

In general, $h = \Theta_E$ is a best approximation of $x - \hat{v}$ in $T(V, \hat{v})$ if $\hat{v}$ is a best approximation of x in V.

By Section III.1.1.1 the statement (III.2.11) is equivalent to the fact that the

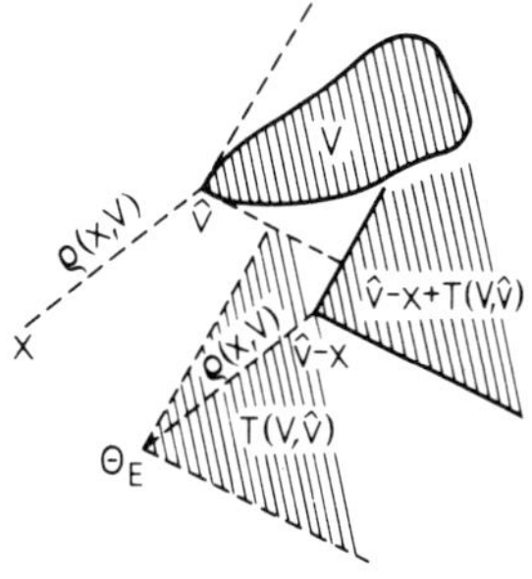

Figure III.2.2 The implication (III.2.10) ⇒ (III.2.11)

$$\rho(x, V) = \rho(\Theta_E, \hat{v} - x + T(V, \hat{v}))$$

pair $(\Theta_E, \|\hat{v} - x\|)$ minimizes the functional $f(h, \gamma) = \gamma$ subject to the side conditions

$$(h, \gamma) \in T(V, \hat{v}) \times \mathbb{R} \qquad \gamma - L(h) - L(\hat{v} - x) \geqslant 0 \qquad \text{for all } L \in B^* \tag{III.2.12}$$

(Proof = Exercise). Here B^* is the unit sphere of the topological dual space E^* of E defined by (II.1.6) (see Section IV.1.2). If one equips E^* with the weak* topology, then B^* is a compact Hausdorff space in E^* and by setting

$$g(y, \gamma, L) = \gamma - L(y) - L(\hat{v} - x) \qquad \text{for } (y, \gamma) \in E \times \mathbb{R}, L \in B^* \tag{III.2.13}$$

an affine linear, and hence concave, mapping $g: E \times \mathbb{R} \to C(B^*)$ is defined (see, for example, Köthe [66] and Section II.2.2), where $C(B^*)$ is the vector space of continuous real-valued functions on B^*. By Lemma II.1.1 we have

$$\delta(\Theta_E, \|\hat{v} - x\|) = \min_{L \in B^*} \{\|\hat{v} - x\| - L(\hat{v} - x)\} = 0.$$

We then set

$$I(\Theta_E, \|\hat{v} - x\|) = \{L \in B^* : g(\Theta_E, \|\hat{v} - x\|, L) = 0\}$$

to obtain

$$I(\Theta_E, \|\hat{v} - x\|) = E_{\hat{v}-x} = \{L \in B^* : L(\hat{v} - x) = \|\hat{v} - x\|\}. \tag{III.2.14}$$

Now $T(V, \hat{v}) \times \mathbb{R}$ is star-shaped with respect to the point $(\Theta_E, \|\hat{v} - x\|)$, the mapping g from $E \times \mathbb{R}$ into $C(B^*)$ defined by (III.2.13) is concave, and $f(h, \gamma) = \gamma$, $(h, \gamma) \in E \times \mathbb{R}$ is linear and hence convex. Therefore, on the basis of the optimality of $(\Theta_E, \|\hat{v} - x\|)$, one may apply Theorem II.6.3, and obtain the implication

$$\gamma - L(h) - L(\hat{v} - x) > 0 \qquad \text{for all } L \in E_{\hat{v}-x} \Rightarrow \|\hat{v} - x\| \leqslant \gamma \tag{III.2.15}$$

for all $(h, \gamma) \in T(V, \hat{v}) \times \mathbb{R}$. From this it follows further that

$$\max_{L \in E_{\hat{v}-x}} L(h) \geqslant 0 \qquad \text{for all } h \in T(V, \hat{v}). \tag{III.2.16}$$

The maximum is attained since $E_{\hat{v}-x}$ is a weakly* closed subset of B^* and hence is weakly* compact (Proof = Exercise) and the linear form $L \to L(h)$ is weakly* continuous on E^* for every fixed $h \in E$. For, if

$$\delta = \max_{L \in E_{\hat{v}-x}} L(h) < 0 \qquad \text{for some } h \in T(V, \hat{v})$$

then for

$$\gamma = \|\hat{v} - x\| + \tfrac{1}{2}\delta \ (< \|\hat{v} - x\|)$$

one would have

$$\gamma - L(h) - L(\hat{v} - x) = \|\hat{v} - x\| + \tfrac{1}{2}\delta - L(h) - \|\hat{v} - x\| \geqslant \tfrac{1}{2}\delta > 0$$

for all $L \in E_{\hat{v}-x}$, which contradicts (III.2.15). As a result we have the next theorem.

Theorem III.2.3 *If $\hat{v} \in V$ is a best approximation of x in V, i.e. if (III.2.10) holds, then (III.2.16) necessarily follows, where $T(V, \hat{v})$ denotes the tangent cone to V at $\hat{v}$ and $E_{\hat{v}-x}$ is defined by (III.2.14).*

If V is a convex subset of E, then (III.2.16) is also sufficient for $\hat{v} \in V$ to be a best approximation of x in V. In general, we have, in fact, our next theorem.

Theorem III.2.4 *If $\hat{v} \in V$ is given and is such that for $E_{\hat{v}-x}$ given by (III.2.14)*

$$\max_{L \in E_{\hat{v}-x}} L(v - \hat{v}) \geqslant 0 \qquad \text{for all } v \in V \tag{III.2.17}$$

then (III.2.10) follows.

Proof For every $v \in V$ we have

$$\| v - x \| - \| \hat{v} - x \| \geqslant L(v - x) - L(\hat{v} - x) = L(v - \hat{v})$$

if one chooses $L \in E_{\hat{v}-k}$ arbitrarily. By (III.2.17) there is an $\hat{L} \in E_{\hat{v}-x}$ with $\hat{L}(v - \hat{v}) \geqslant 0$, which implies $\| v - x \| \geqslant \| \hat{v} - x \|$. Since $v \in V$ is arbitrary, (III.2.10) is proven.

For convex sets V, the conditions (III.2.16) and (III.2.17) are equivalent (Proof = Exercise).

One can sharpen the conditions (III.2.16) and (III.2.17) still further by showing that they also hold, if one replaces $E_{\hat{v}-x}$ by suitable subsets (for this, see, for example Brosowski [69a]). In particular, this is true for the intersection of $E_{\hat{v}-x}$ with the set of extremal points of B^*, which is in the special case of the uniform approximation of functions leads to a well known criterion of Kolmogoroff. We shall not carry this out any further here, however (see Section III.2.2.3).

III.2.2.3 Application to uniform approximation of functions

We take as the basis for our discussion the same situation as in Section III.1.1.2. Thus we wish to minimize the continuous convex functional $f(y) = \| y - x \|$, $x \in E = C(M)$, on the non-empty subset V of $C(M)$, where again $x \in C(M)$ is given and fixed and $\| \ \|$ denotes the maximum norm (I.2.1). Again using Theorem III.2.2, for every best approximation $\hat{v} \in V$ of x in V we conclude the statement (III.2.11) which says that $(\Theta_E, \| \hat{v} - x \|^2)$ minimizes the functional $f(h, \gamma) = \gamma$ subject to the side conditions

$$(h, \gamma) \in T(V, \hat{v}) \times \mathrm{IR}$$

$$\gamma - [\hat{v}(t) - x(t) + h(t)]^2 \geqslant 0 \qquad \text{for all } t \in M \tag{III.2.18}$$

(Proof = Exercise). By setting

$$g(y, \gamma, t) = \gamma - [\hat{v}(t) - x(t) + y(t)]^2$$

$$(y, \gamma) \in C(M) \times \mathrm{IR} \qquad \text{for } t \in M$$

a concave mapping $g : C(M) \times \mathrm{IR} \to C(M)$ is defined. Further

$$\delta(\Theta_E, \| \hat{v} - x \|^2) = \min_{t \in M} \{ \| \hat{v} - x \|^2 - [\hat{v}(t) - x(t)]^2 \} = 0$$

and

$$I(\Theta_E, \| \hat{v} - x \|^2) = \hat{E}_{\hat{v}-x} = \{ t \in M : | \hat{v}(t) - x(t) | = \| \hat{v} - x \| \}. \qquad \text{(III.2.19)}$$

Again Theorem II.6.3 is applicable and yields the implication

$$\gamma - [\hat{v}(t) - x(t) + h(t)]^2 > 0 \qquad \text{for all } t \in \hat{E}_{\hat{v}-x} \Rightarrow \| \hat{v} - x \|^2 \leqslant \gamma \qquad \text{(II.2.20)}$$

for all $(h, \gamma) \in T(V, \hat{v}) \times \mathbb{R}$. From this it follows further that

$$\max_{t \in \hat{E}_{\hat{v}-x}} [\hat{v}(t) - x(t)] h(t) \geqslant 0 \qquad \text{for all } h \in T(V, \hat{v}). \qquad \text{(III.2.21)}$$

For if

$$\max_{t \in E_{\hat{v}-x}} [\hat{v}(t) - x(t)] h(t) < 0 \qquad \text{for some } h \in T(V, \hat{v})$$

then

$$\| h \| > 0 \text{ and } \gamma = \| \hat{v} - x \|^2 + \| h \|^2 \lambda^2 + \delta\lambda < \| \hat{v} - x \|^2 \qquad \text{for } 0 < \lambda < -\delta/\| h \|^2$$

and

$$\gamma - [\hat{v}(t) - x(t) + \lambda h(t)]^2 \geqslant -\delta\lambda > 0 \qquad \text{for all } t \in \hat{E}_{\hat{v}-x}$$

which contradicts (III.2.20).

Therefore, the condition (III.2.21) is necessary for $\hat{v} \in V$ to be a best approximation of x in V. If V is convex, then (III.2.21) is also sufficient for (III.2.10), for in general the following statement is true

Assertion *If $\hat{v} \in V$ is given and is such that*

$$\max_{t \in \hat{E}_{\hat{v}-x}} [\hat{v}(t) - x(t)] [v(t) - \hat{v}(t)] \geqslant 0 \qquad \textit{for all } v \in V \qquad \text{(III.2.22)}$$

with $\hat{E}_{\hat{v}-x}$ given by (III.2.19), then $\hat{v}$ is a best approximation of x in V.

For every $v \in V$ we have, in fact, for all $t \in \hat{E}_{\hat{v}-x}$

$$\begin{aligned} \| v - x \|^2 - \| \hat{v} - x \|^2 &\geqslant [v(t) - x(t)]^2 - [\hat{v}(t) - x(t)]^2 \\ &= [v(t) - \hat{v}(t) + \hat{v}(t) - x(t)]^2 - [\hat{v}(t) - x(t)]^2 \\ &= [v(t) - \hat{v}(t)]^2 + 2[\hat{v}(t) - x(t)] [v(t) - \hat{v}(t)] \end{aligned}$$

so that (III.2.22) yields $\| v - x \|^2 \geqslant \| \hat{v} - x \|^2$, which proves (III.2.10).

For convex sets V, the statements (III.2.21) and (III.2.22) are equivalent.

The condition (III.2.22) is a generalization of a criterion of Kolmogoroff [48] due to Meinardus and Schwedt [64]. Kolmogoroff showed that in the case of linear approximation it is necessary and sufficient for a best approximation.

The question concerning the necessity of (III.2.22) for best approximations led Meinardus to the concept of asymptotic convexity (see, for example, Meinardus

[64] or also Meinardus and Schwedt [64]). We shall not go into this any further, and consider as a representative for an asymptotically convex family of functions V in $C(M)$, the set V of quotients of continuous functions defined by (III.1.5), for finite dimensional subspaces U and W of $C(M)$, with a positive function for the denominator.

Assertion *For every function* $\hat{v} = \hat{u}/\hat{w} \in V$

$$\left\{ \frac{u\hat{w} - w\hat{u}}{\hat{w}^2} : u \in U, w \in W \right\} \subseteq T(V, \hat{v}). \tag{III.2.23}$$

Proof For $k \geqslant 1$ let

$$u^k = \hat{u} + (1/k)u \qquad w^k = \hat{w} + (1/k)w$$

where $u \in U$ and $w \in W$ are arbitrary but fixed. Then it follows that

$$u^k \in U \qquad \text{and} \qquad w^k \in W^+ \qquad \text{for all sufficiently large } k.$$

Furthermore, for all sufficiently large k,

$$\frac{u^k}{w^k} - \frac{\hat{u}}{\hat{w}} = \frac{(1/k)(\hat{w}u - \hat{u}w)}{[\hat{w} + (1/k)w]\hat{w}}$$

whence

$$\lim_{k \to \infty} \|(u^k/w^k) - (\hat{u}/\hat{w})\| = 0.$$

Finally, for sufficiently large k,

$$k\left(\frac{u^k}{w^k} - \frac{\hat{u}}{\hat{w}}\right) = \frac{\hat{w}u - \hat{u}w}{[\hat{w} + (1/k)w]\hat{w}}$$

whence

$$\lim_{k \to \infty} \left\| k\left(\frac{u^k}{w^k} - \frac{\hat{u}}{\hat{w}}\right) - \frac{u\hat{w} - w\hat{u}}{\hat{w}^2} \right\| = 0$$

which proves $(u\hat{w} - w\hat{u})/\hat{w}^2 \in T(V, \hat{v})$ and hence (III.2.23) is established. Taking into account

$$\frac{w}{\hat{w}}\left(\frac{u}{w} - \frac{\hat{u}}{\hat{w}}\right) = \frac{u\hat{w} - w\hat{u}}{\hat{w}^2} \qquad \text{for all } u \in U \text{ and } w \in W^+$$

we obtain from (III.2.21) the necessary condition for a best approximation $\hat{v} = \hat{u}/\hat{w} \in V$ of x in V:

$$\max_{t \in \hat{E}_{\hat{v}-x}} \left(\frac{\hat{u}(t)}{\hat{w}(t)} - x(t)\right)\left(\frac{u(t)}{w(t)} - \frac{\hat{u}(t)}{\hat{w}(t)}\right) \geqslant 0 \qquad \text{for all } u \in U \text{ and } w \in W^+ \tag{III.2.24}$$

which, as we have seen above, is also sufficient (see III.2.22).

III.2.2.4 Application to L_1 approximation

One can also apply Theorem III.2.2 directly to obtain necessary conditions for best approximations. That now will be demonstrated in the case of the approximation of a function in the sense of the L_1 norm. We consider for this purpose the vector space $C[a, b]$ of continuous real-valued functions defined on a finite closed interval $[a, b]$ equipped with the L_1 norm

$$\| g \|_1 = \int_a^b | g(t) | \, dt \qquad \text{for } g \in C[a, b]. \tag{III.2.25}$$

Approximating a function $x \in C[a, b]$ by functions v from a subset V of $C[a, b]$ as well as possible in the sense of the L_1 norm, means, therefore minimizing the functional $f(y) = \| y - x \|_1$ on V. Again f is continuous and convex so that Theorem III.2.2 can be applied.

If $\hat{v} \in V$ is a best approximation of x in V, then one has

$$\| \hat{v} - x \|_1 \leqslant \| \hat{v} - x + h \|_1 \qquad \text{for all } h \in T(V, \hat{v}) \tag{III.2.26}$$

where $T(V, \hat{v})$ is the tangent cone to V at $\hat{v}$. We set

$$N_0(\hat{v} - x) = \{ t \in [a, b] : \hat{v}(t) - x(t) = 0 \}. \tag{III.2.27}$$

Then we have the following theorem.

Theorem III.2.5 If $\hat{v} \in V$ is a best approximation of x in V in the sense of the L_1 norm (III.2.25), then

$$\int_a^b h(t) \operatorname{sgn}[\hat{v}(t) - x(t)] \, dt + \int_{N_0(\hat{v}-x)} | h(t) | \, dt \geqslant 0 \qquad \textit{for all } h \in T(V, \hat{v}) \tag{III.2.28}$$

where $\operatorname{sgn}(0) = 0$.

Proof We assume there to be an $\hat{h} \in T(V, \hat{v})$ with

$$\int_a^b \hat{h}(t) \operatorname{sgn}[\hat{v}(t) - x(t)] \, dt + \int_{N_0(\hat{v}-x)} | \hat{h}(t) | \, dt < 0 \tag{III.2.29}$$

For every positive number λ we define with $\| \hat{h} \| = \max_{a \leqslant t \leqslant b} | \hat{h}(t) | \, (> 0)$ the set

$$M_\lambda = \{ t \in [a, b] : | \hat{v}(t) - x(t) + \lambda \| h \| \}.$$

Then for all $t \in \complement M_\lambda = [a, b] \backslash M$

$$| \hat{v}(t) - x(t) + \lambda \hat{h}(t) | = [\hat{v}(t) - x(t) + \lambda \hat{h}(t)] \operatorname{sgn}[\hat{v}(t) - x(t)]$$

and consequently

$$\int_a^b | \hat{v}(t) - x(t) + \lambda \hat{h}(t) | \, dt = \int_{\complement M_\lambda} [\hat{v}(t) - x(t) + \lambda \hat{h}(t)] \operatorname{sgn}[\hat{v}(t) - x(t)] \, dt$$

$$+ \int_{M_\lambda} | \hat{v}(t) - x(t) + \lambda \hat{h}(t) | \, dt$$

$$= \int_{\complement M_\lambda} |\hat{v}(t) - x(t)|\,\mathrm{d}t + \lambda \int_{\complement M_\lambda} \hat{h}(t)\operatorname{sgn}[\hat{v}(t) - x(t)]\,\mathrm{d}t$$

$$+ \int_{M_\lambda} |\hat{v}(t) - x(t)|\,\mathrm{d}t - \int_{M_\lambda} |\hat{v}(t) - x(t)|\,\mathrm{d}t$$

$$+ \int_{M_\lambda} \lambda\hat{h}(t)\operatorname{sgn}[\hat{v}(t) - x(t)]\,\mathrm{d}t$$

$$- \int_{M_\lambda} \lambda\hat{h}(t)\operatorname{sgn}[\hat{v}(t) - x(t)]\,\mathrm{d}t + \int_{M_\lambda} |\hat{v}(t) - x(t) + \lambda\hat{h}(t)|\,\mathrm{d}t$$

$$= \int_a^b |\hat{v}(t) - x(t)|\,\mathrm{d}t + \lambda \int_a^b \hat{h}(t)\operatorname{sgn}[\hat{v}(t) - x(t)]\,\mathrm{d}t$$

$$+ \int_{M_\lambda} [\hat{v}(t) - x(t) + \lambda\hat{h}(t)]\{\operatorname{sgn}[\hat{v}(t) - x(t)$$

$$+ \lambda\hat{h}(t)] - \operatorname{sgn}[\hat{v}(t) - x(t)]\}\,\mathrm{d}t.$$

Because

$$|\hat{v}(t) - x(t) + \lambda\hat{h}(t)| \leqslant 2\lambda \,\|h\| \qquad \text{for all } t \in M_\lambda$$

it follows that

$$\int_a^b |\hat{v}(t) - x(t) + \lambda\hat{h}(t)|\,\mathrm{d}t - \int_a^b |\hat{v}(t) - x(t)|\,\mathrm{d}t$$

$$\leqslant \lambda \left(\int_{N_0(\hat{v}-x)} |\hat{h}(t)|\,\mathrm{d}t + \int_a^b \hat{h}(t)\operatorname{sgn}[\hat{v}(t) - x(t)]\,\mathrm{d}t + 4\,\|\hat{h}\|\,\mu[M_\lambda \backslash N_0(\hat{v} - x)] \right)$$

where μ denotes the Lebesgue measure. Since $\lim_{\lambda\to 0} \mu[M_\lambda \backslash N(\hat{v} - x)] = 0$, assumption (III.2.29) yields the existence of a number $\lambda > 0$ satisfying

$$\|\hat{v} - x + \hat{\lambda}\hat{h}\|_1 - \|\hat{v} - x\|_1 < 0.$$

Since $\hat{\lambda}\hat{h} \in T(V, \hat{v})$, this contradicts the necessary condition (III.2.26) for the optimality of $\hat{v}$. Therefore the assumption (III.2.29) is false and the assertion (III.2.28) is proven.

Problem III.2.2 Let $E = \mathbb{R}^m$ be equipped with the L_1 norm:

$$\|g\|_1 = \sum_{i=1}^{m} |g_i|, \qquad g \in E.$$

Show that for a given non-empty subset V of E and $x \in E$, if a vector $\hat{v} \in V$ is a best approximation of x in V, then

$$\sum_{i=1}^{m} h_i \operatorname{sgn}(\hat{v}_i - x_i) + \sum_{\hat{v}_i - x_i = 0} |h_i| \geqslant 0 \qquad \text{for all } h \in T(V, \hat{v})$$

where $\operatorname{sgn}(0) = 0$.

For a linear subspace V of $C[a, b]$, under the additional assumption that $N_0(\hat{v} - x)$ (III.2.27) is a finite set, Cheney [66] proves the necessity and sufficiency of (III.2.28) for $\hat{v}$ to be best approximation of x in V. The condition (III.2.28) in this special case goes over into

$$\int_a^b h(t)\operatorname{sgn}[\hat{v}(t) - x(t)]\,\mathrm{d}t = 0 \qquad \text{for all } h \in V.$$

The idea for the proof of Theorem III.2.5 was taken from the paper of Carroll and McLaughlin [73].

III.2.3 Bibliographical remarks

The idea of the tangent cone introduced in Section III.2.1 was taken over from Hestenes [66], who used it systematically in the treatment of optimization and control problems. Abadie [67], Varaiya [67], and Guignard [69] also use this concept in order to obtain necessary conditions for minimal points in the case of finite and infinite nonlinear optimization problems. Dubovitskii and Miljutin [63, 65] were the first to use a definition of the tangent cone which has been adapted by numerous authors (see, for example, Laurent [67, 72], and Lobry [67]). In the recent article by Bazaraa, Goode, and Nashed [74] on tangent cones with applications to optimization, three equivalent definitions of the tangent cone were used. Brosowski [69a] refers to the definition of Dubovitskii and Miljutin and gives the implication (III.2.10) ⇒ (III.2.11). As general sufficient conditions for the fact that an element $\hat{v}$ from a non-empty subset V of a normed vector space E over $\mathbb{R}$ is a best approximation of an element $x \in E$ in V, the following two equivalent statements were given by Brosowski [69a]

$$\max_{L \in E_{\hat{v}-x} \cap \Gamma} L(\hat{v} - v) \leqslant 0 \qquad \text{for all } v \in V \tag{III.2.17a}$$

and

$$\min_{L \in E_{\hat{v}-x} \cap Ep(\Gamma)} L(\hat{v} - v) \leqslant 0 \qquad \text{for all } v \in V. \tag{III.2.17b}$$

Here $E_{\hat{v}-x}$ is defined by (III.2.14), Γ is a (suitably defined) fundamental system of continuous linear forms on E, and $Ep(A)$ denotes the set of extremal points of a set A in a vector space (see, for example, Köthe [66]). Important examples for Γ are the unit ball B^* of the dual space E^* of E and the weakly* closed hull $\overline{Ep(B^*)}$ of the extremal points of B^*. The concept of the fundamental system goes back to Nikolski [65], who [61, 65] also proves that (III.2.17a) is generally sufficient and, in the case of a convex set, also necessary for a best approximation. In particular, if $\Gamma = B^*$ then, since $E_{\hat{v}-x} \cap Ep(\Gamma) = Ep(E_{\hat{v}-x})$, the condition (III.2.17b) goes over into

$$\min_{L \in Ep(E_{\hat{v}-x})} L(\hat{v} - v) \leqslant 0 \qquad \text{for all } v \in V. \tag{III.2.17$'$}$$

Singer [62] showed, for certain normed spaces E, and Choquet [63] for arbitrary normed spaces E, that in the case of a linear subspace V of E, (III.2.17$'$) is characteristic for a best approximation. These results were extended by Garkavi [64], Havinson [67] and Deutsch and Maserick [67] to convex sets V.

The question of the necessity of the so-called global Kolmogoroff criterion (III.2.17*a*) or (III.2.17*b*) for best approximations was answered positively by Brosowski [69a] even for a more general class of subsets V of E, and in fact for so-called α-suns (which had been introduced by Klee [65]). Conversley, these may be characterized precisely by the necessity of the global Kolmogoroff criterion for best approximations for every $x \in E$. Brosowski calls such subsets V of E Kolmogoroff sets of the first kind. As a sufficient condition for the presence of such a set, Brosowski [69a] gives the validity of the implication (III.2.11) ⇒ (III.2.10), for all $x \in E$. One can now easily convince oneself that the necessary condition (III.2.16) derived in Section III.2.2.2 is also sufficient for (III.2.11) and with the same arguments as in the proof of the equivalence of (III.2.17*a*) and (III.2.17*b*) one can also see that (III.2.16) is in turn equivalent to the so-called local Kolmogoroff criterion

$$\min_{L \in Ep(E_{\hat{v}-x})} -L(h) \leqslant 0 \qquad \text{for all } h \in T(V, \hat{v}). \tag{III.2.16$'$}$$

Consequently this is equivalent to (III.2.11) and necessary for best approximations. If it is also sufficient for every $x \in E$ that $\hat{v} \in V$ is a best approximation of x in V, then Brosowski calls V a Kolmogoroff set of the second kind. On the basis of the above considerations, every Kolmogoroff set of the second kind is also one of the first kind.

Brosowski and Wegmann [69] give conditions characterizing Kolmogoroff sets of the first and second kind. For the characterization of Kolmogoroff sets of the first kind, conditions were also given in earlier papers of Brosowski [69a,b] and by Breckner and Kolumban [68a,b], which were however weaker in that they did not include corresponding conditions in special cases.

III.3 NONLINEAR OPTIMIZATION WITH AN INFINITE NUMBER OF SIDE CONDITIONS

III.3.1 Necessary conditions for minimal points

III.3.1.1 The general case

We start with the same situation as in Section II.6.1, and consider a normed vector space E (there E was in fact only a vector space), a non-empty subset X of E, a functional $f : X \to \mathbb{R}$ and a mapping $g : X \to C(T)$, where $C(T)$ is the vector space of continuous real-valued functions on T and T is a compact Hausdorff space.

Thus we consider the problem of minimizing the functional $f = f(x)$ subject to the side conditions

$$x \in X, g(x, t) \geqslant 0 \qquad \text{for all } t \in T. \tag{III.3.1}$$

If one denotes the natural positive cone of $C(T)$ by Y (see Section IV.1.1) and sets

$$S = \{x \in X : g(x) \in Y\} \tag{III.3.2}$$

then we are faced with the problem of minimizing f on S. Therefore we seek an $\hat{x} \in S$ satisfying

$$f(\hat{x}) \leqslant f(x) \qquad \text{for all } x \in S. \tag{III.3.3}$$

We call every such $\hat{x} \in S$ a minimal point of f on S. We assume the existence of such minimal points and first look for necessary conditions for these. Here assumptions on X, f, g are naturally necessary. In Section II.6.1, we occupied ourselves in detail with the case that X is a convex set (Section IV.3), f is a convex functional (Section II.2.1), and g is a concave mapping (Section II.2.2).

If f is defined on all of E, is continuous and convex with respect to a minimal point $\hat{x} \in S$, and X and g are arbitrary, then Theorem III.2.2 furnishes a general necessary condition for a minimal point $\hat{x} \in S$ in the form

$$f(\hat{x}) \leqslant f(\hat{x} + h) \qquad \text{for all } h \in T(S, \hat{x}) \tag{III.3.4}$$

where $T(S, \hat{x})$ denotes the tangent cone to S at $\hat{x}$ (see Section III.2.1). The statement (III.3.4) is very coarse and does not take into account the finer structure of the set S. In this connection, the relation of the two tangent cones $T(X, \hat{x})$ and $T(S, \hat{x})$ to each other is interesting. On the basis of (III.2.5) we have in the first place $T(S, \hat{x}) \subseteq T(X, \hat{x})$, and in general the containment is strict. Under certain circumstances, however, the two sets coincide. In order to see that, we associate with each $x \in S$ the set

$$I(x) = \{t \in T : g(x, t) = 0\}. \tag{III.3.5}$$

Then we have the following theorem.

Theorem III.3.1 *If for some $x \in S$ the set $I(x)$ defined by (III.3.5) is empty, i.e. if*

$$g(x, t) > 0 \qquad \text{for all } t \in T \tag{III.3.6}$$

and if the mapping $g : X \to C(T)$ is continuous with respect to the maximum norm $\| \ \|$ of $C(T)$, then

$$T(S, x) = T(X, x).$$

Proof Let $h \in T(X, x)$ be given. Then there is a sequence $\{x_k\}$ of elements $x_k \in X$ and a sequence $\{\lambda_k\}$ of positive numbers λ_k with

$$x = \lim_{k \to \infty} x_k \qquad \text{and} \qquad h = \lim_{k \to \infty} \lambda_k (x_k - x).$$

From (III.3.6) it follows because of the continuity of $g : X \to C(T)$ that, for all sufficiently large k, one has

$$g(x_k, t) > 0 \qquad \text{for all } t \in T$$

i.e.

$$x_k \in S$$

which implies $h \in T(S, x)$.

A simple consequence of Theorem III.3.1 is the next theorem.

Theorem III.3.2 *Let $f: E \to \mathbb{R}$ and $g: X \to C(t)$ be continuous. Also let $\hat{x} \in S$ be a minimal point of f on S such that the set $I(\hat{x})$ defined by (III.3.5) is empty. If f is convex on E and X is star-shaped with respect to $\hat{x}$, then $\hat{x}$ is also a minimal point of f on X, i.e.*

$$f(\hat{x}) \leqslant f(x) \qquad \text{for all } x \in X. \tag{III.3.7}$$

Proof By Theorem III.2.2 and Theorem III.3.1, one has

$$f(\hat{x}) \leqslant f(\hat{x} + h) \qquad \text{for all } h \in T(S, \hat{x}) = T(X, \hat{x}). \tag{III.3.8}$$

Since X is star-shaped with respect to $\hat{x}$, $X - \hat{x} \subseteq T(X, \hat{x})$ (see the proof of (III.2.3)) and hence (III.3.7) is a direct consequence of (III.3.8).

By Theorem II.6.7, the same statement holds if one replaces the continuity of g by the concavity with respect to $\hat{x}$. Then f has only to be defined on X and does not necessarily have to be continuous.

For the case that the set $I(\hat{x})$ defined by (III.3.5) of a minimal point of f on S is not empty, we also derived in Section II.6.1 necessary conditions for x for the case that f, or g, is convex, or concave, respectively, with respect to $\hat{x}$. If one omits this assumption, then in order to be able to draw conclusions, in general, one has to make suitable differentiability conditions.

III.3.1.2 The differentiable case

In the sequel, suppose X is a subset of an open set $\hat{X}$. Further, let f be defined on $\hat{X}$ and be Fréchet differentiable, i.e. for every $x \in \hat{X}$ there exists a continuous linear functional (see Section IV.1.2) $f_x' : E \to \mathbb{R}$, the so-called Fréchet derivative of f in x, satisfying

$$| f(x + h) - f(x) - f_x'(h) | \leqslant \| h \| \epsilon(\| h \|)$$

for all $h \in E$ such that $x + h \in \hat{X}$, where

$$\lim_{\| h \| \to 0} \epsilon(\| h \|) = 0.$$

Then we have at first the theorem below.

Theorem III.3.3 *If f is defined on $\hat{X}$ and is Fréchet differentiable and if $\hat{x} \in S$ is a minimal point f on S, then it follows that*

$$f_x'(h) \geqslant 0 \qquad \text{for all } h \in T(S, \hat{x}). \tag{III.3.9}$$

Proof We make the assumption

$$f_x'(h) < 0 \qquad \text{for some } h \in T(S, \hat{x}).$$

Let $\{x_k\}$ be a sequence from S and $\{\lambda_k\}$ a sequence of positive numbers with

$$\hat{x} = \lim_{k \to \infty} x_k \qquad \text{and} \qquad h = \lim_{k \to \infty} \lambda_k (x_k - \hat{x}).$$

Then it follows that

$$\lambda_k [f(x_k) - f(\hat{x})] \leqslant f_x' [\lambda_k (x_k - \hat{x})] + \| \lambda_k (x_k - \hat{x}) \| \epsilon(\| x_k - \hat{x} \|)$$

with

$$\lim_{k\to\infty} \epsilon(\| x_k - \hat{x} \|) = 0.$$

However, this implies $f(x_k) < f(\hat{x})$ for sufficiently large k, which contradicts the optimality of $\hat{x}$.

If, in addition to the assumption of Theorem III.3.3, the mapping $g : X \to C(T)$ is continuous and the set $I(\hat{x})$ defined by (III.3.5) is empty, then the optimality of $\hat{x} \in S$ using Theorem III.3.1 necessarily yields

$$f_x'(h) \geqslant 0 \qquad \text{for all } h \in T(X, \hat{x}). \tag{III.3.10}$$

In order now to draw conclusions for the case when $I(\hat{x})$ is not empty, we assume for the following that g is also defined on $\hat{X}$ and Fréchet differentiable, i.e. that for every $x \in \hat{X}$ there exists a continuous linear mapping $g_x' : E \to C(T)$ equipped with the maximum norm $\| \;\; \|_\infty$ from (I.2.1) (see Section IV.1.3), the Fréchet derivative of g in x, with

$$\| g(x+h) - g(x) - g_x'(h) \|_\infty \leqslant \| h \| \alpha(\| h \|) \tag{III.3.11}$$

for all $h \in E$ with $x + h \in \hat{X}$, where

$$\lim_{\| h \| \to 0} \alpha(\| h \|) = 0. \tag{III.3.12}$$

Lemma III.3.4 *The mapping $g : \hat{X} \to C(T)$ is Fréchet differentiable if and only if to each $x \in \hat{X}$ and $t \in T$ there exists a linear form $g_{t,x}' : E \to \mathbb{R}$ such that*

$$\max_{t \in T} | g(x+h, t) - g(x, t) - g_{t,x}'(h) | \leqslant \| h \| \alpha(\| h \|) \tag{III.3.13}$$

for all $h \in E$ with $x + h \in \hat{X}$, where again (III.3.12) is satisfied and for every fixed $x \in \hat{X}$ the mapping $(t, h) \to g_{t,x}'(h)$ of $T \times E \to \mathbb{R}$ is continuous.

Proof 1. Suppose $g : \hat{X} \to C(T)$ is Fréchet differentiable. If one then defines for every $x \in \hat{X}$ and $t \in T$

$$g_{t,x}'(h) = g_x'(h, t) \qquad \text{for } h \in E \tag{III.3.14}$$

where $g_x' : E \to C(T)$ is the Fréchet derivative of g in x, then (III.3.13) follows from (III.3.11). Further from

$$| g_{t,x}'(h) - g_{\hat{t},x}'(\hat{h}) | \leqslant | g_{t,x}'(h) - g_{\hat{t},x}'(h) | + | g_{t,x}'(h) - g_{\hat{t},x}'(\hat{h}) |$$
$$\leqslant | g_x'(h, t) - g_x'(h, \hat{t}) | + \| g_x' \| \, \| h - \hat{h} \|$$

with $(t, h), (\hat{t}, \hat{h}) \in T \times E, x \times \hat{X}$, we conclude the continuity of the mapping $(t, h) \to g_{t,x}'(h)$.

2. Conversely, suppose for every $x \in \hat{X}$ and $t \in T$ there is a linear functional $g_{t,x}' : E \to \mathbb{R}$ with the properties given in the lemma. We then define a mapping $g_x' : E \to C(T)$ by

$$g_x'(h, t) = g_{t,x}'(h) \qquad \text{for all } h \in E, t \in T.$$

Certainly g_x' is linear and satisfies (III.3.11). We must show the continuity, i.e.

$$\| g'_x(h, \cdot) \|_\infty \leqslant \epsilon \qquad \text{for all } h \in E \text{ with } \| h \| \leqslant \delta(\epsilon). \tag{III.3.15}$$

By assumption there is for every $t \in T$ a $\delta_t(\epsilon)$ with

$$| g'_x(h, t) | \leqslant \epsilon/2 \qquad \text{for all } h \in E \text{ with } \| h \| \leqslant \delta_t(\epsilon)$$

whence

$$| g'_x(h, \tau) | \leqslant \epsilon \qquad \text{for all } h \in E \text{ with } \| h \| \leqslant \delta_t(\epsilon)$$

and all τ from a suitable neighbourhood V_t of $t \in T$. Since T is a compact Hausdorff space, T may be covered by finitely many V_{t_i}, $i = 1, \ldots, r$. If we choose

$$\delta(\epsilon) = \min_{i=1,\ldots,r} \delta_{t_i}(\epsilon),$$

then (III.3.15) is satisfied.

For every $x \in S$ we now define the set

$$\mathfrak{L}(S, x) = \begin{cases} T(X, x) & \text{if } I(x) \text{ (by (III.3.5)) is empty} \\ \bigcap\limits_{t \in I(x)} \{h \in T(X, x) : g'_{t,x}(h) > 0\} & \text{otherwise} \end{cases} \tag{III.3.16}$$

where again g is assumed to be Fréchet differentiable on $\hat{X}$ and $g'_{t,\,x}(h)$ is defined for every $t \in T$ and $h \in E$ by (III.3.14). Generalizing the statement (III.3.10) we have the following.

Theorem III.3.5 *If f and g are Fréchet differentiable on $\hat{X}$, then it necessarily follows for every minimal point $\hat{x} \in S$ of f on S, that*

$$f'_{\hat{x}}(h) \geqslant 0 \qquad \text{for all } h \in \mathfrak{L}(S, \hat{x}) \tag{III.3.17}$$

with $l(S, x)$ defined by (III.3.16).

Proof If $I(\hat{x})$ is empty, then because of the continuity of g on X, which follows from the Fréchet differentiability of g on $\hat{X}$ (Proof = Exercise), the assertion (III.3.17) is identical with the statement (III.3.10) already proven.

If $I(\hat{x})$ is not empty, then for the case that $\mathfrak{L}(S, \hat{x})$ is empty, there is nothing to show. On the other hand (III.3.17) follows from Theorem III.3.3 together with the next theorem.

Theorem III.3.6 *For every $x \in S$ such that the set $\mathfrak{L}(S, x)$ defined by (III.3.16) is not empty, we have*

$$\mathfrak{L}(S, x) \subseteq T(S, x) \tag{III.3.18}$$

and there is a sequence $\{x_k\}$ of points $x_k \in S$ with $x = \lim\limits_{k\to\infty} x_k$ such that the associated sets $I(x_k)$ (defined by (III.3.5)) are empty, i.e.

$$g(x_k, t) > 0 \qquad \textit{for all } t \in T \textit{ and for all } k.$$

Proof If $I(x)$ is empty, then by Theorem III.3.1 there is nothing to show. Suppose therefore that $I(x)$ is not empty. Let $h \in \mathfrak{L}(S, x)$ be given. Then we must have $h \neq \Theta_E$ = null element of E. Further, there is a sequence $\{x_k\}$ in X and a sequence $\{\lambda_k\}$ of positive numbers such that

$$x = \lim_{k\to\infty} x_k \qquad h = \lim_{k\to\infty} \lambda_k (x_k - x) \qquad g'_{t,x}(h) > 0 \qquad \text{for all } t \in I(x).$$

Setting $h_k = \lambda_k (x_k - x)$ yields

$$x_k = x + \frac{1}{\lambda_k} h_k \qquad \lim_{k\to\infty} \frac{1}{\lambda_k} h_k = \Theta_E \qquad \text{and} \qquad \lim_{k\to\infty} \lambda_k = \infty.$$

Since $I(x)$ is compact, there is a δ with

$$g'_{t,x}(h) \geqslant \delta > 0 \qquad \text{for all } t \in I(x).$$

Thus we set

$$I(x, h) = \left\{ t \in T : g'_{t,x}(h) > \frac{\delta}{2} \right\}.$$

For every sufficiently large k the Fréchet differentiability of g on $\hat{X}$ yields

$$g(x_k, t) \geqslant g(x, t) + \frac{1}{\lambda_k} \left[g'_{t,x}(h_k) - \| h_k \| \, \alpha\left(\frac{1}{\lambda_k} \| h_k \| \right) \right] \tag{III.3.19}$$

for all $t \in T$, where

$$\lim_{k\to\infty} \alpha\left(\frac{1}{\lambda_k} \| h_k \| \right) = 0.$$

Because of the continuity of $g'_x : E \to C(T)$ there is a k_0 with

$$g'_{t,x}(h_k) > \delta/4 \qquad \text{for all } t \in I(x, h) \text{ and } k \geqslant k_0$$

and we can further assume that

$$\| h_k \| \, \alpha\left(\frac{1}{\lambda_k} \| h_k \| \right) \leqslant \delta/4 \qquad \text{for all } k \geqslant k_0.$$

Thus it follows from (III.3.19) for all $k \geqslant k_0$ that

$$g(x_k, t) > 0 \qquad \text{for all } t \in I(x, h).$$

If $I(x, h) = T$, then $x_k \in S$ and $I(x_k) = \phi$ for all $k \geqslant k_0$, which implies $h \in T(S, x)$ and everything is proven. If $I(x, h) \neq T$, then $T - I(x, h)$ is compact and there are numbers $\beta, \gamma > 0$ with

$$g(x, t) \geqslant \beta \qquad \text{for all } t \in T - I(x, h)$$

and

$$g'_{t,x}(h_k) - \| h_k \| \, \alpha\left(\frac{1}{\lambda_k} \| h_k \| \right) \geqslant -\gamma \qquad \text{for all } t \in T - I(x, h) \text{ and all } k.$$

Therefore it follows from (III.3.19) for all sufficiently large $k \geqslant k_0$ that

$$g(x_k, t) \geqslant \beta - \frac{1}{\lambda_k} \gamma > 0 \qquad \text{for all } t \in T - I(x, h).$$

In conclusion, therefore, $x_k \in S$ and $I(x_k) = \phi$, which again implies $h \in T(S, x)$.

If one defines a functional $g_0 : \hat{X} \to \mathbb{R}$ by

$$g_0(x) = -f(x) \qquad \text{for } x \in \hat{X} \tag{III.3.20}$$

then g_0 together with f is also Fréchet differentiable and the Fréchet derivative is

$$g'_{0,x}(h) = -f'_x(h) \qquad \text{for } h \in E. \tag{III.3.21}$$

If one defines for every $x \in S$

$$I_0(x) = I(x) \cup \{0\} \tag{III.3.22}$$

with $I(x)$ given by (III.3.5), then (III.3.17) becomes

$$\min_{t \in I_0(\hat{x})} g'_{t,\hat{x}}(h) \leqslant 0 \qquad \text{for all } h \in T(X, \hat{x}). \tag{III.3.23}$$

Obviously (III.3.17) is a direct consequence of (III.3.23). Conversely if (III.3.17) is satisfied and $h \in T(X, \hat{x})$ is given, then (III.3.23) follows directly if $h \in \mathfrak{L}(S, \hat{x})$. However, if $h \in \mathfrak{L}(S, \hat{x})$, then $g'_{t,\hat{x}}(h) \leqslant 0$ for at least one $t \in I(\hat{x})$, which also inplies (III.3.23). Subject to the requirements of Theorem III.3.5, (III.3.23) is, therefore, a necessary condition that $\hat{x} \in S$ is a minimal point of f on S.

III.3.2 Sufficient conditions for minimal points

For each $t \in T$ we define a functional $g_t : \hat{X} \to \mathbb{R}$ by

$$g_t(x) = g(x, t) \qquad \text{for } x \in \hat{X} \tag{III.3.24}$$

By Lemma III.3.4 each g_t is Fréchet differentiable on $\hat{X}$ and the mapping $(t, h) \to g'_{t,x}(h)$ is continuous if g is Fréchet differentiable on $\hat{X}$. The Fréchet derivatives $g'_{t,x}$ are given by (III.3.14).

By analogy with Theorems II.6.I and II.6.4, we have the next theorem.

Theorem III.3.7 *Let $f : X \to \mathbb{R}$ be continuous. Further, let the set*

$$S_0 = \{x \in X : g_t(x) = g(x, t) > 0 \qquad \text{for all } t \in T\} \tag{III.3.25}$$

be non-empty, and suppose

$$S \subseteq \bar{S}_0 = \text{closure of } S_0 \tag{III.3.26}$$

with S given by (III.3.2). If $\hat{x} \in S$ is given and satisfies

$$\min_{t \in I_0(\hat{x})} \{g_t(x) - g_t(\hat{x})\} \leqslant 0 \qquad \text{for all } x \in X \tag{III.3.27}$$

where $I_0(\hat{x})$ is defined by (III.3.22) and g_0 by (III.3.20), then $\hat{x}$ is a minimal point of f on S.

Proof Let $x \in S$ be given. If $I(\hat{x})$ is empty, then $f(x) \geqslant f(\hat{x})$ follows directly from (III.3.27). If $I(\hat{x})$ is not empty, then we consider a sequence $\{x_k\}$ of points

$x_k \in X$ with $x = \lim_{k\to\infty} x_k$ and

$$g(x_k, t) > 0 \qquad \text{for all } t \in I(\hat{x}) \text{ and all } k$$

which exists by (III.3.26). However it follows from (III.3.27) that

$$f(x_k) \geqslant f(\hat{x}) \qquad \text{for all } k$$

whence $f(x) \geqslant f(\hat{x})$, because of the continuity of f.

With Theorem III.3.6, we have immediately the following lemma.

Lemma III.3.8 *If g is Fréchet differentiable on $\hat{X}$ (open, $\supseteq X$), S defined by (III.3.2) is not empty and if for every $x \in S$ the set $\mathfrak{L}(S, x)$ defined by (III.3.16) is not empty, then the set S_0 defined by (III.3.25) is not empty and the assertion (III.3.26) holds.*

In the sequel we call a point $x \in S$ regular if $\mathfrak{L}(S, x)$ (see (III.3.16)) is not empty. If f and g are Fréchet differentiable on $\hat{X}$, then by definition the mapping

$$\hat{g}(x) = (g_0(x), g(x)) \qquad \text{for } x \in \hat{X} \tag{III.3.28}$$

(with g_0 given by (III.3.20)) of $\hat{X}$ into $\mathbb{R} \times C(T)$ is called almost pseudo-concave on X if to every closed subset Q of T and to every pair $x, \hat{x} \in X$ with

$$\min_{t \in Q \cup \{0\}} \{g_t(x) - g_t(\hat{x})\} > 0$$

(g_t given by (III.3.24) for $t \in T$) there exists an $h \in T(X, \hat{x})$ such that

$$\min_{t \in Q \cup \{0\}} g'_{t,\hat{x}}(h) > 0.$$

From Theorems III.3.5, III.3.7, and Lemma III.3.8, we have in summary the following theorem.

Theorem III.3.9 *Let f and g be Fréchet differentiable on X. Further let all points $x \in S$ be regular, let the mapping $\hat{g}$ given by (III.3.28) of $\hat{X}$ into $\mathbb{R} \times C(T)$ be almost pseudo-concave on X. Then the conditions (III.3.23) and (III.3.27) are both necessary and sufficient for $\hat{x} \in S$ to be a minimal point of f on S.*

The proof is an immediate consequence of the implication (III.3.23) $\Rightarrow$ (III.3.27) which follows directly from the almost pseudo-concavity of $\hat{g}$ (III.3.28).

The almost pseudo-concavity of g is implied by the Fréchet differentiability of f and g and the following conditions: suppose the set X is convex, f is pseudo-convex on X and every g_t, $t \in T$, defined by (III.3.24) is pseudo-concave on X, i.e. for every pair of points $x, \hat{x} \in X$ and for every $t \in T$, the implications

$$f'_{\hat{x}}(x - \hat{x}) \geqslant 0 \Rightarrow f(x) - f(\hat{x}) \geqslant 0$$

$$g'_{t,\hat{x}}(x - \hat{x}) \leqslant 0 \Rightarrow g_t(x) - g_t(\hat{x}) \leqslant 0.$$

hold (see Mangasarian [69]).

Problem III.3.1 Prove the sufficiency of these assumptions for the almost pseudo-convexity of $\mathfrak{g}$ (III.3.28).

Lemma III.3.10 Suppose g is Fréchet differentiable on $\hat{X}$ and pseudo-concave on X. If the set S_0 defined by (III.3.25) is non-empty, then all the points $x \in S$ are regular.

Proof Let $x \in S$ be given. If $I(x)$ is empty, then there is nothing to show. If $I(x)$ is not empty, then we choose an $x_0 \in S_0$. Then it follows that $g_t(x_0) - g_t(x) > 0$ for all $t \in I(x)$ and hence $g'_{t,x}(x_0 - x) > 0$ for all $t \in I(x)$. Because of (III.2.3), $x_0 - x \in T(X, x)$ and consequently $\mathfrak{L}(S, x)$ defined by (III.3.16) is not empty.

III.3.3 Application to nonlinear uniform approximation

We shall take here as the basis for our considerations an important special case of the situation described in Section III.1.1.2. For this purpose, we consider a non-empty subset Y of $\mathbb{R}^n$ and an open set $\hat{Y}$ containing Y in $\mathbb{R}^n$. Further, let a mapping F of $\hat{Y}$ into the vector space $C(T)$ of continuous real-valued functions defined on the compact Hausdorff space T be given. Assume that for every $t \in T$ and $y \in \hat{Y}$ the partial derivatives $\partial F(y, t)/\partial x_j$, for $j = 1, \ldots, n$, exist and the mapping

$$(y, t) \to \text{grad } F(y, t) = \left\{\frac{\partial F}{\partial y_1}(y, t), \ldots, \frac{\partial F}{\partial y_n}(y, t)\right\}$$

of $\hat{Y} \times T$ into $\mathbb{R}^n$ is continuous.

If one again takes in $C(T)$ the maximum norm $\| \ \|_\infty$ (see (I.2.1)), then by Theorem II.4.6 of Collatz and Krabs [73] the mapping $F: \hat{Y} \to C(T)$ is Fréchet differentiable and the Fréchet derivative is given by

$$F'_y(k, t) = \langle \text{grad } F(y, t), k \rangle = \sum_{j=1}^{n} \frac{\partial F}{\partial y_j}(y, t) k_j \qquad \text{(III.3.29)}$$

for all $y \in \hat{Y}$ and all $k \in \mathbb{R}^n$.

Now let a function $b \in C(T)$ be given. Then we consider the approximation problem of finding a $\hat{y} \in Y$ such that

$$\|F(\hat{y}) - b\|_\infty \leqslant \|F(y) - b\|_\infty \qquad \text{for all } y \in Y \qquad \text{(III.3.30)}$$

i.e. the function b is to be approximated as well as possible by a function $F(\hat{y})$ from the family $V = \{F(y) : y \in Y\} \subseteq C(T)$.

If one defines $\hat{X} = \hat{Y} \times \mathbb{R}$, a mapping $g : \hat{X} \to C(T)$ by

$$g(y, \gamma, t) = \gamma - [F(y, t) - b(t)]^2 \qquad \text{for } t \in T \ (y, \gamma) \in \hat{X} \qquad \text{(III.3.31)}$$

and a functional $f: \hat{X} \to \mathbb{R}$ by

$$f(y, \gamma) = \gamma \qquad \text{for } (y, \gamma) \in \hat{X} \qquad \text{(III.3.32)}$$

then the problem (III.3.30) is equivalent to the optimization problem (as in Section III.3.1) of minimizing the functional $f = f(y, \gamma)$ subject to the side conditions

$$(y, \gamma) \in X = Y \times \mathbb{R} \qquad g(y, \gamma, t) \geq 0 \qquad \text{for all } t \in T.$$

The mapping g defined by (III.3.31) and the functional f defined by (III.3.32) are Fréchet differentiable on $\hat{X}$ and the Fréchet differentiable on $\hat{X}$ and the Fréchet derivatives are (Proof = Exercise)

$$g'_{(y,\gamma)}(k, \lambda, t) = \lambda - 2\,[F(y,t) - b(t)]\,F'_y(k, t) \tag{III.3.33}$$

$t \in T$, $k \in \mathbb{R}^n$, $\lambda \in \mathbb{R}$, with $F'_y(k, t)$ by (III.3.29) and

$$f'_{(y,\gamma)}(k, \lambda) = \lambda \qquad \text{for } (k, \lambda) \in \mathbb{R}^{n+1}. \tag{III.3.34}$$

Problem III.3.2 Show that for every $x = (y, \gamma) \in X = Y \times \mathbb{R}$ the tangent cone $T(X, x)$ is given by (see problem III.2.1*b*)

$$T(X, x) = T(Y, y) \times \mathbb{R}. \tag{III.3.35}$$

Lemma III.3.11 *Every point x of the set S defined by (III.3.2) with $X = Y \times \mathbb{R}$ and g given by (III.3.31) is regular.*

Proof Let $x = (y, \gamma) \in S$ be given. Then there are two possible cases.

1. $\gamma > \| F(y) - b \|_\infty^2$. Then the set

$$I(x) = I(y, \gamma) = \{t \in T : \gamma = [F(y,t) - b(t)]^2\} \tag{III.3.36}$$

is empty and there is nothing to prove.

2. $\gamma = \| F(y) - b \|_\infty^2$. Then $I(y, \gamma)$ is not empty and to every $k \in T(Y, y)$ there exists a sufficiently large $\lambda \in \mathbb{R}$, i.e. $(k, \lambda) \in T(X, x)$ (on the basis of (III.3.35)) with

$$g'_{(y,\lambda)}(k, \lambda, t) = \lambda - 2\,[F(y, t) - b(t)]\,F'_y(k, t) > 0 \qquad \text{for all } t \in I(y, \gamma)$$

(see (III.3.29)), i.e. $\mathfrak{L}(S, x)$ given by (III.3.16) is not empty.

For every $y \in Y$ we define

$$I^*(y) = \{t \in T : |F(y, t) - b(t)| = \|F(y) - b\|_\infty\}. \tag{III.3.37}$$

Obviously we have, with $I(y, \gamma)$ given by (III.3.36),

$$I^*(y) = I(y, \gamma) \qquad \text{for } \gamma = \| F(y) - b \|_\infty\}. \tag{III.3.38}$$

If $\hat{y} \in Y$ is a solution of the approximation problem (III.3.30) and $\hat{\gamma} = \|F(\hat{y}) - b\|_\infty$, then $(\hat{y}, \gamma) \in S$ is a minimal point of f given by (III.3.32) on S, $I(\hat{y}, \hat{\gamma}) = I^*(\hat{y})$ is not empty, and we obtain the next theorem.

Theorem III.3.12 *If $\hat{y} \in Y$ is a solution of the approximation problem (III.3.30), then it follows, with $I^*(\hat{y})$ defined by (III.3.37), that*

$$\min_{t \in I^*(\hat{y})} [b(t) - F(\hat{y}, t)]\,\langle \operatorname{grad} F(\hat{y}, t), k\rangle \leq 0 \qquad \text{for all } k \in T(Y, \hat{y}). \tag{III.3.39}$$

Proof From Theorem III.3.5 it follows using the equivalence of (III.3.17) with (III.3.23) that

$$\min\{\min_{t\in I^*(\hat{y})}\{\lambda - 2\,[F(\hat{y},t)-b(t)]\,\langle \operatorname{grad} F(\hat{y},t),k\rangle\},-\lambda\}\leqslant 0 \qquad \text{(III.3.40)}$$

for all $k\in T(Y,\hat{y})$ and all $\lambda\in \mathbb{R}$, if one uses the Fréchet derivatives given by (III.3.33) and (III.3.34) and takes into account (III.3.35) as well as (III.3.38). If now (III.3.39) were not true for some $k\in T(Y,\hat{y})$ then for

$$-\lambda=\min_{t\in I^*(\hat{y})}[b(t)-F(\hat{y},t)]\,\langle \operatorname{grad} F(\hat{y},t),k\rangle>0$$

(III.3.40) would also be false. Therefore, we conclude (III.3.40) ⇒ (III.3.39).

The converse implication (III.3.39) ⇒ (III.3.40) is trivial. Hence the condition (III.3.39) is equivalent to condition (III.3.23) for the special case at hand. The condition (III.3.39) is the well known local Kolmogoroff condition for best approximations $F(\hat{y})$ of b in $V=F(Y)$ (see, for example, Collatz and Krabs [73]). In Section III.2.2.3 we obtained a sufficient condition (III.2.22) for best approximations $F(\hat{y})$ of b in $V=F(Y)$, which in the notation of this section reads as follows:

$$\min_{t\in I^*(\hat{y})}[b(t)-F(\hat{y},t)]\;[F(y,t)-F(\hat{y},t)]\leqslant 0 \qquad \text{for all } y\in Y \qquad \text{(III.3.41)}$$

with $I^*(\hat{y})$ given by (III.3.37). This statement implies the condition (III.3.27), which here reads

$$\min\{\min_{t\in I^*(\hat{y})}\{\gamma-\hat{\gamma}-[F(y,t)-b(t)]^2+[F(\hat{y},t)-b(t)]^2\},\hat{\gamma}-\gamma\}\leqslant 0$$

$$\text{for all } y\in Y \text{ and all } \gamma\in\mathbb{R}.$$

In order to see that, we remark first that (III.3.42) is equivalent to

$$\min_{t\in I^*(\hat{y})}\{F(\hat{y},t)-b(t)]^2-[F(y,t)-b(t)]^2\}\leqslant 0 \qquad \text{for all } y\in Y \qquad \text{(III.3.43)}$$

which can be proved as the equivalence of (III.3.39) and (III.3.40). Taking into account the identity

$$[F(\hat{y},t)-b(t)]^2-[F(y,t)-b(t)]^2=\{2[b(t)-F(\hat{y},t)]-[F(\hat{y},t)-F(\hat{y},t)]\}\,[F(y,t)-F(\hat{y},t)] \qquad \text{(III.3.44)}$$

one sees finally that (III.3.43) and hence also (III.3.42) is a consequence of (III.3.41).

In Collatz and Krabs [73] a sign condition is given which assures that the implication (III.3.39) ⇒ (III.3.41) holds, so that subject to this sign condition both statements are necessary and sufficient for the fact that $\hat{y}\in Y$ solves the approximation problem (III.3.30).

Definition The mapping $F:\hat{X}\to C(T)$ satisfies the sign condition on X if for each closed subset Q of T and for every pair $y, \hat{y} \in Y$ with

$$\min_{t \in Q} |F(y,t) - F(\hat{y},t)| > 0 \tag{III.3.45}$$

there exists a $k \in T(Y, \hat{y})$ such that

$$\min_{t \in Q} [F(y,t) - F(\hat{y},t)] \langle \text{grad}\, F(\hat{y},t), k\rangle > 0. \tag{III.3.46}$$

In order to see that the sign condition yields the implication (III.3.39) ⇒ (III.3.41), we assume that (III.3.41) does not hold for some $y \in Y$. Then certainly (III.3.45) holds and hence (III.3.46) with $Q = I^*(\hat{y})$ holds for some $k \in T(Y, \hat{y})$. Since

$$\min_{t \in I^*(\hat{y})} [b(t) - F(\hat{y},t)] \langle \text{grade}\, F(\hat{y},t), k\rangle > 0$$

it follows further from (III.3.46) with $Q = I^*(\hat{y})$ that

$$\min_{t \in I^*(\hat{y})} [b(t) - F(\hat{y},t)]\ [F(y,t) - F(\hat{y},t)] > 0$$

i.e. (III.3.39) does not hold, which completes the proof.

Finally we have the following lemma.

Lemma III.3.13 *If F satisfies the sign condition on X, then the mapping $\hat{g}(y,\gamma) = (-\gamma, g(y,\gamma))$ with g defined by (III.3.31) (see (III.3.28)) is almost pseudo-concave.*

Proof Let Q be a closed subset of T and $(y,\gamma), (\hat{y},\hat{\gamma}) \in X = Y \times \mathbb{R}$ be a pair with

$$\min\left\{\min_{t \in Q}\{\gamma - \hat{\gamma} - [F(y,t) - b(t)]^2 + [F(\hat{y},t) - b(t)]^2\}, \hat{\gamma} - \gamma\right\} > 0. \tag{III.3.47}$$

We must show the existence of an $h = (k, \lambda) \in T(X, \hat{x}) = T(Y, \hat{y}) \times \mathbb{R}$ with

$$\min\left\{\min_{t \in Q}\{\lambda - 2\,[F(\hat{y},t) - b(t)]\ \langle \text{grad}\, F(\hat{y},t), k\rangle\}, -\lambda\right\} > 0. \tag{III.3.48}$$

From (III.3.47) follows

$$\min_{t \in Q}\{[F(\hat{y},t) - b(t)]^2 - [F(y,t) - b(t)]^2\} > 0 \tag{III.3.49}$$

and from (III.3.44) we obtain

$$\min_{t \in Q} [b(t) - F(\hat{y},t)]\ [F(y,t) - F(\hat{y},t)] > 0 \tag{III.3.50}$$

which implies (III.3.45). On the basis of the sign condition, (III.3.45) implies (III.3.46) for some $k \in T(Y, \hat{y})$. Now (III.3.46) and (III.3.50) together yield

$$-\lambda = \min_{t \in Q} [b(t) - F(\hat{y},t)]\ \langle \text{grad}\, F(\hat{y},t), k\rangle > 0 \tag{III.3.51}$$

so that for this λ also (III.3.48) is satisfied.

The sign condition encompasses a whole series of special approximation problems, for example, linear, general rational (see Sections III.1.1.3 and III.2.2.3) and exponential approximation. It is characteristic of the fact that the local Kolmogoroff condition (III.3.39) is sufficient for all $b \in C(T)$ that $\hat{y} \in Y$ solves the approximation problem (III.3.30) (see Collatz and Krabs [73]).

Also, the almost pseudo-concavity of the mapping $\hat{g}$ defined in (III.3.28) may be characterized by the fact that the condition (III.3.23) for a certain class of optimization problems (III.3.3) is sufficient for $\hat{x} \in S$ (III.3.2) to solve the problem (see Krabs [73]). For the case of the approximation problem (III.3.30) at hand, it says that for every closed subset Q of T and for every pair $y, \hat{y} \in Y$ defined by (III.3.49) there exists a $k \in T(Y, \hat{y})$ satisfying (III.3.51) (Proof = Exercise).

III.3.4 Semi-infinite nonlinear optimization

The nonlinear approximation problem (III.3.30), treated in Section III.3.3 in its equivalent form as an optimization problem (III.3.3) with g given by (III.3.31) and f by (III.3.32), is a special case of the general problem of semi-fininite nonlinear optimization. This fact one obtains by basing the problem (III.3.3) in Section III.3.1.1 on $\mathbb{R}^n$ for the normed vector space E. We assume further that $\hat{X}$ is an open set in $\mathbb{R}^n$ containing X and $g: \hat{X} \to C(T)$ is a mapping such that for every $t \in T$ and $x \in \hat{X}$ the gradient, grad $g(x, t)$, exists and the mapping $(x, t) \to$ grad $g(x, t)$ of $\hat{X} \times T$ into $\mathbb{R}^n$ is continuous. Then g is Fréchet differentiable on $\hat{X}$ and the Fréchet derivative is given by

$$g'_x(h, t) = \langle \text{grad}\, g(x, t), h \rangle \tag{III.3.52}$$

for all $x \in \hat{X}$, $h \in \mathbb{R}^n$ and $t \in T$ (again see Collatz and Krabs [73], Theorem II.4.6). The functional $f: \hat{X} \to \mathbb{R}$ is assumed to possess for every $x \in \hat{X}$ a gradient, grad $f(x)$, which depends continuously on x. Then f is also Fréchet differentiable on X and the Fréchet derivative is given by

$$f'_x(h) = \langle \text{grad}\, f(x), h \rangle \tag{III.3.53}$$

for all $x \in \hat{X}$ and $h \in \mathbb{R}^n$.

Under these assumptions, we again consider the optimization problem (III.3.3) with the goal of supplementing the two equivalent necessary conditions (III.3.17) and (III.3.23) for minimal points $\hat{x} \in S$ of f on S by a further one, which for regular minimal points (i.e. for those for which $\mathfrak{L}(S, \hat{x})$ given by (III.3.16) is not empty) is equivalent to these two and can be looked upon as a generalized multiplier rule.

First we define for every $x \in S$ the so-called linearizing cone

$$L(S, x) = \bigcap_{t \in I(x)} \{h \in T(X, x) : \langle \text{grad}\, g(x, t), h \rangle \geqslant 0\} \tag{III.3.54}$$

with $I(x)$ defined by (III.3.5). Then we have the next lemma.

Lemma III.3.14 *For every $x \in S$*

$$T(S, x) \subseteq L(S, x) \tag{III.3.55}$$

where $T(S, x)$ denotes the tangent cone to S at x.

Proof

1. Let $I(x)$ be empty. Then (III.2.5) implies

$$T(S, x) \subseteq T(X, x) = L(S, x)$$

2. Suppose $I(x)$ is not empty.

Assumption There is an $h \in T(S, x)$ with

$$\langle \text{grad}\, g(x, t), h \rangle < 0 \qquad \text{for some } t \in I(x). \tag{III.3.56}$$

Now let $\{x_k\}$ be a sequence in S and $\{\lambda_k\}$ be a sequence of positive numbers with $x = \lim_{k \to \infty} x_k$ and $h = \lim_{k \to \infty} \lambda_k (x_k - x)$.

Then for $t \in I(x)$ defined by (III.3.56) and for all sufficiently large k, one has

$$\begin{aligned}\lambda_k g(x_k, t) &= \lambda_k \;[g(x_k, t) \quad g(x, t)]\\ &\leqslant \langle \text{grad}\, g(x, t), \lambda_k(x_k - x) \rangle + \lambda_k \,\| x_k - x \|\, \epsilon\, (\| x_k - x \|)\end{aligned}$$

with $\lim_{k \to \infty} \epsilon\, (\| x_k - x \|) = 0$ (where $\| \;\|$ denotes any norm in $\mathbb{R}^n$).

This implies $g(x_k, t) < 0$, a contradiction to $x_k \in S$ for all k.

Lemma III.3.15 *If $x \in S$ is regular, i.e. the set $\ell(S, x)$ defined by (III.3.16) is not empty, and if the tangent cone $T(X, x)$* to X at x *is convex (for example, if X is convex or open), as long as $I(x)$ given by (III.3.5) is not empty, then it follows that*

$$\overline{\ell(S, x)} = T(S, x) = L(S, x). \tag{III.3.57}$$

Proof By Theorem III.3.6 and Lemma III.3.14 it suffices to show that $\overline{\ell(S, x)} = L(S, x)$. If $I(x)$ is empty, then there is nothing to prove. If $I(x)$ is not empty, then there is nothing to prove. If $I(x)$ is not empty, then we suppose a given $h \in L(S, x)$ and choose an $h_0 \in \ell(S, x)$ arbitrarily. Then for all $\lambda \in [0, 1]$

$$h_\lambda = \lambda h + (1 - \lambda)\, h_0 \in T(X, x)$$

and

$$\langle \text{grad}\, g(x, t), h_\lambda \rangle > 0 \qquad \text{for all } t \in I(x)$$

i.e.

$$h_\lambda \in \ell(S, x) \text{ and } \lim_{\lambda \to 1-0} h_\lambda = h \Rightarrow h \in \overline{\ell(S, x)}.$$

Thus $L(S, x) \subseteq \overline{\ell(S, x)}$. The converse inclusion follows from the fact that $L(S, x)$ is closed and $\ell(S, x) \subseteq L(S, x)$.

The main result of this section is the theorem now given.

Theorem III.3.16 *Let $x \in S$ be a regular point and for the case that $I(x)$ (defined by (III.3.5)) is not empty, let the tangent cone $T(X, x)$ to X at x be convex. Then the two conditions (III.3.17) and (III.3.23) are equivalent to the statement*

$$\langle \operatorname{grad} f(x) - \sum_{t \in I_e(x)} \lambda_t \operatorname{grad} g(x, t), h \rangle \geqslant 0 \quad \text{for all } h \in T(X, x). \qquad \text{(III.3.58)}$$

Here $I_e(x)$ is a finite subset of $I(x)$, the λ_t *are certain non-negative numbers and for the case that $I(x)$ is empty, the sum in (III.3.58) is to be replaced by Θ_n = null element of $\mathbb{R}^n$. ($\hat{x}$ in (III.3.17) and (III.3.23) is to be replaced by x).*

Proof For an empty set $I(x)$ there is nothing to prove. Suppose therefore that $I(x)$ is not empty.

1. We assume that (III.3.58) is satisfied. Then there follows immediately

$$\langle \operatorname{grad} f(x), h \rangle \geqslant 0 \qquad \text{for all } h \in L(S, x) \qquad \text{(III.3.59)}$$

with $L(S, x)$ given by (III.3.54) and since $\mathfrak{L}(S, x) \subseteq L(S, x)$, we therefore also have (III.3.17) $\Leftrightarrow$ (III.3.23).

2. We assume that (III.3.17) is satisfied. By Lemma III.3.15 then (III.3.59) follows (Proof = Exercise).

If one defines

$$Q_x = \{ \operatorname{grad} g(x, t) : t \in I(x) \}$$

then Q_x is a non-empty compact subset of $\mathbb{R}^n$ and hence the convex hull $H(Q_x)$ of Q_x is also compact by Theorem IV.3.2. Furthermore $\Theta_n \notin H(Q_x)$. For otherwise there would be finitely many points $t_1, \ldots, t_r \in I(x)$ and numbers $\lambda_1 \geqslant 0, \ldots, \lambda_r \geqslant 0$ with

$$\sum_{1=1}^{r} \lambda_i = 1 \qquad \text{and} \qquad \Theta_n = \sum_{i=1}^{r} \lambda_i \operatorname{grad} g(x, t_i).$$

For every $h \in \mathfrak{L}(S, x)$ one would then have, however,

$$0 = \langle \Theta_n, h \rangle = \sum_{i=1}^{r} \lambda_i \langle \operatorname{grad} g(x, t_i), h \rangle > 0$$

which is a contradiction. Let $K(Q_x)$ be the cone generated by $H(Q_x)$

$$K(Q_x) = \{ \lambda k : k \in H(Q_x), \lambda \geqslant 0 \}.$$

Then $(-T(X, x)^0) \cap K(Q_x) = \{\Theta_n\}$, where for any cone K in $\mathbb{R}^n$ the cone $\mathring{K}$ is defined by (IV.2.14). For every vector $y \in (-T(X, x)^0) \cap K(Q_x)$ we have

$$y = \sum_{i=1}^{s} \lambda_{t_i} \operatorname{grad} g(x, t_i)$$

and

$$\langle y, h \rangle = \sum_{i=1}^{s} \lambda_{t_i} \langle \operatorname{grad} g(x, t_i), h \rangle \leqslant 0 \qquad \text{for all } h \in T(X, x)$$

where $t_1, \ldots, t_s$ are finitely many points of $I(x)$ and $\lambda_{t_1}, \ldots, \lambda_{t_s}$ are non-negative numbers. Since for every $h \in \mathfrak{L}(S, x)$ one has

$$\langle \text{grad}\, g(x, t_i), h \rangle > 0 \qquad \text{for all } i = 1, \ldots, s$$

all $\lambda_{t_i} = 0$ and, hence $y = \Theta_n$.

By Theorem IV.2.1, therefore the cone $T(X, x)^0 + K(Q_x)$ is closed. The condition (III.3.59) following from (III.3.17) now says, by the definition (III.3.54) of $L(S, x)$,

$$\langle \text{grad}\, f(x), h \rangle \geqslant 0 \qquad \text{for all } h \in T(X, x) \cap K(Q_x)^\circ \tag{III.3.60}$$

(Proof = Exercise). Since by Lemma III.2.1, $T(X, x)$ is closed, it follows from (IV.2.12′) that $T(X, x) = T(X, x)^{\circ\circ}$ and further from (IV.2.9′) that

$$T(X, x) \cap K(Q_x)^\circ = (T(X, x)^\circ)^\circ \cap K(Q_x)^\circ = (T(X, x)^\circ + K(Q_x))^\circ.$$

By again using (IV.2.14) one can therefore write for (III.3.60)

$$\text{grad}\, f(x) \in (T(X, x)^\circ + K(Q_x))^{\circ\circ}$$

and since $T(X, x)^\circ + K(Q_x)$ is closed we conclude from (IV.2.12′) that

$$\text{grad}\, f(x) \in T(X, x)^\circ + K(Q_x) \tag{III.3.61}$$

which is equivalent to (III.3.58).

If in particular, X is an open subset of $\mathbb{R}^n$, then by (III.2.2), $T(X, x) = \mathbb{R}^n$ and $T(X, x)^\circ = \{\Theta_n\}$, so that (III.3.61) goes over into

$$\text{grad}\, f(x) \in K(Q_x)$$

i.e.

$$\text{grad}\, f(x) = \sum_{i=1}^{r} \lambda_{t_i} \text{grad}\, g(x, t_i) \tag{III.3.62}$$

for finitely many $t_1, \ldots, t_r \in I(x)$ and $\lambda_{t_1} \geqslant 0, \ldots, \lambda_{t_r} \geqslant 0$.

Summarizing we have the following result.

Result If $\hat{x} \in S$ is a regular minimal point of f on S such that in the case of a non-empty set $I(\hat{x})$ the tangent cone $T(X, \hat{x})$ at $\hat{x}$ on X is convex, then the condition (III.3.58) necessarily holds with $x = \hat{x}$ or (III.3.62) holds for the case that X is open.

For a finite set T the statement (III.3.62) is the well known Lagrange multiplier rule.

III.3.5 Bibliographical remarks

For the case of finite optimization problems, i.e. for finite T, and under the various assumptions at the beginning of Section III.3.4, John [48] set up a somewhat weaker multiplier rule under the additional assumption $X = \mathbb{R}^n$ and without any regularity requirements. Under the same conditions and with the addition of a so-called constraint qualification, Kuhn and Tucker proved the multiplier rule

(III.3.62) in their fundamental paper [51]. This constraint qualification was later weakened by numerous authors and extended to problems with inequalities and equalities as side conditions, for example, by Arrow, Hurwicz and Uzawa [61], who have given the weakest regularity condition for the case of a convex set S (III.3.2), by Abadie [67], Evans [69], and Mangasarian and Fromovitz [67] (see also Mangasarian [69]). The weakest regularity condition for the general nonlinear case stems from Gould and Tolle [71]. The first general survey of constraint qualification and related conditions are found in Elster and Götz [69]. Recently Peterson [73] in a survey article gave a complete presentation of all results up to 1973.

A generalization of the theorem of John in the form of a maximum principle for nonlinear infinite optimization problems was given by Lempio [72a] (in this connection, see also Lempio [71c, 72b, 73, 74]).

Very general multiplier rules without regularity conditions were derived by Cannon, Cullum and Polak [66, 70] and Hestenes [66] and used to obtain maximum principles in the case of discrete and continuous control problems. Other papers in this connection are Neustadt [66, 67], Halkin [66a, 66b, 67], and Halkin and Neustadt [66].

The regularity condition used in Section III.3.2 is a direct generalization of the sufficient condition given by Abadie [67], Theorem 3, for finite optimization problems for the Kuhn–Tucker constraint qualification in a slightly weakened form. It was used by Krabs [73] to obtain necessary and sufficient conditions for minimal points in the case of infinite optimization problems. Sections III.3.1 and III.3.2, and III.3.3 are based on this paper.

The basic reason for the validity of the multiplier rule (III.3.58) is the fact that the linearizing cone $L(S, x)$ defined by (III.3.54) coincides at a minimal point $\hat{x} \in S$ with the tangent cone $T(S, \hat{x})$ at $\hat{x}$ on S. In the case where these two coincide, Abadie [67] calls the point $\hat{x}$ sequentially qualified and proves for such minimal points a multiplier rule of the form (III.3.62) for $x = \hat{x}$.

In the sixties one began to seek necessary and sufficient conditions for minimal points in the case of infinite nonlinear optimization problems. Besides the above-mentioned contributions by Neustadt and Halkin (which represent only a selection), numerous further papers were written, of which the one by Dubovitskii and Miljutin [65] turned out to be fundamental. In the meantime the topic has been taken up by text books and forms, for example, the nucleus of the book of Pschenitschny [72] on 'Necessary Optimality Conditions', which also contains a series of applications. In Holmes [72] there is also a section dedicated to the Dubovitskii–Miljutin theory with an application to the classical variational problem. It also forms the basis of the book of Girsanov [72] and is applied to control problems. Finally, it is presented in the first chapter of the book of Laurent [72], who in this connection goes back to the doctoral dissertation of Lobry [67].

Halkin [70] again goes into this theory. Besides the Dubovitskii–Miljutin theory, there are numerous other approaches to necessary and sufficient conditions for minimal points, in which to some extent similar aids are used. We name here only the following papers which are representative: Bazaraa and Goode [73], Dem'yanov and Malozemov [71], Gittleman [71], Guignard [69], Guinn et al. [69], Hoffmann [71], Hoffmann and Kolumban [74], Nagahisa and Sakawa [69], Raffin [68], Ritter [67], Rubinov [66], Russell [66], and Varaiya [67].

SOLUTIONS AND HINTS TO THE PROBLEMS

Problem III.2.1a Let $x \in X \subseteq Y$; if $h \in T(X, x)$, then there is a sequence (x_k) in X and a sequence (λ_k) of positive numbers such that

$$x = \lim_{k\to\infty} x_k \qquad \text{and} \qquad h = \lim_{k\to\infty} \lambda_k(x_k - x).$$

Since (x_k) also belongs to Y and $x \in Y$ we have $h \in T(Y, x)$, hence $T(X, x) \subseteq T(Y, x)$.

The proof of the second assertion is similar.

Problem III.2.1b Let $(h, l) \in T(X \times Y, (x, y))$. Then there exist sequences (x_k, y_k) in $X \times Y$ and (λ_k), $\lambda_k > 0$ such that

$$\left.\begin{aligned} x &= \lim_{k\to\infty} x_k, \quad h = \lim_{k\to\infty} \lambda_k(x_k - x) \\ y &= \lim_{k\to\infty} y_k, \quad h = \lim_{k\to\infty} \lambda_k(y_k - y) \end{aligned}\right\} \Rightarrow (h, l) \in T(X, x) \times T(Y, y).$$

Assume X to be open in E. Then $T(X, x) = E$. Let $(h, l) \in E \times T(Y, y)$ be given. Then there is a sequence (y_k) in Y and a sequence (λ_k) of positive numbers such that

$$y = \lim_{k\to\infty} y_k \qquad l = \lim_{k\to\infty} \lambda_k(y_k - y).$$

Put $x_k = x + (1/\lambda_k)h$ then $x = \lim_{k\to\infty} x_k$ and $x_k \in X$ for k sufficiently large. Further, $h = \lim_{k\to\infty} \lambda_k(x_k - x)$, hence $(h, l) \in T(X \times Y, (x, y))$ which proves $T(X, x) \times T(Y, y) \subseteq T(X \times Y, (x, y))$.

Problem III.2.2 Assume that there exists an $h \in T(V, v)$ such that

$$\sum_{i=1}^{m} h_i \operatorname{sgn}(\hat{v}_i - x_i) + \sum_{\hat{v}_i - x_i = 0} |h_i| < 0 \qquad \text{where } \operatorname{sgn}(0) = 0.$$

For sufficiently small $\lambda > 0$ we then obtain

$$\begin{aligned} \sum_{i=1}^{m} |\hat{v}_i - x_i + \lambda h_i| &= \sum_{\hat{v}_i - x_i \neq 0} (\hat{v}_i - x_i + \lambda h_i)\operatorname{sgn}(\hat{v}_i - x_i) + \sum_{\hat{v}_i - x_i = 0} |h_i| \\ &= \sum_{i=1}^{m} |\hat{v}_i - x_i| \\ &\quad + \left(\sum_{i=1}^{m} h_i \operatorname{sgn}(\hat{v}_i - x_i) + \sum_{\hat{v}_i - x_i = 0} |h_i| \right) < \sum_{i=1}^{m} |\hat{v} - x_i|. \end{aligned}$$

By Theorem III.2.2 it follows that $\hat{v} \in V$ is not optimal which complete the proof.

Problem III.3.1 Let Q be a closed subset of T and $x, \hat{x} \in X$ be such that

$$\min_{t \in Q \cup \ 0} \{g_t(x) - g_t(\hat{x})\} > 0.$$

Then

$$f(x) - f(\hat{x}) < 0 \qquad \text{and} \qquad g_t(x) - g_t(\hat{x}) > 0 \qquad \text{for all } t \in Q.$$

Hence

$$f'_x(x - \hat{x}) < 0 \qquad \text{and} \qquad g'_{t,\hat{x}}(x - \hat{x}) > 0 \qquad \text{for all } t \in Q$$

which implies

$$\min_{t \in Q \cup \{0} g'_{t,\hat{x}}(x - \hat{x}) > 0$$

where $x - \hat{x} \in T(X, \hat{x})$. This completes the proof.

Problem III.3.2 The assertion is an immediate consequence of Problem III.2.1*b*.

CHAPTER IV

Appendix: Mathematical Aids

Many concepts and results from functional analysis, including properties of convex sets and representations of linear functionals, are used in the book. They are assembled here. We assume certain basic facts about normed spaces, such as convergence, completeness, and topological properties to be known since they are readily available in standard textbooks. For the same reason, we shall not give proofs of all the relevant theorems, but instead merely give references.

IV.1. CONVEX CONES AND LINEAR MAPPINGS

Basic concepts for the linear optimization theory developed in Section I are partially ordered vector spaces, continuous linear mappings, and adjoint mappings. Their connections will now be investigated.

IV.1.1 Convex cones and partial orderings

Let E be a linear vector space over $\mathbb{R}$. A subset C of E is said to be convex if

$$x, y \in C, 0 \leqslant \lambda \leqslant 1 \Rightarrow \lambda x + (1-\lambda)y \in C.$$

A subset K of E is called a cone (with the vertex Θ_E, the zero element of E) if

$$x \in K, \lambda \geqslant 0 \Rightarrow \lambda x \in K.$$

A subset K of E is called a convex cone if K is a cone and is also convex.

Lemma IV.1.1 *A cone K is convex if and only if*

$$x, y \in K \Rightarrow x + y \in K. \tag{IV.1.1}$$

Proof

1. Suppose K is a convex cone. Given $x, y \in K$, it follows that $\frac{1}{2}(x+y) = \frac{1}{2}x + \frac{1}{2}y \in K$, since K is convex, and $x + y = 2[\frac{1}{2}(x+y)] \in K$, since K is a cone.
2. Suppose K is a cone satisfying (IV.1.1). If $x, y \in K$ and $\lambda \in [0, 1]$ are given, then it follows that $\lambda x \in K$ and $(1-\lambda)y \in K$, since K is a cone. Now (IV.1.1) implies $\lambda x + (1-\lambda)y \in K$.

If K is a non-empty convex cone in E, then we define an ordering relation in E by

$$x \geqslant y (\Leftrightarrow y \leqslant x) \Leftrightarrow x - y \in K. \tag{IV.1.2}$$

Since $\Theta_E \in K$ we obviously have

$$x \geqslant x \qquad \text{for all } x \in E \text{ (reflexivity).} \tag{IV.1.3}$$

Furthermore

$$x \geqslant y, y \geqslant z \Rightarrow x \geqslant z \text{ (transitivity)} \tag{IV.1.4}$$

since (IV.1.2) gives $x - y \in K$ and $y - z \in K$ and, hence, (IV.1.1) implies $x - z = (x - y) + (y - z) \in K$.

The order relation (IV.1.2) is also compatible with the algebraic structure of E, i.e. we have

$$x \geqslant \Theta_E, \lambda \geqslant 0 \Rightarrow \lambda x \geqslant \Theta_E, \tag{IV.1.5}$$

$$x \geqslant y, z \in E \Rightarrow x + z \geqslant y + z. \tag{IV.1.6}$$

Statement (IV.1.5) follows directly from

$$K = \{x \in E : x \geqslant \Theta_E\} \tag{IV.1.7}$$

and (IV.1.6) is an immediate consequence of (IV.1.2).

Conversely, if E has an order relation with the properties (IV.1.3)–(IV.1.6) and if K is defined by (IV.1.7), then K is a non-empty convex cone, and (IV.1.2) holds (Proof = Exercise).

Definition IV.1.1 A partially ordered vector space E over $\mathbb{R}$ is a vector space with an order relation, having the properties (IV.1.3)–(IV.1.6). The order relation may be defined in terms of a non-empty convex cone K by (IV.1.2). Then (IV.1.7) holds, and K is called the associated positive cone.

A cone is said to be pointed if

$$x \in K \text{ and } -x \in K \Rightarrow x = \Theta_E.$$

If K is a non-empty convex cone in E and the order relation is given by (IV.1.2), then K is pointed if and only if the following implication holds: $x \geqslant y$ and $y \geqslant x \Rightarrow x = y$ (Proof = Exercise).

Example 1 $E = \mathbb{R}^n$; let $0 \leqslant r \leqslant n$. For $x = (x_1, \ldots, x_n)^{\mathrm{T}}$ and $y = (y_1, \ldots, y_n)^{\mathrm{T}}$ we define

$$x \geqslant y \Leftrightarrow x_i \geqslant y_i \qquad \text{for } 1 \leqslant i \leqslant r. \tag{IV.1.2$'$}$$

The associated positive cone defined by (IV.1.7) is

$$K = K_r^n = \{x = (x_1, \ldots, x_n)^{\mathrm{T}} : x_i \geqslant 0, i = 1, \ldots, r\}. \tag{IV.1.7$'$}$$

In particular, $K_0^n = \mathbb{R}^n$ and K_n^n is the non-negative quadrant of $\mathbb{R}^n$; K_n^n is pointed. However, K_r^n is not pointed for $0 \leqslant r < n$.

Example 2 E = vector space of real-valued functions defined on a set B. If Q is any subset of B (perhaps empty), then for $x, y \in E$ we define

$$x \geqslant y \Leftrightarrow x(t) \geqslant y(t) \qquad \text{for all } t \in Q. \tag{IV.1.2$''$}$$

The associated positive cone defined by (IV.1.7) is given by

$$K = K_Q^B = \{x \in E : x(t) \geqslant 0 \text{ for all } t \in Q\}. \tag{IV.1.7$''$}$$

In particular, if $K_\phi^B = E$ (where ϕ = empty set), K_B^B is pointed. However, K_Q^B is not pointed if Q is a genuine subset of B.

Example 3 Suppose E is a unitary space with scalar product denoted by $\langle \cdot , \cdot \rangle$ and suppose $\{a_i\}_{i \in I}$ is a non-empty family of elements $a_i \in E$. Then we define an order relation in E by

$$x \geqslant y \Leftrightarrow \langle a_i, x \rangle \geqslant \langle a_i, y \rangle \qquad \text{for all } i \in I. \tag{IV.1.2'''}$$

Problem IV.1.1 Show that (IV.1.2‴) is an order relation with the properties (IV.1.3)–(IV.1.6). How can (IV.1.2′) be obtained from (IV.1.2‴)?

IV.1.2 The (topological) dual space

Suppose E is a normed vector space over $\mathbb{R}$. A real-valued function g on E is said to be continuous if

$$x_n \to x, \quad x_n, x \in E \Rightarrow \lim_{n \to \infty} g(x_n) = g(x).$$

A real-valued function c on E is called a linear functional (or also a linear form) on E if

$$c(\lambda x + \mu y) = \lambda c(x) + \mu c(y) \qquad \text{for all } x, y \in E, \lambda, \mu \in \mathbb{R}.$$

Theorem IV.1.2 *A linear form c on E is continuous if and only if there is a constant $\alpha \geqslant 0$ such that*

$$| c(x) | \leqslant \alpha \| x \| \qquad \textit{for all } x \in X. \tag{IV.1.8}$$

For the proof see, for example, Köthe [66], page 132.

Let E^* be the vector space of continuous linear forms on E. If one associates with every $c \in E^*$ the number

$$\| c \| = \sup_{\| x \| \leqslant 1} | c(x) | \tag{IV.1.9}$$

then E^* becomes a normed vector space with $\| c \|$ the smallest number α satisfying (IV.1.8) (Proof = Exercise).

In the sequel we shall always regard E^* as normed in this way, whereby in fact E^* becomes a Banach space. Convergence and topological concepts in E are always understood in the sense of the norm (IV.1.9).

Example 1 Let E be an n-dimensional normed vector space over $\mathbb{R}$. Then E has a basis $\{e_1, \ldots, e_n\} \subseteq E$ such that each $x \in E$ can be represented uniquely as

$$x = \sum_{i=1}^{n} x_j e_j \qquad \text{for } x_j \in \mathbb{R}, j = 1, \ldots, n.$$

In this case E^* consists of all linear forms on E (Proof = Exercise), and E and E^* are isomorphic to each other. An isomorphism is given, for example, by associating with every $c \in E^*$ the $x \in E$ which has the representation

$$x = \sum_{j=1}^{n} c(e_j)e_j$$

(Proof = Exercise).

Problem IV.1.2 Show that this mapping is continuous in both directions, i.e. that

$$c_n, c \in E^*, c_n \to c \text{ (in the sense of norm (IV.1.9))} \Leftrightarrow x_n \to x.$$

Example 2 Suppose E is a unitary space. If $y \in E$ is fixed, then by setting $c(x) = \langle y, x \rangle$ for $x \in E$, a linear form is defined on E which satisfies

$$|c(x)| \leqslant \|y\| \, \|x\| \qquad \text{for all } x \in E$$

by the Cauchy–Schwartz inequality. Thus c is continuous and $\|c\| \leqslant \|y\|$. If $y = \Theta_E$, then c is the zero mapping and $\|c\| = \|y\| = 0$. Otherwise $\|y\| > 0$ and $|c(z)| \leqslant \|c\|$ for $z = y/\|y\|$, which implies $\|y\| \leqslant \|c\|$. Hence, $\|y\| = \|c\|$.

In general, there are other continuous linear forms in a unitary space. If E is complete, then one has the following theorem.

Theorem IV.1.3 *If E is a Hilbert space, then for each $c \in E^*$ there is precisely one $y \in E$ satisfying*

$$c(x) = \langle y, x \rangle \qquad \text{for all } x \in E \text{ (and } \|c\| = \|y\|). \tag{IV.1.10}$$

For the proof see, for example, Collatz [68], page 83.

By the mapping $y \to c$ satisfying (IV.1.10) a norm preserving isomorphism of E onto E^* is defined (which is continuous in both directions).

Example 3 We consider further the important special case $E = \mathbb{R}^n$ with the orthonormal basis $\{e_1, \ldots, e_n\}$, $e_j^i = \delta_{ij}$ = Kronecker delta for $i, j = 1, \ldots, n$. Every continuous linear form c on E is then given by

$$c(x) = \langle \hat{c}, x \rangle \qquad \text{for } x \in \mathbb{R}^n \tag{IV.1.11}$$

with $\hat{c} = (c(e_1), \ldots, c(e_n))^{\mathrm{T}}$. If $E = \mathbb{R}^n$ is given the Euclidean norm, then $\|c\| = \|\hat{c}\|_2 = \langle \hat{c}, \hat{c} \rangle^{1/2}$.

For subsequent considerations, let E again be an arbitrary normed vector space and E^* the topological dual space of E, equipped with the norm (IV.1.9). By E^{**} we denote the topological dual space of E^*. E^{**} is a Banach space with the norm

$$\|x^{**}\| = \sup_{\substack{\|x^*\| \leqslant 1 \\ x^* \in E^*}} x^{**}(x^*). \tag{IV.1.12}$$

If one associates with every $x \in E$ a linear form Φ_x on E^* by

$$\Phi_x(x^*) = x^*(x) \qquad \text{for } x^* \in E^* \tag{IV.1.13}$$

then there follows

$$|\Phi_x(x^*)| \leqslant \|x\| \, \|x^*\| \qquad \text{for } x^* \in E$$

i.e. Φ_x is continuous ($\Phi_x \in E^{**}$) and $\| \Phi_x \| \leqslant \| x \|$.

Theorem IV.1.4 *For every $x \neq \Theta_E$ there is an $x^* \in E^*$ satisfying*

$$x^*(x) = \| x \| \qquad \text{and} \qquad \| x^* \| = 1.$$

For the proof see, for example, Köthe [66], page 199.

Corollary *For the continuous linear form Φ_x on E^* defined by (IV.1.13) one has, in fact, $\| \Phi_x \| = \| x \|$.*

Since the mapping $x \to \Phi_x$ for $x \in E$ is linear, a norm preserving isomorphism of E into E^{**} is defined by (IV.1.13).

The image $\Phi(E)$ may be a proper subset of E^{**}.

A normed vector space E is called reflexive if $\Phi(E) = E^{**}$, i.e. if the norm preserving isomorphism Φ defined by (IV.1.13) is a mapping of E onto E^{**}. Since E^{**} is a Banach space, E is necessarily also a Banach space if E is reflexive.

Examples of reflexive Banach spaces are Hilbert spaces and finite dimensional normed vector spaces.

Now let E be a partially ordered normed vector space with the non-empty positive cone K. Then we define an order relation in E^* by

$$x^* \geqslant \hat{x}^*, x^*, \hat{x}^* \in E^* \Leftrightarrow x^*(x) - \hat{x}^*(x) \geqslant 0 \qquad \text{for all } x \in K. \qquad \text{(IV.1.14)}$$

Problem IV.1.3 Prove that this order relation satisfies the properties (IV.1.3)–(IV.1.6).

The order relation defined by (IV.1.14) is called the induced ordering in E^*. The associated positive cone is given by

$$K^* = \{x^* \in E^* : x^*(x) \geqslant 0 \qquad \text{for all } x \in K\}.$$

K^* is closed in E^* (Proof = Exercise).

If $K = E$ then $K^* = \{\Theta_{E^*}\}$, where Θ_{E^*} is the zero mapping, and if $K = \{\Theta_E\}$, then $K^* = E^*$.

Example For $K = K_r^n$ defined by (IV.1.7′) we have

$$K^* = \{x \in \mathbb{R}^n : x_i \geqslant 0 \text{ for } i = 1, \ldots, r, x_i = 0 \text{ for } i = (r+1), \ldots, n\}$$

if E^* is identified with $\mathbb{R}^n$ (Proof = Exercise).

Theorem IV.1.5 *Suppose E is a partially ordered normed vector space such that the associated positive cone K has a non-empty interior $\mathring{K}$. For some $x^* \in E^*$, if*

$$x^* \geqslant \Theta_{E^*} \qquad \text{and} \qquad x^*(x_0) \leqslant 0 \qquad \text{for some } x_0 \in \mathring{K}$$

then $x^* = \Theta_{E^*}$.

Proof We assume there is an $x \in E$ with $x^*(x) \neq 0$ and, without loss of generality, that $x^*(x) < 0$. Let $x_\lambda = \lambda x + (1 - \lambda)x_0$, $\lambda \in [0, 1]$. Then there exists $\lambda_0 \in [0, 1]$ with $x_\lambda \in K$ for all $\lambda \in [0, \lambda_0]$, which implies $x^*(x_\lambda) \geqslant 0$ for all $\lambda \in [0, \lambda_0]$.

On the other hand, however, by assumption $x^*(x_\lambda) < 0$ for all $\lambda \in [0, \lambda_0]$, which is a contradiction. Thus $x^*(x) = 0$ for all $x \in E$, i.e. $x^* = \Theta_{E^*}$.

IV.1.3 Linear mappings

Let E and F be two normed vector spaces. A mapping $A : E \to F$ is said to be linear if

$$A(\lambda x + \mu y) = \lambda A(x) + \mu A(y) \qquad \text{for all } x, y \in E \text{ and } \lambda, \mu \in \mathbb{R}.$$

A mapping is $A : E \to F$ is called continuous if

$$x_n \to x,\, x_n, x \in E \Rightarrow A(x_n) \to A(x).$$

Theorem IV.1.6 *A linear mapping A of E into F is continuous if and only if there is a constant $\alpha \geqslant 0$ with*

$$\| A(x) \|_F \leqslant \alpha \| x \|_E \qquad \text{for all } x \in E.$$

For the proof see, for example, Köthe [66], pages 131–2.

Introducing the definition

$$\| A \| = \sup_{\| x \|_E \leqslant 1} \| A(x) \|_F$$

the vector space $L(E, F)$ of continuous linear mappings of E into F becomes a normed vector space. Let E^* and F^* be the topological dual spaces of E and F. For each $A \in L(E, F)$, the adjoint mapping $A^* : F^* \to E^*$ is defined for $y^* \in F^*$ by

$$A^*(y^*)(x) = y^*(A(x)) \qquad \text{for all } x \in E. \qquad \text{(IV.1.15)}$$

Obviously we have for all $x \in E$, $y^*, z^* \in F^*$ and $\lambda, \mu \in \mathbb{R}$,

$$\begin{aligned} A^*(\lambda y^* + \mu z^*)(x) &= (\lambda y^* + \mu z^*)(A(x)) \\ &= \lambda y^*(A(x)) + \mu z^*(A(x)) = \lambda A^*(y^*)(x) + \mu A^*(z^*)(x) \\ &= (\lambda A^*(y^*) + \mu A^*(z^*))(x). \end{aligned}$$

Thus A^* is linear. Since A is continuous, Theorem IV.1.6 gives an $\alpha \geqslant 0$ with

$$\| A(x) \|_F \leqslant \alpha \| x \|_E \qquad \text{for all } x \in E.$$

From this, for all $x \in E$ and all $y^* \in F^*$ it follows that

$$\begin{aligned} | A^*(y^*)(x) | &= | y^*(A(x)) | \\ &\leqslant \| y^* \|_{F^*} \| A(x) \|_F \leqslant \| y^* \|_{F^*} \alpha \| x \|_E \\ &\Rightarrow \| A^*(y^*) \|_{E^*} \leqslant \alpha \| y^* \|_{F^*}. \end{aligned}$$

Thus A^* is also continuous, and $\| A^* \| \leqslant \alpha$ (which implies $\| A^* \| \leqslant \| A \|$).

Example 1 $E = \mathbb{R}^n$, $F = \mathbb{R}^m$. If A is a real $m \times n$ matrix, then a mapping $A \in L(E, F)$ is defined by

$$A(x) = Ax \qquad \text{for } x \in \mathbb{R}^n.$$

If E^* is identified with $\mathbb{R}^n$ and F^* with $\mathbb{R}^m$, then the adjoint mapping is given by

$$A^*(y^*) = A^{\mathrm{T}}y \qquad \text{for all } y^* \in \mathbb{R}^m$$

(Proof = Exercise).

Example 2 Let E and F be two Hilbert spaces and $A \in L(E, F)$. If for each fixed $y^* \in F^*$ one defines

$$h^*(x) = y^*(A(x)) \qquad \text{for all } x \in E$$

then $h^* \in E^*$. By Theorem IV.1.3 there are unique elements $y \in F$ and $h \in E$ with

$$\langle h, x \rangle_E = h^*(x) = y^*(A(x)) = \langle y, A(x) \rangle_F. \tag{IV.1.16}$$

Since the mappings $y \to y^*, A$ and $h^* \to h$ are continuous and linear, a continuous linear mapping $A' : F \to E, y \to h$, is defined by (IV.1.16) with

$$A^*(y^*)(x) = y^*(A(x)) = \langle y, A(x) \rangle_F = \langle A'(y), x \rangle_E$$

for all $x \in E$.

Suppose E, F, and G are linear vector spaces. The Cartesian product $E \times F$ is then also a linear vector space, with componentwise addition and scalar multiplication.

Theorem IV.1.7 *For each linear mapping $A : E \times F \to G$ there are unique linear mappings $A_1 : E \to G$ and $A_2 : F \to G$ satisfying*

$$A(x, y) = A_1(x) + A_2(y) \qquad \textit{for all } x \in E \textit{ and } y \in F. \tag{IV.1.17}$$

Proof If $A_1(x) = A(x, \Theta_F)$ for all $x \in E$ and $A_2(y) = A(\Theta_E, y)$ for all $y \in F$, then A_1 and A_2 are linear mappings and (IV.1.17) holds. Conversely, if $A_1 : E \to G$ and $A_2 : F \to G$ are two linear mappings satisfying (IV.1.17), then it necessarily follows that $A_1(x) = A(x, \Theta_F)$ for all $x \in E$ and $A_2(y) = A(\Theta_E, y)$ for all $y \in F$.

Additional remark If E, F, and G are normed vector spaces, then $E \times F$ also becomes a normed vector space if, for example, one defines

$$\| (x, y) \|_{E \times F} = \| x \|_E + \| y \|_F \qquad \text{for all } x \in E, y \in F.$$

Then A is continuous if and only if A_1 and A_2 are continuous. In particular, every $z^* \in (E \times F)^*$ is represented uniquely in the form

$$z^*(x, y) = x^*(x) + y^*(y) \qquad \text{for all } (x, y) \in E \times F \tag{IV.1.18}$$

with $x^* \in E^*$ and $y^* \in F^*$.

IV.2 PROPERTIES OF CONVEX CONES AND REPRESENTATION OF POSITIVE LINEAR FORMS

IV.2.1 Closedness of convex cones

The duality theory of linear optimization problems developed in Section I makes essential use of the fact that certain cones are closed in normed vector spaces and that they may be geometrically characterized. We now take up these topics.

Let E be a normed vector space and K a convex cone in E (see Section IV.1.1). Then the closure $\bar{K}$ of K is a necessarily closed convex cone. This was used in Section I, in order to obtain far-reaching existence and duality theorems for suitable generalized linear optimization problems.

In order now to obtain sufficient conditions for K itself to be closed, we assume that K is generated by a non-empty convex set Q in E (see Section IV.1.1), i.e. it is of the form

$$K = K(Q) = \{\lambda q : \lambda \geqslant 0, q \in Q\}. \tag{IV.2.1}$$

(It should be verified that $K(Q)$ is a convex cone).

If Q is the convex hull (see Section IV.3.1) of finitely many points of E, then $K(Q)$ is closed (for the proof see, for example, Hestenes [66], page 15). This fact is the reason why it is unnecessary to make topological assumptions in the theory of ordinary linear optimization. For optimization problems in normed vector spaces (for example, in function spaces), finitely generated cones $K(Q)$ do not, in general, suffice. Here the following theorem turns out to be useful.

Theorem IV.2.1. *Let Q be a non-empty, compact convex subset of E with $\Theta_E \notin Q$. Further, let K be a closed convex cone in E such that*

$$K(Q) \cap K = \{\Theta_E\} \tag{IV.2.2}$$

where $K(Q)$ is given by (IV.2.1). Then

$$K(Q) - K = \{\lambda q - k : \lambda \geqslant 0, q \in Q, k \in K\} \tag{IV.2.3}$$

is a non-empty closed convex cone in E.

Proof First, $K(Q) - K$ is a non-empty convex cone in E (Proof = Exercise).

For the proof that $K(Q) - K$ is closed, we consider an $x \in \overline{K(Q) - K}$, the closure of $K(Q) - K$. Then there are sequences

$$\{\lambda_i\} \text{ with } \lambda_i \geqslant 0, \{q_i\} \text{ in } Q, \text{ and } \{k_i\} \text{ in } K, \text{ with } x = \lim_{i\to\infty} \lambda_i q_i - k_i.$$

Since Q is compact, there is a $q \in Q$ and a subsequence of $\{q_i\}$, which we again denote by $\{q_i\}$, such that $q = \lim_{i\to\infty} q_i$.

We distinguish two cases.

(α) The associated sequence $\{\lambda_i\}$ is bounded. Then there is a $\lambda \geqslant 0$ and a subsequence, which we again denote by $\{\lambda_i\}$ with $\lambda = \lim_{i\to\infty} \lambda_i$.

If we define $x_i = \lambda_i q_i - k_i$, then

$$k = \lim_{i\to\infty} k_i = -x + \lambda q$$

exists and $k \in K$, since K is closed. Consequently

$$x = \lambda q - k \in K(Q) - K.$$

(β) The associated sequence $\{\lambda_i\}$ is not bounded. Then there is a subsequence again denoted by $\{\lambda_i\}$ such that $\lim_{i\to\infty} \lambda_i = +\infty$, and all $\lambda_i > 0$.

If again $x_i = \lambda_i q_i - k_i$, then there follows

$$q_i = \frac{1}{\lambda_i} x_i + \frac{1}{\lambda_i} k_i \to q = \lim_{i\to\infty} \frac{1}{\lambda_i} k_i \text{ because of } \frac{1}{\lambda_i} x_i \to \Theta_E .$$

Since K is closed and $(1/\lambda_i)\, k_i \in K$ for all i, we conclude that $q \in K$. Now (IV.2.2) implies $q = \Theta_E$, which contradicts $\Theta_E \notin Q$. Thus the case (β) cannot occur.

If E is finite dimensional, then Theorem IV.2.1 can be sharpened to read as follows.

Theorem IV.2.2 *Suppose K_1 and K_2 are two non-empty closed convex cones in E, with* $\dim E < +\infty$, *such that*

$$K_1 \cap K_2 = \{\Theta_E\}. \tag{IV.2.4}$$

Then

$$K_1 - K_2 = \{k_1 - k_2 : k_1 \in K_1 \text{ and } k_2 \in K_2\} \tag{IV.2.5}$$

is a non-empty closed convex cone in E.

Proof That $K_1 - K_2$ is a convex cone is again easy to see. For the proof that $K_1 - K_2$ is closed, consider an element $k \in \overline{K_1 - K_2}$, the closure of $K_1 - K_2$, and two sequences $\{k_i^1\} \subset K_1$ and $\{k_i^2\} \subset K$ with $k = \lim_{i\to\infty} k_i^1 - k_i^2$.

Depending on whether the sequence $\{k_i^1\}$ is bounded or not, the proof follows the same lines as the proof of Theorem IV.2.1 (Exercise).

IV.2.2 Adjoint cones

Let E be a partially ordered normed vector space with K as the non-empty positive cone. As in Section IV.1.2 an order relation is induced in E^* by means of definition (IV.1.14), with ordering cone given by

$$K^* = \{x^* \in E^* : x^*(x) \geqslant 0 \text{ for all } x \in K\} \tag{IV.2.6}$$

and is called the cone adjoint to K. The following theorem plays a central roll in the duality theory of linear optimization problems (see Section I.4).

Theorem IV.2.3 *Let K be a non-empty closed convex cone in E and $x_0 \in E$ an arbitrary point. Then*

$$x_0 \in K \Leftrightarrow x^*(x_0) \geqslant 0 \qquad \textit{for all } x^* \in K^*. \tag{IV.2.7}$$

Proof The implication '$\Rightarrow$' is a direct consequence of the definition (IV.2.6) of K^*. For the proof of '$\Leftarrow$' we assume there is an $x_0 \notin K$. Then by the separation theorem 2 in Section IV.3.2 there is an $x_0^* \in E^*$ with

$$0 \leqslant \alpha = \sup_{x\in K} x_0^*(x) < x_0^*(x_0).$$

If $x_0^*(x) > 0$ for some $x \in K$, then $\alpha = +\infty$, which is impossible. Therefore $\alpha = 0$. If one sets $x_1^* = -x_0^*$, then $x_1^*(x) \geqslant 0$ for all $x \in K$, whence $x_1^* \in K^*$. However,

$x_1^*(x_0) < 0$, i.e. the right-hand statement in (IV.2.7) does not hold. The implication '$\Leftarrow$' therefore follows by contraposition.

Problem IV.2.1 For two non-empty convex cones K_1 and K_2 in E prove the statements

$$K_1 \supseteq K_2 \quad \Rightarrow K_1^* \subseteq K_2^* \tag{IV.2.8}$$

$$(K_1 + K_2)^* = K_1^* \cap K_2^* \tag{IV.2.9}$$

$$(K_1 \cap K_2)^* \supseteq K_1^* + K_2^*. \tag{IV.2.10}$$

From the partial ordering (IV.1.14) in E^* a partial ordering in E^{**} is again induced, with ordering cone given by

$$K^{**} = \{x^{**} \in E^{**} : x^{**}(x^*) \geqslant 0 \text{ for all } x^* \in K^*\}. \tag{IV.2.11}$$

If one again denotes by Φ the canonical isomorphism of E into E^{**} (see Section IV.1.2), then there follows $\Phi(K) \subseteq K^{**}$. For each $x \in K$ we have in fact

$$\Phi_x(x^*) = x^*(x) \geqslant 0 \qquad \text{for all } x^* \in K^*$$

i.e.

$$\Phi_x \in K^{**}.$$

Problem IV.2.2 Let E be reflexive and partially ordered with the convex ordering cone K. By using Theorem IV.2.3 show

$$K \text{ closed} \Leftrightarrow \Phi(K) = K^{**}. \tag{IV.2.12}$$

Remark The implication '$\Rightarrow$' was used implicitly in the double dualization of a linear optimization problem in Section I.4.1.

Problem IV.2.3 Let E be finite dimensional and let K_1 and K_2 be two non-empty closed convex cones in E with $K_1 \cap (-K_2) = \{\Theta_E\}$. By using (IV.2.9), (IV.2.10), (IV.2.12) and Theorem IV.2.2 show

$$\Phi(K_1 + K_2) = (K_1^* \cap K_2^*)^* = K_1^{**} + K_2^{**}. \tag{IV.2.13}$$

Now let $E = \mathbb{R}^n$, equipped with the Euclidean norm $\| x \|_2 = \langle x, x \rangle^{1/2}$. By Section IV.1.2 there is for every $x^* \in E^*$ precisely one $\hat{x}^* \in \mathbb{R}^n$ with $x^*(x) = \langle \hat{x}^*, x \rangle$ for all $x \in \mathbb{R}^n$, and the mapping $x^* \to \hat{x}^*$ is a norm preserving isomorphism of E^* onto $\mathbb{R}^n$. If K is a non-empty convex cone in E and K^* the cone adjoint to K defined by (IV.2.6), then $x^* \to \hat{x}^*$ is a one-to-one mapping of K^* onto the cone

$$\mathring{K} = \{\hat{k}^* \in \mathbb{R}^n : \langle \hat{k}^*, k \rangle \geqslant 0 \text{ for all } k \in K\} \tag{IV.2.14}$$

in $\mathbb{R}^n$.

The cone

$$-\mathring{K} = \{\hat{k}^* \in \mathbb{R}^n : \langle \hat{k}^*, k \rangle \leqslant 0 \qquad \text{for all } k \in K\}$$

is denoted as the polar cone to K (see Figure IV.2.1).

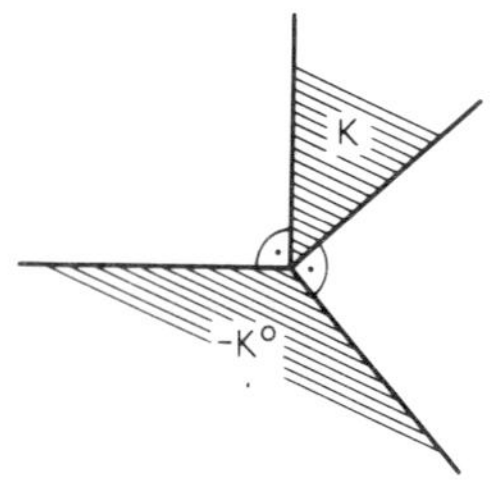

Figure IV.2.1 The polar cone $-K^\circ$ of K (in the plane)

Analogous to (IV.2.8), (IV.2.9), (IV.2.10) the following statements hold

$$K_1 \supseteq K_2 \quad \Rightarrow \mathring{K}_1 \subseteq \mathring{K}_2 \tag{IV.2.8'}$$

$$(K_1 + K_2)^\circ = \mathring{K}_1 \cap \mathring{K}_2 \tag{IV.2.9'}$$

$$(K_1 \cap K_2)^\circ \supseteq \mathring{K}_1 + \mathring{K}_2 \tag{IV.2.10'}$$

if K_1 and K_2 are two non-empty convex cones in $E = \mathbb{R}^n$. For every convex cone K in $\mathbb{R}^n$ we have, by analogy to (IV.2.12), the implication (Proof = Exercise)

$$K \text{ closed} \Leftrightarrow K = K^{\circ\circ}. \tag{IV.2.12'}$$

IV.2.3 Representation of positive linear forms on vector spaces of continuous functions

Let $C(M)$ be the vector space of continuous real-valued functions on a compact subset M of a normed vector space. Let $F = C(M)$ be equipped with the maximum norm

$$\| g \|_\infty = \max_{t \in M} | g(t) | \qquad \text{for } g \in F$$

and partially ordered in the natural way by

$$f, g \in F, f \geqslant g \Leftrightarrow f(t) \geqslant g(t) \qquad \text{for all } t \in M. \tag{IV.2.15}$$

A linear form $L : F \to \mathbb{R}$ (see Section IV.1.2) is called positive if

$$f \in F, f \geqslant \Theta_F \text{ (= null function)} \Rightarrow L(f) \geqslant 0.$$

Lemma IV.2.4 *Every positive linear form L on $F = C(M)$ is continuous, i.e. $L \geqslant \Theta_{F^*}$ in the sense of the partial ordering defined by (IV.1.14) of F^*, the topological dual space of F.*

Proof Let $e \equiv 1$ on M. Then, for every $f \in F$, one has

$$-\| f \|_\infty e \leqslant f \leqslant \| f \|_\infty e \Rightarrow -\| f \|_\infty L(e) \leqslant L(f) \leqslant \| f \|_\infty L(e).$$

Thus, $\| L(f) \| \leqslant L(e) \| f \|_\infty$. By Theorem IV.1.2, L is continuous. Furthermore, the norm of L (see (IV.1.9)) satisfies (Proof = Exercise)

$$\| L \| = \sup_{\|f\|_\infty \leqslant 1} | L(f) | = L(e).$$

For every positive linear form L on F there is precisely one regular Borel measure μ

such that

$$L(f) = \int_M f(t)\,\mathrm{d}\mu(t) \qquad \text{for } f \in F$$

(for the proof see, for example, Halmos [51]).

If M is a finite real interval $[a, b]$ $(a < b)$, then by the theorem of Riesz (see, for example, Ljusternik and Sobolew [68]) there is, for every positive linear form L on F, a bounded monotonically non-decreasing function g on $[a, b]$, such that for every every $f \in F$ the value $L(f)$ is given by the Riemann–Stieltjes integral

$$L(f) = \int_a^b f(t)\,\mathrm{d}g(t).$$

Now let V be a finite dimensional linear subspace of $C(M)$, which is spanned by the functions $v_1, \ldots, v_n \in C(M)$. A linear form $L : V \to \mathbb{R}$ is called positive if

$$v \in V, v \geqslant \Theta_F \Rightarrow L(v) \geqslant 0.$$

The aim of the following considerations is a representation theorem for positive linear forms on V. For this we need the following assumption.

Assumption There exists a $\hat{v} \in V$ with $\hat{v}(t) > 0$ for all $t \in M$. (IV.2.16)

Lemma IV.2.5 *Under the assumption (IV.2.16) every positive linear form $L : V \to \mathbb{R}$ is continuous.*

Proof If one sets $\hat{\rho} = \min_{t \in M} \hat{v}(t)$, then $\hat{\rho} > 0$ and for every $v \in V$ we have

$$-\| v \|_\infty \hat{v}/\hat{\rho} \leqslant v \leqslant \| v \|_\infty \hat{v}/\hat{\rho} \Rightarrow -\| v \|_\infty L(\hat{v})/\hat{\rho} \leqslant L(v) \leqslant \| v \|_\infty L(\hat{v})/\hat{\rho}$$

i.e.

$$|L(v)| \leqslant \| v \|_\infty L(\hat{v})/\hat{\rho}$$

whence, by Theorem IV.1.2, the continuity of L again follows.

To obtain the desired representation theorem, we start with the set

$$C = \{(v_1(t), \ldots, v_n(t))^{\mathrm{T}} : t \in M\} \subseteq \mathbb{R}^n$$

which is compact since it is the image of M under the continuous mapping $t \to (v_1(t), \ldots, v_n(t))^{\mathrm{T}}$ of M into $\mathbb{R}^n$. By Theorem IV.3.2 the convex hull $Q = H(C)$ of C is also compact in $\mathbb{R}^n$. By Theorem IV.2.1 the convex cone $K(Q) = \{\lambda q : \lambda \geqslant 0, q \in Q\}$ generated by Q is closed since $\Theta_n \notin Q$. If in fact $\Theta_n \in Q$, then there would exist numbers

$$\lambda_1 \geqslant 0, \ldots, \lambda_m \geqslant 0 \qquad \text{with} \qquad \sum_{i=1}^m \lambda_i = 1$$

and points $t_1, \ldots, t_m \in M$ with

$$\left.\begin{array}{l} \sum_{i=1}^m v_j(t_i)\lambda_i = 0 \\ j = 1, \ldots, n \end{array}\right\} \Rightarrow \sum_{i=1}^m v(t_i)\lambda_i = 0 \qquad \text{for all } v \in V.$$

If we choose $v = \hat{v}$ satisfying (IV.2.16), then necessarily all $\lambda_i = 0$, a contradiction to $\sum_{i=1}^{m} \lambda_i = 1$. Thus $\Theta_n \notin Q$.

By Theorem IV.2.3 and Section IV.1.2, Example 3, we have the equivalence relation

$$h \in K(Q) \Leftrightarrow \langle x, h \rangle = \sum_{j=1}^{n} x_j h_j \geqslant 0 \qquad \text{for all } x \in \mathbb{R}^n$$
$$\text{with } \langle x, k \rangle \geqslant 0 \qquad \text{for all } k \in K(Q). \tag{IV.2.17}$$

Problem IV.2.4 Prove the equivalence

$$\langle x, k \rangle \geqslant 0 \text{ for all } k \in K(Q) \Leftrightarrow \langle x, c \rangle \geqslant 0 \text{ for all } c \in C. \tag{IV.2.18}$$

From (IV.2.17) and (IV.2.18) we therefore obtain

$$h \in K(Q) \Leftrightarrow \{\langle x, c \rangle \geqslant 0 \text{ for all } c \in C \Rightarrow \langle x, h \rangle \geqslant 0\}. \tag{IV.2.19}$$

Now let $L : V \to \mathbb{R}$ be a positive linear form. Then

$$\sum_{j=1}^{n} v_j(t) x_j \geqslant 0 \text{ for all } t \in M \Rightarrow \sum_{j=1}^{n} L(v_j) x_j \geqslant 0. \tag{IV.2.20}$$

If one defines $h = (L(v_1), \ldots, L(v_n))$ then, by the definition of C, the statement (IV.2.20) is equivalent to the implication

$$\langle x, c \rangle \geqslant 0 \text{ for all } c \in C \Rightarrow \langle x, h \rangle \geqslant 0.$$

From this follows $h \in K(Q)$, i.e. there exist numbers $\lambda_1 \geqslant 0, \ldots, \lambda_m \geqslant 0$ and points $t_1, \ldots, t_m \in M$ with

$$\left.\begin{array}{l} L(v_j) = \sum_{i=1}^{m} \lambda_i v_j(t_i) \\ \text{for } j = 1, \ldots, n \end{array}\right\} \Rightarrow L(v) = \sum_{i=1}^{m} \lambda_i v(t_i) \qquad \text{for all } v \in V.$$

Conversely, if one chooses points $t_1, \ldots, t_m \in M$ and numbers $\lambda_1 \geqslant 0, \ldots, \lambda_m \geqslant 0$ and defines

$$L(v) = \sum_{i=1}^{m} \lambda_i v(t_i) \qquad \text{for all } v \in V \tag{IV.2.21}$$

then $L : V \to \mathbb{R}$ is a positive linear form. Altogether therefore we obtain the following representation theorem and corollary.

Theorem IV.2.6 *Under the assumption (IV.2.16), every positive linear form L on V can be represented in the form (IV.2.21) where $t_1, \ldots, t_m \in M$ are distinct points and $\lambda_1, \ldots, \lambda_m$ are non-negative numbers.*

Corollary *Under the assumption (IV.2.16) every $L \in V^*$ (the topological dual space of V) such that $L \geqslant \Theta_{V^*}$ may be represented in the form (IV.2.21) with distinct points $t_1, \ldots, t_m \in M$ and non-negative numbers $\lambda_1, \ldots, \lambda_m$.*

IV.2.4 Representation of continuous linear forms on vector spaces of continuous functions

We again take as our starting point the vector space $C(M)$ of continuous real-valued functions defined on a compact subset M of a normed vector space. Let $F = C(M)$ again be equipped with the maximum norm and partially ordered by (IV.2.15) in the natural way.

For the continuous linear forms on F (see Section IV.1.2) the following decomposition holds (see Royden [71]).

Theorem IV.2.7 *For every $L \in F^*$ (the topological dual space of F) there are two positive (hence continuous by Lemma IV.2.4) linear forms L^+ and L^- with*

$$L(f) = L^+(f) - L^-(f) \qquad \text{for all } f \in F = C(M) \tag{IV.2.22}$$

and

$$\| L \| = \sup_{\|f\|_\infty \leq 1} | L(f) | = L^+(e) + L^-(e) \tag{IV.2.23}$$

with $e = 1$.

Conversely, if one has two positive linear forms L_1, L_2 on F, then by Lemma IV.2.4 there is a continuous linear form on F given by $L = L_1 - L_2$ and we have

$$\| L \| \leq \| L_1 \| + \| L_2 \| = L_1(e) + L_2(e). \tag{IV.2.24}$$

The representation (IV.2.22) of L as the difference of two positive linear forms L^+ and L^- on F is therefore minimal in the sense that $L^+(e) + L^-(e) = \| L^+ \| + \| L^- \|$ is minimized.

A representation theorem for positive linear forms on F induces one for continuous linear forms.

If, for example, M is a real interval, then by Section IV.2.4 every positive linear form L on F may be represented as a Reimann–Stieltjes integral, i.e. we have

$$L(f) = \int_a^b f(t)\, dg(t) \qquad \text{for all } f \in F$$

where g is a bounded, monotonically non-decreasing function. Using Theorem IV.2.7 we have further the following theorem.

Riesz representation theorem *To every $L \in F^*$ there are two bounded, monotonically non-decreasing functions $g^+ = g^+(t)$ and $g^- = g^-(t)$ such that*

$$L(f) = \int_a^b f(t) dg(t) \qquad \text{for all } f \in F$$

with

$$g(t) = g^+(t) - g^-(t) \qquad \text{for } t \in [a, b]$$

and furthermore

$$\| L \| = \sup_{\|f\| \leq 1} | L(f) | = \int_a^b dg^+(t) + \int_a^b dg^-(t)$$
$$= g^+(b) + g^-(b) - g^+(a) - g^-(a).$$

Representation theorems for continuous linear forms on linear subspaces of F may be obtained from Theorem IV.2.7 from those for positive linear forms, which is particularly important for finite dimensional subspaces of F. Here we have first the next theorem.

Theorem IV.2.8 *Let V be an r-dimensional linear subspace of $F = C(M)$ and V^* the topological dual space of V (see Section IV.1.2). Then for every $y^* \in V^*$ with $y^* \neq \Theta_{V^*}$, there are distinct points $t_1, \ldots, t_s \in M$ and numbers $y_1, \ldots, y_s \in \mathbb{R}$ with*

$$y^*(v) = \sum_{i=1}^{s} y_i v(t_i) \qquad \text{for all } v \in V \tag{IV.2.25}$$

and

$$\| y^* \|_v = \sup_{\substack{v \in V \\ \| v \| \leq 1}} | y^*(v) | = \sum_{i=1}^{s} | y_i | > 0. \tag{IV.2.26}$$

Proof By the Hahn–Banach theorem (see, for example, Köthe [66]), every $y^* \in V^*$ has a norm-preserving extension to all of F, i.e. there is an $L \in F^*$ with $L(v) = y^*(v)$ for all $v \in V$ and $\| L \| = \| y^* \|_V$. From this it also follows, for every linear subspace W of F which contains V, that

$$\| L \| = \| L \|_W = \sup_{\substack{w \in W \\ \| w \| \leq 1}} | L(w) | \tag{IV.2.27}$$

(Proof = Exercise). In particular, let W be the linear subspace of F, which is spanned by V and $e \equiv 1$. By Theorem IV.2.7 there are then two positive linear forms L^+ and L^- on F satisfying (IV.2.22) and (IV.2.23). Then these are also positive on W. So by Theorem IV.2.6 there are distinct points $t_1^k, \ldots, t_{s_k}^k \in M$ for $k = +, -$ and positive numbers $\lambda_1^k, \ldots, \lambda_{s_k}^k$ such that

$$L^k(w) = \sum_{i=1}^{s_k} \lambda_i^k w(t_i^k) \qquad \text{for all } w \in W.$$

Using (IV.2.23) and (IV.2.27) we have further that

$$\sum_{i=1}^{s_+} \lambda_i^+ + \sum_{i=1}^{s_-} \lambda_i^- = L^+(e) + L^-(e) = \sup_{\substack{w \in W \\ \| w \| \leq 1}} \left| \sum_{i=1}^{s_+} \lambda_i^+ w(t_i^+) - \sum_{i=1}^{s_-} \lambda_i^- w(t_i^-) \right|. \tag{IV.2.28}$$

Since the unit sphere $\{w \in W : \| w \| \leq 1\}$ is compact, the supremum is assumed for some $\hat{w} \in W$ with $\| \hat{w} \| \leq 1$, which is only possible if

$$\hat{w}(t_i^+) = 1 \qquad \text{for all } i = 1, \ldots, s_+$$

and

$$\hat{w}(t_i^-) = -1 \qquad \text{for all } i = 1, \ldots, s_-.$$

From this it follows that all t_i^k are distinct from each other. If one defines

$$t_i = t_i^+ \text{ and } y_i = \lambda_i^+ \qquad \text{for } i = 1, \ldots, s_+$$

as well as

$$t_{i+s_+} = t_i^- \text{ and } y_{i+s_+} = -\lambda_i^- \qquad \text{for } i = 1, \ldots, s_-$$

then (IV.2.25) and (IV.2.26) are satisfied for $s = s_+ + s_-$.

The statement of Theorem IV.2.8 can be still further sharpened.

Corollary *In Theorem IV.2.8 one may choose* $s \leqslant r$.

Proof Let $y^* \in V^*$ with $y^* \neq \Theta_{V^*}$ be given. Then there is a representation (IV.2.25) which is equivalent to the statement

$$\sum_{i=1}^{s} v(t_i)y_i = y^*(v_j) \qquad \text{for } j = 1, \ldots, r \tag{IV.2.29}$$

where $v_1, \ldots, v_r$ comprise a basis of V. With y^* given, the right-hand sides of (IV.2.29) are determined and this system of linear equations has a non-trivial solution $y_1, \ldots, y_s$. We assume without loss of generality that all $y_i \neq 0$. If the column vectors $(v_1(t_i), \ldots, v_r(t_i))^{\mathrm{T}}$, $i = 1, \ldots, s$, are linearly independent, then necessarily $s \leqslant r$ and there is nothing more to prove. If not, there are numbers $z_1, \ldots, z_s \in \mathbb{R}$, not all zero, such that

$$\sum_{i=1}^{s} v_j(t_i)z_i = 0 \qquad \text{for } j = 1, \ldots, r.$$

We can assume further without loss of generality that there is a $z_i \neq 0$ with $y_i z_i < 0$. If one then chooses

$$\lambda = \min_{z_i \neq 0} |y_i/z_i|$$

then

$$y_i > 0 \Rightarrow \begin{cases} y_i + \lambda z_i > 0 & \text{for } z_i \geqslant 0 \\ y_i + \lambda z_i = y_i - \lambda |z_i| \geqslant y_i - (y_i/|z_i|)|z_i| = 0 & \text{for } z_i < 0 \end{cases}$$

and

$$y_i < 0 \Rightarrow \begin{cases} y_i + \lambda z_i < 0 & \text{for } z_i \leqslant 0 \\ y_i + \lambda z_i = -|y_i| + \lambda z_i \leqslant -|y_i| + (|y_i|/z_i)z_i = 0 & \text{for } z_i > 0. \end{cases}$$

Furthermore, for at least one $i = 1, \ldots, s$, there is a $y_i + \lambda z_i = 0$. Then the $y_i^* = y_i + \lambda z_i$ that are different from zero also form a solution of (IV.2.29) with $\operatorname{sgn} y_i^* = \operatorname{sgn} y_i$, which has less than s components different from zero.

This process can eventually be continued, in fact as long as the associated column vectors $(v_1(t_i), \ldots, v_r(t_i))^{\mathrm{T}}$ are linearly independent. It breaks off after finitely many steps with a vector $(y_{i_1}^*, \ldots, y_{i_p}^*)$, with $p \leqslant r$ all of whose components are different from zero and satisfy the following conditions:

$$\operatorname{sgn} y_{i_k}^* = \operatorname{sgn} y_{i_k} \qquad \text{for } k = 1, \ldots, p$$

where the y_i satisfy the conditions (IV.2.25), (IV.2.26) and

$$y^*(v) = \sum_{k=1}^{p} y^*_{i_k} v(t_{i_k}) \qquad \text{for all } v \in V.$$

Since $\{v \in V : \| v \| \leqslant 1\}$ is compact, the supremum in (IV.2.26) is assumed for some $\hat{v} \in V$ with $\| \hat{v} \| \leqslant 1$. This is possible only if

$$\hat{v}(t_i) = \operatorname{sgn} y_i \qquad \text{for } i = 1, \ldots, r$$

which implies

$$\hat{v}(t_{i_k}) = \operatorname{sgn} y^*_{i_k} \qquad \text{for } k = 1, \ldots, p$$

and hence

$$\| y^* \|_V = \sum_{k=1}^{p} | y^*_{i_k} |.$$

IV.3 CONVEX SETS

IV.3.1 Algebraic and topological properties

If E is a vector space over $\mathbb{R}$, then according to Section IV.1.1 a subset C of E is said to be convex if

$$x, y \in C, 0 \leqslant \lambda \leqslant 1 \Rightarrow \lambda x + (1 - \lambda) y \in C \tag{IV.3.1}$$

i.e. if the closed line segment connecting any two points in C also belongs to C. We denote this line segment by

$$[x, y] = \{\lambda x + (1 - \lambda) y : 0 \leqslant \lambda \leqslant 1\}. \tag{IV.3.2}$$

Example 1 The empty set is convex.

Example 2 For any two points $x, y \in E$, the closed segment $[x, y]$ and, for $x \neq y$, the open line segment

$$]x, y[= \{\lambda x + (1 - \lambda) y : 0 < \lambda < 1\}$$

connecting x and y is convex.

Example 3 If E is a normed vector space, $x \in E$, and $\rho > 0$, then the open, or closed ball

$$\mathring{K}(x, \rho) = \{y \in E : \| y - x \| < \rho\}$$

and

$$K(x, \rho) = \{y \in E : \| y - x \| \leqslant \rho\} \tag{IV.3.3}$$

respectively, centred at x of radius ρ is convex.

Example 4 If $x^* : E \to \mathbb{R}$ is a non-trivial linear form on the vector space E (see Section IV.1.2) and if $\alpha \in \mathbb{R}$, then the hyperplane

$$H = \{x \in E : x^*(x) = \alpha\}$$

is convex. More generally every linear manifold in E (see Section IV.3.2) as well as the half spaces defined by H

$$R^+ = \{x \in E : x^*(x) \geqslant \alpha\} \qquad \text{and} \qquad R^- = \{x \in E : x^*(x) \leqslant \alpha\}$$

are convex.

Let A be a non-empty subset of E, and $H(A)$ the set of all elements $x \in E$ which may be represented in the form

$$x = \sum_{i \in I} \lambda_i a_i \qquad \lambda_i \geqslant 0, a_i \in A \text{ for all } i \in I, \sum_{i \in I} \lambda_i = 1 \tag{IV.3.4}$$

where I is a suitable, finite index set depending on x. Every such $x \in H(A)$ is called a convex combination of elements of A.

Assertion *$H(A)$ is the smallest convex set containing A (and therefore is called the convex hull of A).*

Proof Obviously $A \subseteq H(A)$. Further, $H(A)$ is convex (Proof = Exercise), and obviously $H(A)$ is contained in every convex set containing A, which completes the proof.

Therefore a subset A of a linear vector space is convex if and only if $A = H(A)$, which is equivalent to the following implication:

$$m \in \mathbb{N}, a_i \in A, \lambda_i \geqslant 0, \quad i = 1, \ldots, m, \quad \sum_{i=1}^{m} \lambda_i = 1 \Rightarrow \sum_{i=1}^{m} \lambda_i a_i \in A. \tag{IV.3.5}$$

Problem IV.3.1a Show directly by induction that A is convex if and only if (IV.3.5) holds.

Problem IV.3.1b Show that every intersection of arbitrarily many convex sets is convex.

Problem IV.3.1c
1. Show that if A and B are convex and $\lambda, \mu \in \mathbb{R}$, then the set $\lambda A + \mu B = \{\lambda x + \mu y : x \in A, y \in B\}$ is convex.
2. Show that if A is convex and $\lambda, \mu \geqslant 0$, then

$$(\lambda + \mu)A = \lambda A + \mu A \qquad \text{with } \rho A = \{\rho x : x \in A\}.$$

Theorem IV.3.1 *The closed hull $\bar{A}$ of a convex subset A of a normed vector space E is also convex.*

Problem IV.3.2 Prove this theorem.

The convex hull $H(A)$ of a closed subset of a normed vector space E is in general not closed, as the following counter-example shows: $E = \mathbb{R}^2$ equipped with any norm. We see that

$$A = \{\Theta_2\} \cup \{(x_1, x_2) : x_1 x_2 \geqslant 1\}$$

is closed, but

$$H(A) = \{\Theta_2\} \cup \{(x_1, x_2) : x_1 > 0, x_2 > 0\}$$

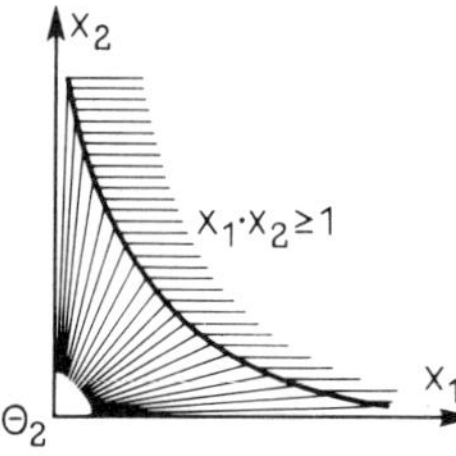

Figure IV.3.1 Example of a non-closed convex hull of a closed set

is not closed (see Figure IV.3.1).

If E is finite dimensional, then we have the following theorem.

Theorem of Carathêodory *If* $\dim E \leqslant n$ *and* A *is a non-empty subset of* E, *then every element of the convex hull* $H(A)$ *may be represented as a convex combination (IV.3.4) of at most* $(n + 1)$ *elements of* A.

For the proof see for example Collatz and Krabs [73].

An important consequence of the theorem of Carathéodory is the next theorem.

Theorem IV.3.2 *Suppose* E *is a finite dimensional normed vector space over* $\mathbb{R}$ *and* A *is a non-empty compact subset of* E. *Then the convex hull* $H(A)$ *of* A *is also compact.*

For the proof see also Collatz and Krabs [73].

If A is a compact subset of an infinite dimensional space, it follows, by a theorem of Mazur, only that $H(A)$ is relatively compact, i.e. $\overline{H(A)}$ is compact (see for this, for example, Collatz [68]).

Theorem IV.3.3 *Suppose* A *is a convex subset of a normed vector space. For every* $x_0 \in \mathring{A}$ *(the interior of* A*) and* $x \in \bar{A}$ *with* $x \neq x_0$, *we have*

$$]x_0, x[= \{\lambda x + (1 - \lambda)x_0 : 0 < \lambda < 1\} \subseteq \mathring{A}. \tag{IV.3.6}$$

For the proof see for example Valentine [68].

As an immediate consequence of Theorem IV.3.3 we have our next theorem.

Theorem IV.3.4 *If* A *is a convex subset of a normed vector space, then the interior* $\mathring{A}$ *of* A *is also convex.*

Corollary *The convex hull of an open subset of a normed vector space is also open.*

Problem IV.3.3 Prove this corollary.

IV.3.2 Separation theorems

Let X be a normed vector space over $\mathbb{R}$. A subset A of X is called a linear manifold if

$$x, y \in A, \lambda \in \mathbb{R} \Rightarrow \lambda x + (1 - \lambda)y \in A.$$

If $x^* : X \to \mathbb{R}$ is a nontrivial linear form and $\alpha \in \mathbb{R}$, then

$$H = \{x \in X : x^*(x) = \alpha\} \tag{IV.3.7}$$

is a linear manifold and is called a hyperplane.

One can show that every hyperplane is a maximal linear manifold, i.e.

$$A = \text{linear manifold},\ A \supseteq H \Rightarrow A = H \text{ or } A = X.$$

Furthermore, a hyperplane H of the form (IV.3.7) is closed if x^* is continuous (Proof = Exercise). Both assertions may be reversed, i.e. a linear manifold H in X is maximal and closed if and only if it is of the form (IV.3.7) with $x^* \in X^*$, the topological dual space of X, and $x^* \neq \Theta_{X^*}$.

Definition Two subsets A and B are separated by a hyperplane H of the form (IV.3.7) if

$$x^*(x) \leqslant \alpha \leqslant x^*(y) \qquad \text{for all } x \in A \text{ and } y \in B. \tag{IV.3.8}$$

Separation Theorem 1 *Suppose A and B are two non-empty disjoint convex subsets. Suppose further that the interior $\mathring{B}$ of B is not empty. Then there is a closed hyperplane H of the form (IV.3.7) (i.e. with $x^* \in X^*$) which separates A and B. Thus (IV.3.8) holds and*

$$\alpha < x^*(y) \qquad \textit{for all } y \in \mathring{B}.$$

For the proof see, for example, Köthe [66].

Separation Theorem 2 *Suppose A is a non-empty convex closed subset of X and $x_0 \in X$ is a point with $x_0 \notin A$. Then there is a closed hyperplane H of the form (IV.3.7) which separates x_0 and A strictly, i.e.*

$$\sup_{x \in A} x^*(x) < x^*(x_0).$$

For the proof also see Köthe [66].

IV.3.3 Weak convergence

In the minimization of convex functionals (see Section II.2.1), the concept of weak convergence plays a significant role in connection with existence statements. Suppose E is a normed vector space and E^* is the topological dual space of E (see Section IV.1.2).

Definition A sequence $\{x_k\}$, $x_k \in E$, is said to be weakly convergent to $x \in E$, denoted by $x_k \rightharpoonup x$, if for all $x^* \in E^*$, $\lim_{k\to\infty} x^*(x_k) = x^*(x)$. Then x is called the weak limit of the sequence $\{x_k\}$.

Properties

(*a*) Every weakly convergent sequence has a unique weak limit.

(*b*) Let the sequences $\{x_k\}$ in E, $\{y_k\}$ in E, $\{\lambda_k\}$ in $\mathbb{R}$, $\{\mu_k\}$ in $\mathbb{R}$ be given as well as the points $x, y \in E$ and the numbers $\lambda, \mu \in \mathbb{R}$. Then

$$\left.\begin{array}{l} x_k \rightharpoonup x,\ y_k \rightharpoonup y \\ \\ \lambda_k \to \lambda,\ \mu_k \to \mu \end{array}\right\} \Rightarrow \lambda_k x_k + \mu_k y_k \rightharpoonup \lambda x + \mu y.$$

(c) Every subsequence of a weakly convergent sequence converges weakly to the same limit (Proof of (a), (b), (c) = Exercise).

Lemma IV.3.5 *Every convergent sequence in a normed vector space E converges weakly to the same limit.* (Proof = Exercise).

The converse of Lemma IV.3.5 is in general false, as the following counterexample shows:

Let $E = l_2$ be the vector space of all real sequences $x = \{x^i\}$ with

$$\sum_{i=1}^{\infty} | x^i |^2 < +\infty$$

and let addition and scalar multiplication be defined componentwise. As is well known, l_2 is a Hilbert space with the scalar product

$$\langle x, y \rangle = \sum_{i=1}^{\infty} x^i y^i$$

and the norm

$$\| x \| = \left(\sum_{i=1}^{\infty} | x^i |^2 \right)^{1/2}.$$

By Theorem IV.1.3 there is for every $x^* \in E^* = l_2^*$ precisely one $x \in E$ with $x^*(y) = \langle x, y \rangle$ for all $y \in l_2$. Now we define in l_2 a sequence

$$e_k = \{\delta_k^i\} \qquad \text{with } \delta_k^i = 0 \text{ for } i \neq k \text{ and } \delta_k^k = 1.$$

Then $\| e_k \| = 1$ for all k, so certainly we do not have $e_k \to \Theta_E$. On the other hand,

$$\lim_{k \to \infty} \langle x, e_k \rangle = \lim_{k \to \infty} x^k = 0 \qquad \text{for every } x \in l_2$$

since $\langle x, e_k \rangle = x^k$. This implies that

$$\lim_{k \to \infty} x^*(e_k) = 0 \qquad \text{for every } x^* \in E^*$$

and hence $e_k \rightharpoonup \Theta_E$. We do have, however, the next lemma.

Lemma IV.3.6 *If E is a finite dimensional normed vector space, then a sequence in E is weakly convergent if and only if it is convergent (Proof = Exercise).*

Definition A subset A of a normed vector space is said to be weakly sequentially closed if

$$x_k \rightharpoonup x,\ x_k \in A \Rightarrow x \in A.$$

By Lemma IV.3.5 a weakly sequentially closed subset of a normed vector space is also closed. The converse is in general false. However, we do have the following theorem.

Theorem IV.3.7 *A convex subset A of a normed vector space E is closed if and only if it is weakly sequentially closed.*

Proof We have already observed that weak sequential closedness implies closedness (even without the convexity). Now let A be closed and convex and $\{x_k\}$ be a sequence in A with $x_k \rightharpoonup x$. If $x \notin A$, then by the separation theorem 2 in Section IV.3.2 there would be an $x^* \in E^*$ with

$$x^*(x_k) \leqslant \sup_{y \in A} x^*(y) < x^*(x) \qquad \text{for all } k$$

a contradiction to the fact that $\lim_{k \to \infty} x^*(x_k) = x^*(x)$. Hence $x \in A$.

Definition A subset of a normed vector space is called weakly sequentially compact if every sequence in A contains a weakly convergent subsequence whose limit belongs to A. If the limit does not belong to A, then A is called relatively weakly sequentially compact.

Every weakly sequentially compact subset A of a normed vector space is weakly sequentially closed (Proof = Exercise).

If A is a compact subset of a normed vector space, then every sequence in A has a subsequence which is convergent and, by Lemma IV.3.5, also weakly convergent to a limit in A. Thus we obtain the following theorem.

Theorem IV.3.8 *Every compact subset of a normed vector space is also weakly sequentially compact.*

For applications (see Sections II.2.1 and II.2.3) the following theorem is important.

Theorem IV.3.9 *Every bounded subset of a reflexive Banach space (see Section IV.1.2) is relatively weakly sequentially compact.*

Here a subset B of a normed vector space E is bounded if a number $b > 0$ exists with $\| x \| \leqslant b$ for all $x \in B$.

For the proof of Theorem IV.3.9, see, for example, Goldstein [66], page 140.

From Theorem IV.3.7 and Theorem IV.3.9 we obtain the following theorem.

Theorem IV.3.10 *Every closed, convex and bounded subset of a reflexive Banach space (for example, of a Hilbert space) is weakly sequentially compact.*

Problem IV.3.4 Show that in a finite dimensional normed vector space, a subset is closed, or compact if and only if it is weakly sequentially closed, or weakly sequentially compact, respectively, so that in Theorem IV.3.8 the converse also holds.

From this, Theorem IV.3.10 follows. The converse holds without the requirement of convexity since, as is well known, in finite dimensional vector spaces, subsets are compact if and only if they are closed and bounded.

SOLUTIONS AND HINTS TO THE PROBLEMS

Problem IV.1.1 The proof of (IV.1.3)–(IV.1.6) is pure verification.

In order to obtain (IV.1.2′) from (IV.1.2‴) one has to choose $E = \mathbb{R}^n$ equipped with the ordinary scalar product and put a_i = unit vector with 1 as ith component for $i \in I = \{1, \ldots, r\}$.

Problem IV.1.2 The inequalities

$$\| x \| \leqslant \sum_{j=1}^{n} | c(e_j) | \, \| e_j \| \leqslant \| c \| \sum_{j=1}^{n} \| e_j \|^2$$

prove the continuity of the mapping

$$c \to x = \sum_{i=1}^{n} c(e_j) e_j$$

from E^* onto E.

In order to prove the continuity of the inverse mapping, we first consider the bi-unique mapping

$$(x_1, \ldots, x_n)^{\mathrm{T}} \to x = \sum_{j=1}^{n} x_j e_j$$

from $\mathbb{R}^n$ onto E which is continuous because

$$\| x \| \leqslant \sum_{j=1}^{n} \| e_j \| \max_j | x_j |.$$

Since

$$S = \left\{ x = \sum_{j=1}^{n} x_j e_j : \max_j | x_j | = 1 \right\}$$

is compact and $x \to \| x \|$ is continuous we conclude that

$$\gamma = \min_{x \in S} \| x \| > 0$$

and

$$\| x \| \geqslant \gamma \max_j | x_j | \qquad \text{for all } x = \sum_j x_j e_j \in E.$$

This implies the continuity of the mapping

$$x = \sum_j x_j e_j \to (x_1, \ldots, x_n)^{\mathrm{T}}$$

and hence the continuity of the mapping

$$x = \sum_j c(e_j) e_j \to (c(e_1), \ldots, c(e_n))^{\mathrm{T}}.$$

Because of

$$| c(x) | = \left| \sum_{j=1}^{n} x_j c(e_j) \right| \leqslant \max_j | x_j | \sum_j | c(e_j) | \leqslant \frac{1}{\gamma} \sum_j | c(e_j) | \, \| x \|$$

we have

$$\| c \| \leqslant \frac{1}{\gamma} \sum_{j=1}^{n} | c(e_j) |$$

which implies the continuity of the mapping

$$x = \sum_{j=1}^{n} c(e_j) e_j \to c.$$

Problem IV.1.3 The proof of (IV.1.3)–(IV.1.6) is again pure verification.

Problem IV.2.1 The proof of (IV.2.8) follows immediately from the definition (IV.2.6).

Let $x^* \in K_1^* \cap K_2^*$ then

$$\left.\begin{array}{l} x^*(x_1) \geqslant 0 \text{ for all } x_1 \in K_1 \text{ and} \\ x^*(x_2) \geqslant 0 \text{ for all } x_2 \in K_2 \end{array}\right\} \Rightarrow \begin{array}{l} x^*(x_1 + x_2) \geqslant 0 \text{ for all } x_1 \in K_1 \text{ and} \\ x_2 \in K_2, \text{ i.e. } x^* \in (K_1 + K_2)^*. \end{array}$$

Let $x^* \in (K_1 + K_2)^*$; then

$$x^*(x_1 + x_2) \geqslant 0 \qquad \text{for all } x_1 \in K_1 \text{ and } x_2 \in K_2$$

which implies

$$x^*(x_1) \geqslant 0 \qquad \text{for all } x_1 \in K_1$$

and

$$x^*(x_2) \geqslant 0 \qquad \text{for all } x_2 \in K_2$$

i.e. $x^* \in K_1^* \cap K_2^*$.

This completes the proof of (IV.2.9).

Let $x^* \in K_1^* + K_2^*$, i.e. $x^* = x_1^* + x_2^*$, $x_1^* \in K_1^*$, $x_2^* \in K_2^*$. Then we have

$$x^*(x) = x_1^*(x) + x_2^*(x) \geqslant 0 \qquad \text{for all } x \in K_1 \cap K_2$$

i.e. $x^* \in (K_1 \cap K_2)^*$.

Problem IV.2.2 We have to show that $K^{**} \subseteq \Phi(K)$.

Let $x^{**} \in K^{**}$. Then $x^{**}(x^*) \geqslant 0$ for all $x^* \in K^*$. Since E is reflexive there exists exactly one $x \in E$ with $x^{**} = \phi_x$ and $x^{**}(x^*) = x^*(x) \geqslant 0$ for all $x \in K$. By Theorem IV.2.3 this implies $x \in K$, hence, $x^{**} \in \Phi(K)$.

Problem IV.2.3 By Theorem IV.2.2 the cone $K_1 + K_2$ is closed. Therefore (IV.2.12) implies $\Phi(K_1 + K_2) = (K_1 + K_2)^{**}$. By (IV.2.9) we obtain $\Phi(K_1 + K_2) = (K_1^* \cap K_2^*)^*$ and by (IV.2.10) it follows that $\Phi(K_1 + K_2) \supseteq K_1^{**} + K_2^{**} = \Phi(K_1) + \Phi(K_2) = \Phi(K_1 + K_2)$.

Problem IV.2.4 The implication '⇒' follows from $C \subseteq K(Q)$. Let $k \in K(Q)$ be given. Then there exist numbers $\lambda_i \geqslant 0$, and elements $c_i \in C, i = 1, \dots, r$, such that $k = \sum_{i=1}^{r} \lambda_i c_i$. Hence $\langle x, c \rangle \geqslant 0$ for all $c \in C \Rightarrow \langle x, c_i \rangle \geqslant 0$ for $i = 1, \dots, r \Rightarrow$ $\langle c, k \rangle \geqslant 0$. This proves '⇐'.

Problem IV.3.1a The statement (IV.3.5) implies convexity of A. Conversely let A be convex. The statement (IV.3.5) is obviously true for $m = 1$. Assume (IV.3.5) to be true for some $m \geqslant 1$. Choose $a_1, \dots, a_{m+1} \in A$ and $\lambda_1 \geqslant 0, \dots, \lambda_{m+1} \geqslant 0$ with $\sum_{i=1}^{m+1} \lambda_i = 1$. If $\lambda_{m+1} = 0$ or 1, then (IV.3.5) holds by assumption. If $\lambda_{m+1} \in (0, 1)$, then $1 - \lambda_{m+1} > 0$ and

$$\sum_{i=1}^{m} \frac{\lambda_i}{1 - \lambda_{m+1}} = 1.$$

Hence, by assumption,

$$\sum_{i=1}^{m} \frac{\lambda_i}{1 - \lambda_{m+1}} a_i \subset A$$

and, by the convexity of A,

$$\sum_{i=1}^{m+1} a_i = (1 - \lambda_{m+1}) \sum_{i=1}^{m} \frac{\lambda_i}{1 - \lambda_{m+1}} a_i + \lambda_{m+1} a_{m+1} \in A.$$

Problem IV.3.1b This assertion is an immediate consequence of the definition of convexity.

Problem IV.3.1c

1. Consider $\lambda x_i + \mu y_i, x_i \in A, y_i \in B, i = 1, 2$, and $\rho \in [0, 1]$. Then

$$\rho(\lambda x_1 + \mu y_1) + (1 - \rho)(\lambda x_2 + \mu y_2) = \lambda[\rho x_1 + (1 - \rho)x_2] + \mu[\rho y_1 + (1 - \rho)y_2] \in \lambda A + \mu B.$$

2. Obviously $(\lambda + \mu)A \subseteq \lambda A + \mu A$. Consider $a_1, a_2 \in A$ and $\lambda > 0, \mu > 0$. Then, by the convexity of A,

$$\frac{\lambda}{\lambda + \mu} a_1 + \frac{\mu}{\lambda + \mu} a_2 \in A \Rightarrow \lambda a_1 + \mu a_2 \in (\lambda + \mu) A.$$

Problem IV.3.2 Let $x, y \in \bar{A}$. Then there exist sequences (x_k) and (y_k) in A such that $x = \lim_{k \to \infty} x_k$ and $y = \lim_{k \to \infty} y_k$. If $\lambda \in [0, 1]$ is given, then $\lambda x + (1 - \lambda)y = \lim_{k \to \infty} [\lambda y_k + (1 - \lambda)y_k] \in \bar{A}$ because $\lambda x_k + (1 - \lambda)y_k \in A$ for all k.

Problem IV.3.3 Let A be open and $H(A)$ be the convex hull of A. Then $A = \mathring{A} \subseteq \widehat{H(A)}^{\circ} \Rightarrow H(A) \subseteq \widehat{H(A)}^{\circ}$ since $\widehat{H(A)}^{\circ}$ is convex by Theorem IV.3.4.

Problem IV.3.4 The assertion follows directly from Lemma IV.3.6.

References

The numbers in square brackets denote the final digits of the year of appearance of the reference.

Abadie, J. [67]. On the Kuhn–Tucker theorem, in *Nonlinear Programming* (ed. by J. Abadie), Amsterdam, 1967, pp. 21–36.

Arndt, D. [75]. Approximation des Extremalwertes bei einem Randkontrollproblem der Warmeleitung, *Bonner Math. Schriften,* **77** (1975), 20–30.

Arrow, K., Hurwicz, L., and Uzawa, H. [61]. Constraint qualifications in maximization problems, *Naval Res. Logist. Quart.,* **8** (1961), 171–91.

Barrodale, J., and Young, A. [70]. Computational experience in solving linear operator equations using the Chebychev norm, in *Numerical Approximation to Functions and Data* (ed. by J. G. Hayes), London, 1970, pp. 115–42.

Bazaraa, M. S., and Goode, J. J. [73]. Necessary optimality criteria in mathematical programming in normed linear spaces, *J. Optim. Th. Appl.,* **11** (1973), 235–44.

Bazaraa, M. S., Goode, J. J., and Nashed, M. Z. [74]. On the cones of tangents with applications to mathematical programming, *J. Optim. Th. Appl.,* **13** (1974), 389–426.

Ben-Israel, A., and Charnes, A. [68]. On the intersection of cones and subspaces, *Bull. Am. Math. Soc.,* **74** (1968), 541–4.

Ben-Israel, A., Charnes, A., and Kortanek, K. [69]. Duality and asymptotic solvability over cones, *Bull. Am. Math. Soc.,* **75** (1969), 318–24.

Bratton, D. [55]. The duality theorem in linear programming, *Cowles Commission Discussion Paper: Mathematics,* No. 427, Jan 6, 1955.

Breckner, W., and Kolumban, J. [68a]. Théorèmes de caractérisation des éléments de la meilleure approximation, *C. R. Acad. Sci., Paris,* **266** (1968), 206–8.

Breckner, W., and Kolumban, J. [68b]. Über die Charakterisierung von Minimallösungen in linearen normierten Räumen, *Matematica* (Cluj), **10** (1968), 33–46.

Brφndsted, A. [64]. Conjugate convex functions in topological vector spaces, *Mat. Fys. Medd. Danske Vid. Selsk.,* **34**, No. 2 (1964).

Brosowski, B. [69a]. Einige Bemerkungen zum verallgemeinerten Kolmogoroffschen Kriterium, in *Funktionalanalytische Methoden der Numerischen Mathematik* (ed. by L. Collatz, and H. Unger), Basel–Stuttgart, 1969. = Internationale Schriftenreihe zur Numerischen Mathematik, Bd. 12, 25–34.

Brosowski, B. [69b]. Nichtlineare Approximation in normierten Vektorräumen, in *Abstract Spaces and Approximation* (ed. by P. L. Butzer, and B. Sz.-Nagy), Basel–Stuttgart, 1969 = Internationale Schriftenreihe zur Numerischen Mathematik, Bd. 10, 140–59.

Brosowski, B., and Wegmann, R. [69]. Charakterisierung bester Approximationen in normierten Vektorräumen, *J. Appr. Theory,* **3** (1970), 369–97.

Butkovskiy, A. G. [69]. *Theory of optimal control of distributed parameter systems,* New York–London–Amsterdam, 1969.

Cannon, M. D., Cullum, C. D., and Polak, E. [66]. Constrained minimization problems in finite dimensional spaces, *SIAM J. Control,* **4** (1966), 528–47.

Cannon, M. D., Cullum, C. D., and Polak, E. [70]. *Theory of optimal control and mathematical programming,* New York, 1970.

Carroll, M. P., and McLaughlin, H. W. [73]. L_1 Approximation of vector valued functions, *J. Appr. Theory,* **7** (1973), 122–31.

Charnes, A., Cooper, W. W., and Kortanek, K. [63]. Duality in semi-infinite programs and some works of Haar and Carathéodory, *Management Sci.,* **9** (1963) 209–28.

Charnes, A., Cooper, W. W., and Kortanek, K. [65]. Semi-infinite programs which have no duality gap, *Management Sci.,* **12** (1965), 113–21.

Cheney, E. W. [66]. *Introduction to approximation theory,* New York, 1966.

Choquet, G. [63]. Sur la meilleure approximation dans les espaces vectoriels normes, *Rev. Roumaine Math. Pures Appl.,* **8** (1963), 1–2.

Coddington, E. A., and Levinson, N. [55]. *Theory of ordinary differential equations,* New York–Toronto–London, 1955.

Collatz, L. [68]. *Funktionalanalysis und Numerische Mathematik, Nachdr. der 1. Aufl.,* Berlin–Heidelberg–New York, 1968. = Grundlehren der mathematischen Wissenschaften, Bd. 120.

Collatz, L., and Krabs, W. [73]. *Approximationstheorie,* Stuttgart, 1973.

Collatz, L., and Wetterling, W. [71]. *Optimierungsaufgaben, 2. Aufl.,* Berlin–Heidelberg–New York, 1971. = Heidelberger Taschenbücher, Bd. 15.

Dem'yanov, V. F., and Malozemov, V. N. [71]. On the theory of non-linear minimax problems, *Russ. Math. Surveys,* **26**, No. 3 (1971), 57–115.

Deutsch, F. R., and Maserick, P. H. [67]. Applications of the Hahn Banach theorem in approximation theory, *SIAM Rev.,* **9** (1967), 516–30.

Dieter, U. [66]. Optimierungsaufgaben in topologischen Vektorräumen. I: Dualitätstheorie, *Z. Wahrscheinlichkeitstheorie verw. Gebiete,* **5** (1966), 89–117.

Dubovitskii, A. J., and Miljutin, A. A. [63]. Extremum problems with constraints, *Sov. Math. Dokl.,* **4** (1963), 452–5.

Dubovitskii, A. J., and Miljutin, A. A. [65]. Extremum problems in the presence of restrictions, *USSR Comp. Math. and Math. Phys.,* **5**, No. 3 (1965), 1–80.

Duffin, R. J. [56]. Infinite programs, in *Kuhn–Tucker, Linear inequalities and related systems,* Princeton, N.J., 1956, pp. 157–71.

Duffin, R. J., and Karlovitz, L. A. [65]. An infinite linear program with a duality gap, *Management Sci.,* **12** (1965), 122–34.

Elster, K.-H., and Götz, R. [69]. Über die 'Constraint Qualification' und damit verwandte Bedingungen, *Wiss. Z. TH Ilmenau,* **15** (1969), 27–35.

Evans, J. [69]. A note on constraint qualifications, *Rep. 6917, Center for Mathematical Studies in Business and Economics, Graduate School of Business,* Chicago, 1969.

Fan, Ky [70, 69]. Asymptotic cones and duality of linear relations, in *Inequalities II* (ed. by Shisha), London, 1970, pp. 179–86; *J. Appr. Theory,* **2** (1969), 152–9.

Farkas, J. [02]. Über die Theorie der einfachen Ungleichungen, *J. Reine Angew. Math.,* **124** (1902), 1–24.

Fenchel, W. [49]. On conjugate convex functions, *Can. J. Math.,* **1** (1949), 73–7.

Fenchel, W. [53]. *Convex cones, sets and functions,* Princeton, N.J., 1953.

Friedman, A. [64]. *Partial differential equations of parabolic type,* Englewood Cliffs, N.J., 1964.

Garkavi, A. L. [64]. Über ein Kriterium für ein Element bester Approximation (Russ.), *Sibirski Mat., Ž.* **5** (1964), 472–6.

Girsanov, J. V. [72]. *Lectures on Mathematical Theory of Extremum Problems,* Berlin–Heidelberg–New York, 1972. = Lecture Notes on Economics and Mathematical Systems, No. 67.

Gittleman, A. [71]. A general multiplier rule, *J. Optim. Th. Appl.,* **7** (1971), 29–38.

Glashoff, K., and Gustafson, S.-Å. [76]. On the numerical treatment of a parabolic boundary value control problem, *J. Optim. Th. Appl.,* **19** (1976), 645–63.

Glashoff, K., and Krabs, W. [75]. Dualität und Bang-Bang-Prinzip bei einem parabolischen Rand-Kontrollproblem, *Bonner Math. Schriften*, **77** (1975), 1–8.
Goldstein, A. A. [66]. *Constructive real analysis*, New York–Evanston–London, 1966.
Gol'stein, E. G. [67]. Dual problems of convex and fractionally-convex programming in functional spaces, *Sov. Math. Dokl.*, **8** (1967), 212–6.
Göpfert, A. [73]. *Mathematische Optimierung in allgemeinen Vektorräumen*, Leipzig, 1973.
Gould, F. J., and Tolle, J. W. [71]. A necessary and sufficient qualification for constrained optimization, *SIAM J. Appl. Math.*, **20** (1971), 164–72.
Guignard, M. [69]. Generalized Kuhn–Tucker conditions for mathematical programming problems in a Banach space, *SIAM J. Control*, **7** (1969), 232–41.
Guinn, T., Landesman, E. M., and Mikami, E. Y. [69]. A Lagrange multiplier rule in Hilbert space, *J. Optim. Th. Appl.*, **4** (1969), 386–93.
Gustafson, S.-Å. [72]. Nonlinear systems in semi-infinite programming, in *Series in Numerical Optimization and Pollution Abatement, Technical Report*, No. 2, 1972.
Gustafson, S.-Å., and Kortanek, K. O. [73a]. Mathematical models for air pollution control: Numerical determination of optimizing abatement policies, in *Models for Environmental Pollution Control* (ed. by R. A. Deininger), Ann Arbor, 1973, pp. 251–65.
Gustafson, S.-Å., and Kortanek, K. O. [73b]. Mathematical models for optimizing air pollution abatement policies: Numerical treatment, *Proceedings of the Bilateral U.S.–Czechoslovakia Environmental Protection Seminar*, 1973.
Gustafson, S.-Å., Kortanek, K. O., and Rom, W. [70]. Non-Chebyshevian moment problems, *SIAM J. Numer. Anal.*, **7** (1970), 335–42.
Haar, A. [24]. Über lineare Ungleichungen, *Acta Math.* (Szeged), **2** (1924), 1–14.
Halkin, H. [66a]. An abstract framework for the theory of process optimization, *Bull. Am. Math. Soc.*, **72** (1966), 677–8.
Halkin, H. [66b]. A maximum principle of the Pontryagin type for systems described by nonlinear difference equations, *SIAM J. Control*, **4** (1966), 90–111.
Halkin, H. [67]. Nonlinear nonconvex programming in an infinite dimensional space, in *Mathematical Theory of Control* (ed. by A. V. Balakrishnan), New York–London, 1967, pp. 10–25.
Halkin, H. [70]. A satisfactory treatment of equality and operator constraints in the Dubovitskii–Miljutin optimization formalism, *J. Optim. Th. Appl.*, **6** (1970), 138–49.
Halkin, H., and Neustadt, L. W. [66]. General necessary conditions for optimization problems, *Proc. Nat. Acad. Sci.*, **56** (1966), 1066–71.
Halmos, P. R. [51]. *Measure Theory*, New York, 1951.
Hart, J. F., *et al.* [68]. *Computer Approximations*, New York–London–Sydney, 1968.
Hastings, C., Jr [55]. *Approximations for digital computers*, Princeton, N.J., 1955.
Havinson, S. J. [67]. Approximation by elements of convex sets, *Sov. Math. Dokl.*, **8** (1967), 98–101.
Hestenes, M. R. [66]. *Calculus of variations and optimal control theory*, New York –London–Sydney, 1966.
Hoffmann, K.-H. [71]. *Nichtlineare Optimierung, Habil. Schrift.*, München, 1971.
Hoffmann, K.-H., and Kolumban, J. [74]. Verallgemeinerte Differenzierbarkeitsbegriffe und ihre Anwendung in der Optimierungstheorie, *Computing*, **12** (1974), 17–41.
Holmes, R. B. [72]. *A course on optimization and best approximation*, Berlin–Heidelberg–New York, 1972. = Lecture Notes in Mathematics, No. 257.
Hurwicz, L. [58]. Programming in linear spaces, in *Studies in linear and nonlinear*

programming (ed. by K. J. Arrow, L. Hurwicz, and H. Uzawa), Stanford, Calif., 1958, pp. 38–102.

Ioffe, A. D., and Tikhomirov, V. M. [68]. Duality of convex functions and extremum problems, *Russ. Math. Surveys,* **23**, No. 6 (1968), 53–124.

John, F. [48]. Extremum problems with inequalities as subsidiary conditions, in *Studies and Essays, Courant Anniversary Volume,* New York, 1948, pp. 187–204.

Joly, J. L., and Laurent, P. J. [71]. Stability and duality in convex minimization problems, *R.I.R.O.,* **2** (1971), 3–42.

Kallina, C., and Williams, A. C. [71]. Linear programming in reflexive spaces, *SIAM Rev.,* **13** (1971), 350–76.

Kantorowitsch, L. W., and Akilov, G. P. [64]. *Funktionalanalysis in normierten Räumen,* Berlin, 1964.

Kantorowitsch, L. W., and Krylow, W. I. [56]. *Näherungsmethoden der höheren Analysis,* Berlin, 1956.

Karlin, S. [59]. *Mathematical methods and theory in games, programming, and economics,* 2 vols, Reading, Mass., 1959.

Klee, V. [65]. Remarks on nearest points in normed linear spaces, *Proc. of a Colloqu. on Convexity.,* Copenhagen, 1965, pp. 168–76.

Klee, V. L. [69]. Separation and support properties of convex sets – a survey, in *Control Theory and the Calculus of Variations* (ed. by A. V. Balakrishnan), New York–London, 1969, pp. 235–303.

Kolmogoroff, A. N. [48]. Eine Bemerkung zu den Polynomen von P. L. Tschebyscheff, die von einer gegebenen Funktion am wenigsten abweichen (Russ.), *Usp. Mat.Nauk,* **3** (1948), 216–21.

Kortanek, K. O. [69]. Compound asymptotic duality classification schemes, *Management Sciences Rep. 185, School of Urban and Public Affairs, Carnegie Mellow Univ.,* Pittsburgh, Pa., 1969.

Köthe, G. [66]. *Topologische lineare Räume I. 2. Aufl.,* Berlin–Heidelberg–New York, 1966.

Krabs, W. [68]. Lineare Optimierung in halbgeordneten Vektorräumen, *Numer. Math.,* **11** (1968), 220–31.

Krabs, W. [71]. Zur Dualitätstheorie bei linearen Optimierungsproblemen in halbgeordneten Vektorräumen, *Math. Z.,* **121** (1971), 320–8.

Krabs, W. [73]. Nonlinear optimization and approximation, *J. Appr. Theory,* **9** (1973), 316–26.

Krabs, W. [75]. Zur Berechnung des Extremalwertes bei einem parabolischen Rand-Kontrollproblem, *Beitr. Numer. Math.,* **4** (1975), 129–39.

Krabs, W. [76]. Lower bounds for the extreme value of a parabolic control problem, in *Dynamical Systems,* vol 1, An International Symposium, 1976, Academic Press, New York–San Francisco–London, pp. 291–5.

Krabs, W., and Weck, N. [74]. Über ein Kontrollproblem in der Wärmeleitung, in *Numerische Methoden bei Optimierungsaufgaben,* Bd. 2 (ed. by L. Collatz, and W. Wetterling), Basel–Stuttgart, 1974. = Internationale Schriftenreihe zur Numerischen Mathematik, Bd. 23, 85–100.

Kretschmer, K. S. [61]. Programs in paired spaces, *Can. J. Math.,* **13** (1961), 221–38.

Kuhn, H., and Tucker, A. [51]. Nonlinear programming, in *Proc. Second Berkeley Symposium on Mathematical Statistics and Probability* (ed. by J. Neymann), Berkeley, Calif., 1951, pp. 481–92.

Laurent, P.-J. [67]. Théorèmes de caractérisation en approximation convexe, *Communications au 'Colloque sur la théorie de l'approximation des fonctions',* Cluj (Roumanie), 15–20 Sept. 1967; *Matematica,* **10** (1968), 95–111.

Laurent, P.-J. [72]. *Approximation et optimisation,* Paris, 1972.

Lempio, F. [71a]. Separation und Optimierung in linearen Räumen, *Dissertation,* Hamburg, 1971.

Lempio, F. [71b]. Lineare Optimierung in unendlich-dimensionalen Vektorräumen, *Computing,* **8** (1971), 284–90.

Lempio, F. [71c]. Differenzierbare Optimierung mit unendlich vielen Nebenbedingungen, in *Operations Research-Verfahren* XIII (1971), 265–73.

Lempio, F. [72a]. Eine Verallgemeinerung des Satzes von Fritz John, in *Operations Research-Verfahren* XVIII (1972), 239–47.

Lempio, F. [72b]. Tangentialmannigfaltigkeiten und infinite Optimierung, *Habilitationsschrift.,* Hamburg, 1972.

Lempio, F. [73]. Positive Lösungen unendlicher Gleichungs- und Ungleichungssysteme und Lagrange-Multiplikatoren für infinite differenzierbare Optimierungsprobleme, *Z. Angew. Math. Mech.,* **53** (1973), 61–2.

Lempio, F. [74]. Anwendungen der Lagrangeschen Multiplikatorenregel auf Approximations-, Variations- und Steuerungsprobleme, in *Numerische Methoden bei Differentialgleichungen und mit funktional-analytischen Hilfsmitteln* (ed. by J. Albrecht, and L. Collatz), Basel–Stuttgart, 1974. = Internationale Schriftenreihe zur Numerischen Mathematik, Bd. 19, 147–57.

Ljusternik, L. A., and Sobolew, W. J. [68]. *Elemente der Funktionalanalysis,* Berlin, 1968.

Lobry, C. [67]. Etude géometrique des problems d'optimisation en présence de contrainte, *Thèse,* Grenoble, 1967.

Luenberger, D. G. [69]. *Optimization by vector space methods,* New York–London –Sydney–Toronto, 1969.

Mangasarian, O. [69]. *Nonlinear programming,* New York, 1968.

Mangasarian, O., and Fromovitz, S. [67]. The Fritz John necessary optimality conditions in the presence of equality and inequality constraints, *J. Math. Anal. Appl.,* **17** (1967), 37–47.

Meinardus, G. [64]. *Approximation von Funktionen und ihre numerische Behandlung,* Berlin–Heidelberg–New York, 1964. = Springer Tracts in Natural Philosophy, No. 4.

Meinardus, G., and Schwedt, D. [64]. Nicht-lineare Approximationen, *Arch. Rat. Mech. Anal.,* **17** (1964), 297–326.

Mikhlin, S. G., and Smolitskiy, K. L. [67]. *Approximate methods for solution of differential and integral equations,* New York–London–Amsterdam, 1967.

Moreau, J.-J. [62]. Fonctions convexes en dualité, *Faculté des Sciences de Montpellier, Seminaire de Math.,* 1962.

Mosco, U. [71]. On the continuity of the Young–Fenchel transform, *J. Math. Anal. Appl.,* **35** (1971), 518–35.

Nagahisa, Y., and Sakawa, Y. [69]. Nonlinear programming in Banach spaces, *J. Optim. Th. Appl.,* **4** (1969), 182–90.

Neustadt, L. W. [66, 67]. An abstract variational theory with applications to a broad class of optimization problems: I General theory; II Applications; *SIAM J. Control,* **4** (1966), 505–27; **5** (1967), 90–137.

Neustadt, L. W. [69]. A general theory of extremals, *J. Comput. System Sci.,* **3** (1969), 57–92.

Neustadt, L. W. [70]. Sufficiency conditions and a duality theory for mathematical programming problems in arbitrary linear spaces, in *Nonlinear Programming* (ed. by J. B. Rosen, O. L. Mangasarian, and K. Ritter), New York–London, 1970, pp. 323–48.

Nikolski, W. N. [61]. Verallgemeinerung eines Satzes von A. N. Kolmogoroff auf Banach–Räume, in *Untersuchungen moderner Probleme der konstruktiven Funktionentheorie* (ed. by V. J. Smirnov) (Russ.), Moskau, 1961, pp. 335–7.

Nikolski, W. N. [65]. Ein charakteristisches Kriterium für die am wenigsten abweichenden Elemente aus konvexen Mengen, in *Untersuchungen moderner Probleme der konstruktiven Funktionentheorie* (Russ.), Aserbeidschan, Baku, 1965, pp. 80–4.

Norris, D. O. [67]. Lagrangian saddle points and optimal control, *SIAM J. Control,* **5** (1967), 594–9.
Peterson, D. W. [73]. A review of constraint qualifications in finite-dimensional spaces, *SIAM Rev.,* **15** (1973), 639–54.
Petrowski, J. G. [54]. *Lectures on partial differential equations,* New York–London, 1954.
Protter, M. H., and Weinberger, H. F. [67]. *Maximum Principles in Differential Equations,* Englewood Cliffs, N.J., 1967.
Pschenitschny, B. N. [72]. *Notwendige Optimalitätsbedingungen*, München–Wien, 1972.
Raffin, C. [68]. Programmes linéaire d'appui d'un programme convexe, application aux conditions d'optimalité et a la dualite, *Rev. Fr. Inf. Rech. Opér.,* **2**, No. 13 (1968), 27–60.
Ritter, K. [67]. Duality for nonlinear programming in a Banach space, *SIAM J. Appl. Math.,* **15** (1967), 294–302.
Rockafellar, R. T. [67]. Duality and stability in extremum problems involving convex functions, *Pacific J. Math.,* **21**(1967), 167–87.
Rockafellar, R. T. [69]. *Convex Analysis,* Princeton, N.J., 1969.
Royden, H. L. [71]. *Real analysis,* New York, 1971.
Rubinov, A. M. [66]. Necessary conditions for an extreme value and their use in the study of certain equations, *Sov. Math. Dokl.,* **7**, No. 4 (1966), 978–80.
Rubinshtein, G. Sh. [70]. Duality in mathematical programming and some problems of convex analysis, *Russ. Math. Surveys,* **25**, No. 5 (1970), 171–200.
Russell, D. L. [66]. The Kuhn–Tucker conditions in Banach space with an application to control theory, *J. Math. Anal. Appl.,* **15** (1966), 200–12.
Sander, H.-J. [73]. *Dualität bei Optimierungsaufgaben*, München–Wien, 1973.
Schechter, M. [72]. Duality in continuous linear programming, *J. Math. Anal. Appl.,* **37** (1972), 130–41.
Schechter, M. [73]. Linear programs in topological vector spaces, *J. Math. Anal. Appl.,* (to appear).
Singer, J. [62]. Choquet spaces and best approximation, *Math. Ann.,* **148** (1962), 330–40.
Sion, M. [58]. On general minimax theorems, *Pacific J. Math.,* **8** (1958), 171–6.
Slater, M. L. [50]. Lagrange multipliers revisited: A contribution to nonlinear programming, *Cowles Commission Discussion Paper: Mathematics,* No. 403, November 1950.
Stiefel, E. [65]. *Einführung in die Numerische Mathematik. 3. Aufl.,* Stuttgart, 1965.
Stoer, J. [63]. Duality in nonlinear programming and the minimax theorem, *Numer. Math.,* **5** (1963), 371–9.
Stoer, J. [64]. Über einen Dualitätssatz der nichtlinearen Programmierung, *Numer. Math.,* **6** (1964), 55–8.
Stoer, J., and Witzgall, Chr. [70]. *Convexity and Optimization in Finite Dimensions I,* Berlin–Heidelberg–New York, 1970. = Die Grundlehren der mathematischen Wissenschaften, Bd. 163.
Suchowitzki, S. J., and Awdejewa, L. J. [69]. *Lineare und konvexe Programmierung,* München–Wien, 1969.
Uzawa, H. [58]. The Kuhn–Tucker Theorem in concave programming, in *Studies in linear and nonlinear programming* (ed. by K. J. Arrow, K. J. Hurwicz, and H. Uzawa), Stanford, Calif., 1958, pp. 33–7.
Valentine, F. A. [68]. *Konvexe Mengen,* Mannheim, 1968.
Van Slyke, R. M., and Wets, R. J.-B. [68]. A duality theory for abstract mathematical programs with applications to optimal control theory, *J. Math. Anal. Appl.,* **22** (1968), 679–706.

Varaiya, P. P. [67]. Nonlinear programming in Banach space, *SIAM J. Appl. Math.*, **15** (1967), 284–93.

Vershik, A. M. [70]. Some remarks on the infinite-dimensional problems of linear programming, *Russ. Math. Surveys,* **25**, No. 5 (1970), 117–24.

Von Neumann, J. [28]. Zur Theorie der Gesellschaftsspiele, *Math. Ann.,* **100** (1928), 295–320.

Weck, N. [75]. Über Existenz, Eindeutigkeit und das 'Bang-Bang-Prinzip' bei Kontrollproblemen aus der Wärmeleitung, *Bonner Math. Schriften,* 77 (1975), 9–19.

Yavin, Y. [71]. Lower bounds on the cost functional for a class of distributed systems, in *Proceedings of the IFAC Symposium on The Control of Distributed Parameter Systems,* vol. I, Banff, Canada, 1971.

Yavin, Y. [73]. Lower Bounds on the cost functional for systems governed by partial differential equations, *J. Optim. Th. Appl.,* **11** (1973), 605–12.

Yegorov, J. V. [63]. Some problems in the theory of optimal control, *USSR Comp. Math.,* **3** (1963), 1209–32.

Index

Abadie 164, 181, 209
air pollution, control of 5, 137
Akilov 80, 212
Albrecht 213
approximation,
  best 7, 49, 53, 57, 78, 124, 144, 157, 162
    characterization of 53, 56, 146, 160
    necessary conditions for 119, 144
    sufficient conditions for 119
  convex 6, 78, 85, 86, 88, 124
    uniform 115
  L_1 1, 162
  linear 78
    discrete 1, 8, 9
    subject to nonlinear side conditions 148
    with an infinite number of linear side conditions 5
  nonlinear in normed vector spaces 144, 157
  polynomial 7, 8
  rational, general 145
    ordinary 146
  uniform 7, 10, 13, 48, 78, 82, 115, 144, 159
    linear 7, 10, 48, 78, 82
    linear with interpolation side conditions 8, 133–136
    nonlinear 145, 151, 173, 177
    one-sided 12, 15, 23, 55–59, 146
    one-sided with side conditions 149
  with a mixed norm 86
Arndt 67, 209
Arrow 181, 209, 212, 214
Awdejewa 2, 9, 214

Balakrishnan 211, 212
bang-bang principle 67
Barrodale 87, 209
Bazaraa 164, 182, 209
Ben-Israel 44, 209
bottle-neck process 45
boundary maximum principle (for the potential equation) 11, 21, 59
boundary value problem,
  for the potential equation 11, 21, 59
  linear 13, 147
  nonlinear 6, 7, 146, 149, 150
  of monotonic type 14, 22
Bratton 43, 209
Breckner 165, 209
Brøndstedt 106, 209
Brosowski 159, 164, 165, 209
Butkovskiy 3, 4, 209

Cannon 181, 210
Carroll 164, 210
Charnes 44, 210
Cheney 146, 163, 210
Choquet 164, 210
Coddington 13, 16, 87, 95, 210
Collatz 2, 9, 11, 14, 22, 64, 89, 145, 146, 173, 175, 177, 187, 202, 209, 210, 212, 213
complementary slackness theorem 17, 47, 52
cone 184
  adjoint 192
  closed (convex) 190, 191, 192, 193
  convex 184
  linearizing 177, 181
  pointed 185
  polar 194
  tangent 154, 155, 156, 164, 181
constraint qualification 180, 181
control-approximation problem 3, 61
  linear 15
control problem 87, 181
  linear 95
convex combination 201
convex hull 201, 202
Cooper 44, 210
Cullum 181, 209

defect estimates, for linear operator
equations 80ff
for nonlinear boundary value
problems 150ff
degenerate kernel 85
Deininger 211
Dem'yanov 182, 210
Deutsch 164, 210
Dieter 45, 106, 210
double dualization 31
duality,
gap 27, 30
theorems for convex optimization
96, 101
theorems for linear optimization 31,
35
dual space 186
Dubovitskii 164, 181, 210, 211
Duffin 30, 43, 44, 106, 210

element,
admissible 17
for the dual problem 17
consistent 96
dually consistent 99
dually optimal 99
optimal 17, 96
Elster 181, 210
error estimate,
for a linear boundary value problem
14, 83
for a linear integral equation 82
for a linear operator equation 81
for a nonlinear boundary value
problem 153
estimate,
defect 80, 150
operator 83
optimal 24, 80, 87
Evans 181, 210
existence,
and duality statements, counter-
examples to 26
statements for linear control problems
94
theorems, for convex approximation
problems 114
for convex optimization problems
96, 101, 104
for linear optimization problems
31, 35, 37
for the dual convex optimization
problem 112
for the dual linear optimization
problem 39, 41

extremal value,
for convex optimization 97, 109
dual problem 99
for linear optimization 33
dual problem 35

Fan 30, 210
Farkas 210
Lemma 44
Fenchel 106, 210
Fréchet derivative 167, 168, 173, 177
Fréchet differentiable 167–172, 177
Friedman 3, 61, 210
Fromovitz 181, 213
functional,
affine linear 88
concave 88
with respect to a point 125
convex 88
with respect to a point 125
linear 186
continuous 186
pseudo-concave 172
pseudo-convex 172
weakly semi-continuous from below
89

Gale 30
Garkavi 164, 210
Girsanov 181, 210
Gittleman 182, 210
Glashoff 67, 210, 211
Goldstein 205, 211
Gol'stein 106, 211
Goode 164, 182, 209
Göpfert 30, 43, 45, 106, 211
Götz 181, 210
Gould 181, 211
Grinold 45
Guignard 164, 182, 211
Guinn 211
Gustafson 5, 44, 67, 211

Haar 44, 211
half space 201
Halkin 181, 211
Halmos 195, 211
Hart 2, 211
Hastings 2, 211
Havinson 164, 211
Hestenes 164, 181, 191, 211
Hoffmann 182, 211
Holmes viii, 123, 181, 211
Hurwicz 43, 114, 181, 209, 211, 214

hyperplane 203
  non-horizontal 99, 100
  separating 203

initial boundary value problem (in heat conduction) 3, 9
integral equation 81, 85
Ioffe 106, 212

John 180, 212
Joly 106, 212

Kallina 44, 212
Kantorowitsch 80, 87, 212
Karlin 105, 114, 212
Karlovitz 30, 44, 210
Klee 45, 212
Kolmogoroff 160, 212, 213
  condition, generalized 154, 160
    local 165, 169
  criterion, local 159, 165
    global 165
  set of the first kind 165
  set of the second kind 165
Kolumban 165, 182, 209, 211
Kortanek 5, 44, 209, 210, 211, 212
Köthe 123, 158, 164, 186, 188, 189, 198, 203, 212
Kretschmer 30, 43, 212
Krylow 87, 212
Kuhn 180, 212

Lagrange,
  functional 107
  multiplier rule 180, 181
Landesman 211
Laurent 106, 164, 181, 212
Lempio 45, 181, 213
Levinson 13, 16, 45, 87, 95, 210
linear form 186
  continuous 186
    decomposition of 197
    representation of 197, 198
  positive 194, 195, 196
    representation of 194, 196
linear manifold 203
linear operator equation 80
Ljusternik 82, 195, 213
Lobry 164, 181, 213
Luenberger viii, 106, 213

Malozemov 182, 210
Mangasarian 89, 172, 181, 213
mapping,
  adjoint 16, 31, 189
  affine-linear 92
  almost pseudo-concave 172, 176
  concave 92
    with respect to a point 125
  continuous 189
    linear 189
  convex 91
    with respect to a point 125
  linear 189
Maserick 164, 210
maximum,
  –minimum principle (for the heat equation) 10
  norm 7
  principle for control problems 181
  principle for linear optimization 45
  principle for nonlinear optimization 181
McLaughlin 164, 210
Meinardus 160, 213
method of successive approximation 80
Mikami 211
Mikhlin 87, 213
Milyutin 164, 181, 210, 211
minimal deviation,
  for a control-approximation problem 63–67
  for a convex approximation 122, 124
  for a general linear approximation 78
  for a one-sided uniform approximation 55, 56
  for a uniform approximation 7
minimal point(s) (optimal element(s)),
  necessary conditions for 125, 144, 154, 156, 165, 166, 172, 181
  sufficient conditions for 125, 171, 172, 181
minimizing sequence 90
minimum of a functional 91
  local 91
Moreau 106, 213
Mosco 106, 213

Nagahisa 182, 213
Nashed 164, 209
Neustadt 114, 181, 211, 213
Neymann 212
Nikolski 164, 213
normality 32
Norris 114, 214

optimization problem,
  completely infinite 3

convex 2, 78, 96, 122
in function spaces 124
without explicit side conditions 79
infinite 2
nonlinear 181
linear 12, 80, 105, 106, 124
convex 129
derived from a boundary value problem 11
general 16ff
ordinary 2, 9
subject to infinitely many side conditions 16
nonlinear 7, 144, 145, 148
with an infinite number of side conditions 165
semi-infinite 2, 5, 6, 18, 40, 137, 149
in function spaces 19, 45
nonlinear 177
order(ing) relation 16, 184, 185
orthonormal system of (eigen-)functions 85

pair,
dually consistent 99
dually optimal 99
Peterson 181, 214
Petrowski 10, 214
Polak 181, 209
problem,
consistent 33, 98
dual 35, 99
dual 16, 35, 99, 108, 112
max–inf 108
min–sup 106
normal 33, 99
of optimal control 87, 181
semi-infinite, unsolvable 26
subconsistent 33, 98
Protter 10, 11, 14, 147, 214
Pschenitschny viii, 181, 214

Raffin 214
regular point 172, 174, 178
Ritter 182, 214
Rockafellar 44, 105, 214
Rom 44, 211
Rosen 213
Royden 197, 214
Rubinov 182, 214
Rubinstein 106, 214
Russell 182, 214

saddle point 110, 113, 114
Sakawa 182, 213
Sander 45, 106, 214
Schechter 45, 214
Schwedt 160, 213
sequentially qualified 181
set,
bounded 205
closed 205
compact 205
convex 184, 200
algebraic properties of 200–202
separation theorems for 202, 203
topological properties of 200–202
star shaped with respect to a point 125
weakly sequentially closed 204
weakly sequentially compact 205
sign condition 175
Singer 164, 214
Sion 114, 214
Slater 114, 214
condition 103, 112, 113, 114
Smolitzky 87, 213
Sobolew 82, 195, 213
stability of convex optimization problems 44
statement,
max–inf 110
min–sup 110
saddle point 110
Stiefel 2, 214
Stoer 105, 114, 214
subconsistency 32
subvalue for a convex optimization problem 98, 102
subvalue for a linear optimization problem 33–37
Suchowitzki 2, 9, 214
superconsistency 43

tangent vector 154
Tchebychev approximation, discrete linear 1, 2
theorem of Carathéodory 202
theorem of Hahn–Banach 198, 210
theorem of Kuhn–Tucker 111
theorem of Riesz 195, 197
theorem of Weierstrass 89
Tikhomirov 106, 212
Tolle 181, 211
Tucker 180, 212
Tyndall 45

Unger 209
Uzawa 114, 181, 209, 212, 214

Valentine 202, 214
van Slyke 30, 106, 214
Varaiya 164, 182, 215
vector space,
  normed, reflexive 188
  partially ordered 16, 185, 205
Vershik 215
von Neumann 114, 215

weak convergence 89, 203
weak duality theorem 16, 17
Weck 4, 67, 212, 215
Wegmann 165, 209
Weinberger 10, 11, 14, 147, 214
Wets 30, 106, 214
Wetterling 2, 9, 30, 89, 210, 212
Williams 44, 212
Witzgall 105, 114, 214

Yavin 67, 215
Yegorov 67, 215
Young 87, 209